Richard Blum
Christine Bresnahan

Sams **Teach Yourself**

# Python Programming for Raspberry Pi®

in **24** Hours

st 96th Street, Indianapolis, Indiana, 46240 USA

# Sams Teach Yourself Python Programming for Raspberry Pi® 24 Hours

ISBN-13: 978-0-7897-5205-5

ISBN-10: 0-7897-5205-0

Library of Congress Control Number: 2013946052

Printed in the United States of America

First Printing: October 2013

## Trademarks

All terms mentioned in this book that are known to be trademarks or service marks have been appropriately capitalized. Sams Publishing cannot attest to the accuracy of this information. Use of a term in this book should not be regarded as affecting the validity of any trademark or service mark.

## Warning and Disclaimer

Every effort has been made to make this book as complete and as accurate as possible, but no warranty or fitness is implied. The information provided is on an "as is" basis. The author(s) and the publisher shall have neither liability nor responsibility to any person or entity with respect to any loss or damages arising from the information contained in this book.

## Bulk Sales

Sams Publishing offers excellent discounts on this book when ordered in quantity for bulk purchases or special sales. For more information, please contact

> U.S. Corporate and Government Sales
> 1-800-382-3419
> corpsales@pearsontechgroup.com

For sales outside of the U.S., please contact

> International Sales
> international@pearsoned.com

**Editor-in-Chief**
Greg Wiegand

**Executive Editor**
Rick Kughen

**Development Editor**
Mark Renfrow

**Managing Editor**
Kristy Hart

**Project Editor**
Andy Beaster

**Copy Editor**
Kitty Wilson

**Indexer**
Tim Wright

**Proofreader**
Sarah Kearns

**Technical Editor**
Jason Foster

**Publishing Coordinator**
Kristen Watterson

**Interior Designer**
Mark Shirar

**Cover Designer**
Mark Shirar

**Compositor**
Nonie Ratcliff

# Contents at a Glance

# Table of Contents

# About the Authors

**Richard Blum** has worked in the IT industry for over 25 years as a network and systems administrator, managing Microsoft, Unix, Linux, and Novell servers for a network with more than 3,500 users. He has developed and teaches programming and Linux courses via the Internet to colleges and universities worldwide. Rich has a master's degree in management information systems from Purdue University and is the author of several Linux books, including *Linux Command Line and Shell Scripting Bible* (coauthored with Christine Bresnahan, 2011, Wiley), *Linux for Dummies*, 9th edition (2009, Wiley), and *Professional Linux Programming* (coauthored with Jon Masters, 2007, Wiley). When he's not busy being a computer nerd, Rich enjoys spending time with his wife, Barbara, and two daughters, Katie Jane and Jessica.

**Christine Bresnahan** started working in the IT industry more than 25 years ago as a system administrator. Christine is currently an adjunct professor at Ivy Tech Community College in Indianapolis, Indiana, teaching Python programming, Linux system administration, and Linux security classes. Christine produces Unix/Linux educational material and is the author of *Linux Bible*, 8th edition (coauthored with Christopher Negus, 2012, Wiley) and *Linux Command Line and Shell Scripting Bible* (coauthored with Richard Blum, 2011, Wiley). She has been an enthusiastic owner of a Raspberry Pi since 2012.

# Dedication

*To the Lord God Almighty.*

*"I am the vine, you are the branches; he who abides in Me and I in him,
he bears much fruit, for apart from Me you can do nothing."*
*—John 15:5*

# Acknowledgments

First, all glory, and praise go to God, who through His Son, Jesus Christ, makes all things possible and gives us the gift of eternal life.

Many thanks go to the fantastic team of people at Sams Publishing, for their outstanding work on this project. Thanks to Rick Kughen, the executive editor, for offering us the opportunity to work on this book and keeping things on track. We are grateful to the development editor, Mark Renfrow, who provided diligence in making our work more presentable. Thanks to the production editor, Andy Beaster, for making sure the book was produced. Many thanks to the copy editor, Kitty Wilson, for her endless patience and diligence in making our work readable. Also, we are indebted to our technical editor, Jason Foster, who put in many long hours double-checking all our work and keeping the book technically accurate.

Thanks to Tonya of Tonya Wittig Photography, who created incredible pictures of our Raspberry Pis and was very patient in taking all the photos we wanted for the book. We would also like to thank Carole Jelen at Waterside Productions, Inc., for arranging this opportunity for us and for helping us out in our writing careers.

Christine would also like to thank her student, Paul Bohall, for introducing her to the Raspberry Pi, and her husband, Timothy, for his encouragement to pursue the "geeky stuff" students introduce her to.

# We Want to Hear from You!

As the reader of this book, you are our most important critic and commentator. We value your opinion and want to know what we're doing right, what we could do better, what areas you'd like to see us publish in, and any other words of wisdom you're willing to pass our way.

We welcome your comments. You can email or write to let us know what you did or didn't like about this book—as well as what we can do to make our books better.

Please note that we cannot help you with technical problems related to the topic of this book.

When you write, please be sure to include this book's title and author as well as your name and email address. We will carefully review your comments and share them with the author and editors who worked on the book.

Email:    consumer@samspublishing.com

Mail:     Sams Publishing
          ATTN: Reader Feedback
          800 East 96th Street
          Indianapolis, IN 46240 USA

# Reader Services

Visit our website and register this book at informit.com/register for convenient access to any updates, downloads, or errata that might be available for this book.

# Introduction

Officially launched in February 2012, the Raspberry Pi personal computer took the world by storm, selling out the 10,000 available units immediately. It is an inexpensive credit card–sized exposed circuit board, a fully programmable PC running the free open source Linux operating system. The Raspberry Pi can connect to the Internet, can be plugged into a TV, and costs around $35.

Originally created to spark schoolchildren's interest in computers, the Raspberry Pi has caught the attention of home hobbyist, entrepreneurs, and educators worldwide. Estimates put the sales figures around 1 million units as of February 2013.

The official programming language of the Raspberry Pi is Python. Python is a flexible programming language that runs on almost any platform. Thus, a program can be created on a Windows PC or Mac and run on the Raspberry Pi and vice versa. Python is an elegant, reliable, powerful, and very popular programming language. Making Python the official programming language of the popular Raspberry Pi was genius.

## Programming with Python

The goal of this book is to help guide both students and hobbyists through using the Python programming language on a Raspberry Pi. You don't need to have any programming experience to benefit from this book; we walk through all the necessary steps in getting your Python programs up and running!

Part I, "The Raspberry Pi Programming Environment," walks through the core Raspberry Pi system and how to use the Python environment that's already installed in it. Hour 1, "Setting Up the Raspberry Pi," demonstrates how to set up a Raspberry Pi system, and then in Hour 2, "Understanding the Raspbian Linux Distribution," we take a closer look at Raspbian, the Linux distribution designed specifically for the Raspberry Pi. Hour 3, "Setting Up a Programming Environment," walks through the different ways you can run your Python programs on the Raspberry Pi, and it goes through some tips on how to build your programs.

Part II, "Python Fundamentals," focuses on the Python 3 programming language. Python v3 is the newest version of Python, and is fully supported in the Raspberry Pi. Hours 4 through 7 take you through the basics of Python programming, from simple assignment statements (Hour 4,

"Understanding Python Basics"), arithmetic (Hour 5, "Using Arithmetic in Your Programs"), and structured commands (Hour 6, "Controlling Your Program"), to complex structured commands (Hour 7, "Learning About Loops").

Hours 8, "Using Lists and Tuples," and 9, "Dictionaries and Sets," kick off Part III, "Advanced Python," showing how to use some of the fancier data structures supported by Python—lists, tuples, dictionaries, and sets. You'll use these a lot in your Python programs, so it helps to know all about them!

In Hour 10, "Working with Strings," we take a little extra time to go over how Python handles text strings. String manipulation is a hallmark of the Python programming language, so we want to make sure you're comfortable with how that all works.

After that primer, we walk through some more complex concepts in Python: using files (Hour 11, "Using Files"), creating your own functions (Hour 12, "Creating Functions"), creating your own modules (Hour 13, "Working with Modules"), object-oriented Python programming (Hour 14, "Exploring the World of Object-Oriented Programming"), inheritance (Hour 15, "Employing Inheritance"), regular expressions (Hour 16, "Regular Expressions"), and working with exceptions (Hour 17, "Exception Handling").

Part IV, "Graphical Programming," is devoted to using Python to create real-world applications. Hour 18, "GUI Programming," discusses GUI programming so you can create your own windows applications, and Hour 19, "Game Programming," introduces you to the world of Python game programming.

Part V, "Business Programming," takes a look at some business-oriented applications that you can create. In Hour 20, "Using the Network," we look at how to incorporate network functions such as email and retrieving data from webpages into your Python programs, Hour 21, "Using Databases in Your Programming," shows how to interact with popular Linux database servers, and Hour 22, "Web Programming," demonstrates how to write Python programs that you can access from across the Web.

Part VI, "Raspberry Pi Python Projects," walks through Python projects that focus specifically on features found on the Raspberry Pi. Hour 23, "Creating Basic Pi/Python Projects," shows how to use the Raspberry Pi video and sound capabilities to create multimedia projects. Hour 24, "Working with Advanced Pi/Python Projects," explores connecting your Raspberry Pi with electronic circuits using the General Purpose Input/Output (GPIO) interface.

# Who Should Read This Book?

This book is aimed at readers interested in getting the most from their Raspberry Pi system by writing their own Python programs, including these three groups:

▶ Students interested in an inexpensive way to learn Python programming.

▶ Hobbyists who want to get the most out of their Raspberry Pi system.

▶ Entrepreneurs looking for an inexpensive Linux platform to use for application deployment.

If you are reading this book, you are not necessarily new to programming but you may be new to using the Python programming

# Conventions Used in This Book

To make your life easier, this book includes various features and conventions that help you get the most out of this book and out of your Raspberry Pi:

| | |
|---|---|
| Steps | Throughout the book, we've broken many coding tasks into easy-to-follow step-by-step procedures. |
| Filenames, folder names, and code | These things appear in a `monospace` font. |
| Commands | Commands and their syntax use **bold**. |
| Menu commands | We use the following style for all application menu commands: *Menu, Command,* where *Menu* is the name of the menu you pull down and *Command* is the name of the command you select. Here's an example: File, Open. This means you select the File menu and then select the Open command. |

This book also uses the following boxes to draw your attention to important or interesting information:

## BY THE WAY

By the Way boxes present asides that give you more information about the current topic. These tidbits provide extra insights that offer better understanding of the task.

## DID YOU KNOW

Did You Know boxes call your attention to suggestions, solutions, or shortcuts that are often hidden, undocumented, or just extra useful.

## WATCH OUT!

Watch Out! boxes provide cautions or warnings about actions or mistakes that bring about data loss or other serious consequences.

# PART I

# The Raspberry Pi Programming Environment

# HOUR 1
# Setting Up the Raspberry Pi

---

**What You'll Learn in This Hour:**

▶ What is the Raspberry Pi?

▶ How to get a Raspberry Pi.

▶ What peripherals you need for the Raspberry Pi.

▶ How to get a Raspberry Pi working.

▶ How to troubleshoot a Raspberry Pi.

This lesson introduces the Raspberry Pi: what it is, its history, and why you should learn how to program in Python on it. By the end of this hour, you will know what peripherals are needed for a Raspberry Pi and how to get one up and running.

# What Is a Raspberry Pi?

A Raspberry Pi is a very inexpensive, fully programmable computer that is small enough to fit into the palm of your hand (see Figure 1.1). While the Raspberry Pi is small in size, it is mighty in potential. You can use it like a regular desktop computer or create a super-cool project with it. For example, you could use a Raspberry Pi to set up your very own home-based cloud storage server.

# Raspberry Pi History

The Raspberry Pi is still a fairly young device. It was created in the United Kingdom by Eben Upton and a few colleagues. The first commercial version, Model A, was officially offered for sale in early 2012 at the low price of $25.

**FIGURE 1.1**
The Raspberry Pi, Model B. Note the paperclip for scale.

BY THE WAY

## Different Raspberry Pi Names

People use a few different names for the Raspberry Pi. You also will see it called names such as
*RPi* and just *Pi*.

Upton created the Raspberry Pi to address a concern that he and others in his field shared: Too
few young people were getting involved in computer science. Offering a cheap, flexible, and
small computing device seemed like a good way to trigger more interest.

Upton formed the Raspberry Pi Foundation, with expected sales around 10,000 units. When the
Raspberry Pi went on sale in February 2012, it sold out immediately. An upgraded model, Model
B, was offered during late summer 2012, and sales continued to skyrocket. While the Pi was
originally created to spark young people's interest in computers, it has also caught the attention
of home hobbyist, entrepreneurs, and educators worldwide. In just one year, the Raspberry Pi
Foundation sold approximately 1 million Raspberry Pi computers!

**Supporting the Raspberry Pi Foundation**

The Raspberry Pi Foundation is a charitable organization. It asks that you help support its cause of sparking young people's interest in computers, by purchasing a Raspberry Pi! www.raspberrypi.org

Raspberry Pi owners have used their devices in a variety of really creative projects. People around the world have used Pi to create fun projects, like voice-controlled garage door openers, weather stations, and pinball machines. Also, business-oriented projects have been done, such as using the Raspberry Pi to demonstrate potential computer security threats.

## Why Learn to Program Python on a Raspberry Pi?

A common thread in Raspberry Pi projects is the use of the Python programming language. Python allows a Raspberry Pi owner to increase the field of project possibilities to an incredible size.

Python is an interpreted object-oriented and cross-platform programming language. It is also one of the most popular programming languages, due to its reliability, clear syntax, and ease of use. Python is an elegant, powerful language.

The Raspberry Pi offers an incredibly cheap development platform for Python programming. Though Python can be considered "educational" because it is easy to learn, by no means is Python wimpy.

Armed with Python and Pi, your only limit is your imagination. You can write games in Python and run them on gaming consoles controlled by your Raspberry Pi. You can write programs to control robots attached to your Raspberry Pi. Or you could be like Dave Akerman and send your Raspberry Pi over 39,000 miles above the earth to take incredible pictures (see www.daveakerman.com/?p=592).

**Raspberry Pi Already Up and Running?**

If you are currently a Pi owner and have your Raspberry Pi up and running, you can skip the rest of this hour.

# Acquiring a Raspberry Pi

Before you purchase a Pi, you need to understand a few things:

▶ What you get when you buy a Raspberry Pi

▶ The different Pi models available

▶ Where you can buy a Raspberry Pi

▶ What peripherals you'll need

When you buy a Raspberry Pi, you get an exposed circuit board about the size of your palm, with a system on a chip (SoC), memory, and ports. Figure 1.2 shows a Raspberry Pi Model B diagram depicting what you receive. It does not come with an internal storage device, a keyboard, or any peripherals—so you will want to acquire a few peripherals to get your Pi up and running.

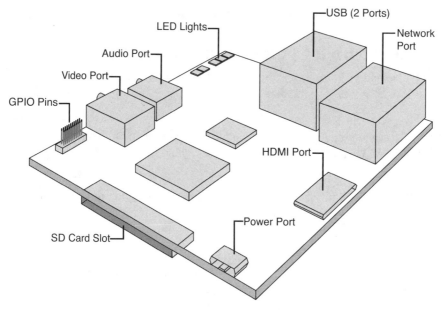

**FIGURE 1.2**
Diagram of the Raspberry Pi Model B.

DID YOU KNOW

### What Is a SoC?

A system on a chip (SoC) is a single microchip or integrated circuit (IC) that contains all the components needed for a system. SoCs are typically found on cell phones and embedded devices. For the Raspberry Pi, the SoC contains both an ARM processor for application processing and a Graphics Processing Unit (GPU) for video processing.

Two models of the Pi currently exist. While the two models are similar, you should review their different features, listed in Table 1.1, to pick which one is right for you.

**TABLE 1.1**   Model A Versus Model B Raspberry Pi

| Model A | Model B |
| --- | --- |
| 256MB RAM | 512MB RAM |
| One USB port | Two USB ports |
| No Ethernet port | One Ethernet port |

Model B has more features, and it costs only $10 more than Model A. The focus in this book is on Model B. However, either model will work fine for learning the Python programming language.

BY THE WAY

**Delay in Receiving Your Pi?**

Demand for the Raspberry Pi has been so great that you might experience a delay in getting your Pi. Don't be surprised if you have to wait from two weeks to nearly two months after purchase!

Where can you buy a Raspberry Pi? When the Raspberry Pi first came out, there were only a few places to buy them. Now the following are just a few of the businesses that sell the Raspberry Pi:

- **Farnell element14**—www.farnell.com

- **RS Components**—uk.rs-online.com

- **Allied Electronics, Inc.**—www.alliedelec.com

- **Amazon**—www.amazon.com

# What Raspberry Pi Peripherals Are Necessary?

At this point, you have a decision to make. You can buy the Raspberry Pi with all its necessary peripherals in a prepackaged kit, or you can buy the Raspberry Pi and its necessary peripherals separately. A prepackaged kit will save you time but cost you more money. Buying everything separately will save you money but cost you time. It's best to look at both options before making your decision.

WATCH OUT!

**Purchasing Peripherals**

Be sure to read the rest of the hour before you purchase a Raspberry Pi and peripherals. There are several important facts you need to know to avoid wasting time and money.

The following sections describe the basic peripherals you need to get your Raspberry Pi up and running:

- An SD card

- Power peripherals

- A television and/or computer monitor with HDMI

- A USB keyboard

The following sections provide more information on these necessary peripherals. Later in this hour, you'll learn about some nice-to-have additional peripherals.

# The SD Card

The Raspberry Pi comes with no internal storage device and no preloaded operating system. The SD card is used to provide the operating system to the Pi, in order for it to run. You must have an SD card in order to boot your Raspberry Pi.

Most prepackaged Raspberry Pi kits come with a supported SD card that is preloaded with the necessary operating system. If you don't buy a prepackaged kit, you have two choices:

- Buy a supported SD card and load the operating system onto it yourself. (You'll learn about that later in this hour.)

- Buy an SD card that has the needed operating system already on it. elinux.org/RPi_Easy_SD_Card_Setup lists companies that sell these preloaded SD cards.

WATCH OUT!

### Getting the Correct SD Card

Spend some time making sure you are purchasing the right SD card for your Raspberry Pi as discussed below. The right SD card can make your Raspberry Pi experience wonderful. The wrong SD card can cause you lots of heartache and pain.

If you decide to get your own SD card and load the operating system yourself, you can't just run out to the store and buy any old SD card. You must get one that works with a Raspberry Pi. So how do you find out which SD card to buy? Fortunately, the good people at the Linux Embedded Wiki page are here to help. On their RPi SD card page, elinux.org/RPi_SD_cards, various Raspberry Pi enthusiasts have listed which SD cards will work and which ones won't. Generally speaking, you need an SDHC card in the standard physical size with at least 4GB of storage (but 8GB is better).

BY THE WAY

### SD Card Size

You are not stuck with only the space on your SD card for storing files and programs. You can also attach storage via the Raspberry Pi's USB port. However, you do need the SD card to boot your Pi.

# Power Supply

The Raspberry Pi does not come with a power cord ready to be plugged into the wall. It simply has a USB micro B female power port. These are the basic power requirements for the Raspberry Pi:

- ▶ 5 volts

- ▶ 700mA

You can go over the 700mA. In fact, it is better to have more power because the more peripherals you add, such as a USB mouse, the more power that will be needed.

You have several options here. The options range from super cheap to very flexible but expensive. Read on to learn more.

## Cheap Power Supplies

If you have a phone charger with a micro B male connector, you may be in luck! Look on the plug end and see if the volts and mA are listed. If your phone charger provides 5 volts 700mA, then you can use it to power your Raspberry Pi. Some people have found that other chargers, such as those to power ebook readers, work as well.

BY THE WAY

### The Longer Cable

When finding a cable for your Raspberry Pi, keep in mind that the longer the cable, the more flexibility you will have. If you use a short cable to connect your Pi to power, then you will have some limitations on where your Pi can move and be set down. In general, longer cables equal greater flexibility.

If you happen to live in a modern apartment or home and your wall sockets have USB A ports in them, you can power your Raspberry Pi through those ports. You will need to purchase a cable with a USB A male connector on one end and a micro B male connector on the other end. If you don't already have these wall sockets, you can have an electrician replace your regular wall sockets with A port sockets or you can use adapters.

## A Middle-of-the-Road Power Supplies

If you do not want to share a charger with your phone or ebook reader, you can buy your Raspberry Pi its own power peripherals. In this case, you will need a USB power plug that plugs into a wall outlet with a USB A port. Also, you will need a cable that has a USB A male connector on one end and a USB micro B male connector on the other. Figure 1.3 shows an example of this.

**FIGURE 1.3**
The Raspberry Pi power port and USB power plug.

The power plug will allow you to plug into any wall socket for power. And you can use the USB power plug to power other USB-compatible devices. If you plan on sticking the Raspberry Pi in a backpack or case for travel, consider getting a USB power plug that has the ability to fold up its power prongs. This will make the power plug into a nice small cube that is compact and easy to carry.

## Portable Power Supplies

A portable power charger is wonderful, basically giving your Raspberry Pi power wherever it goes—but it is not cheap! A portable power charger typically contains a lithium ion battery and can be charged either via a wall socket at home or a USB cable connected to a computer. You can charge your portable power charger and carry it with you to power your Raspberry Pi when other power is not available. To be able to power a Raspberry Pi, a portable power charger must be able to provide the necessary 5 volts and 700mA. More expensive portable power chargers can be powered by multiple sources, such as your car's 12-volt power port as well as wall sockets.

You will still need to purchase a cable that has a USB A male connector on one end and a USB micro B male connector on the other end, in order to connect the Pi to the portable power charger. The nice thing about this is that you can charge your portable power at the same time you are powering your Raspberry Pi at home. Just don't forget to unplug your portable power charger when you remove or insert peripherals on your Pi!

# Output Display

For a very small device, the Raspberry Pi has the ability to display incredible images. It sports an HDMI port for output and enables Blu-ray-quality playback. The Raspberry Pi also provides composite output, allowing you the flexibility of using older equipment for output display. Once again, you get a choice of what you use to get your Raspberry Pi functional.

## Working with Older Display Equipment

If you have an old analog television, you can display your Raspberry Pi's output to it. All you need is a video composite cable with an RCA connector, typically yellow in color. The Raspberry Pi's composite output port is conveniently colored in a matching yellow color.

The composite output port is for video only. To get sound as well, you need an audio cable to plug into the audio-out port on the Raspberry Pi. The audio cable's other end is then connected to your chosen sound output device (for example, external speakers).

WATCH OUT!
_____

### No VGA Support

The Raspberry Pi does not provide VGA support. You can use a DVI-to-VGA converter with the DVI connection described below. However, this can add an additional point of failure to your Raspberry Pi setup.
_____

You can also hook up a computer monitor with a DVI port on it. In this case, you need an adapter to go from HDMI to DVI output. Also, like a composite video cable, DVI does not carry an audio signal. Thus, you also need an audio cable for your Raspberry Pi's sound output.

### Working with Modern Display Equipment

Using modern display equipment is the easiest way to capture a Raspberry Pi's video and audio output. To use this method, purchase an HDMI male–to-male cable, as shown in Figure 1.4. Plug one end into your Raspberry Pi's HDMI output port and the other end into either the HDMI input port on a computer monitor or television. Of course, you should make sure you purchase an HDMI cable that is long enough to accommodate your needs. HDMI handles both video and audio signals, so you need just the one cord.

**FIGURE 1.4**
The Raspberry Pi HDMI port and HDMI cable.

# Keyboard

What keyboard to use is the easiest decision you will have to make about Raspberry Pi peripherals. In order to type in your Python programs and try out various Python commands, you need a keyboard. The Raspberry Pi Model B has two USB A ports (Model A has one USB port), and you can use one of them for any USB-connected keyboard. Keep in mind that most prepackaged Raspberry Pi kits do not include a USB keyboard—but you probably already have one or two of them lying around.

# Nice Additional Peripherals

Now that you know what peripherals you absolutely have to have to run your Raspberry Pi, you can think about a few additional peripherals that will make your life with the Raspberry Pi easier. While not absolutely necessary, these peripherals are helpful:

▶ A case for the Raspberry Pi

▶ A USB mouse

▶ A self-powered USB hub

▶ Networking peripherals

## Choosing a Case

Your Raspberry Pi will come as an exposed single circuit board in an antistatic bag for protection. You don't have to have a case to protect your Pi, but having one is a good idea. Cases for the Raspberry Pi come in all kinds of shape, sizes, and colors. Figure 1.5 shows a very professional-looking case with all the ports nicely labeled.

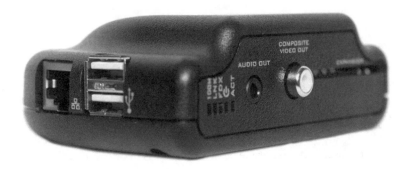

**FIGURE 1.5**
A professional Raspberry Pi case.

Many Raspberry Pi enthusiasts like using a clear case to protect the Pi's circuit board but allow it to be proudly displayed. You need to decide which kind of case meets your needs. You can easily switch your Raspberry Pi to a different case if you change your mind later!

### Static Electricity

Static electricity and circuit boards do not mix! A small spark from your hand on the exposed circuit board could permanently damage your Raspberry Pi. This is a good reason for keeping your Pi in a case.

## Using a USB Mouse

If you plan on using the Pi's graphical user interface, a USB mouse will be very handy. Keep in mind that a USB mouse and a USB keyboard will draw between 50 and 100mA as well as consume both of your Model B's USB ports.

## Looking at a Self-Powered USB Hub

If you want to connect a USB keyboard and a USB mouse, how will you connect your other USB devices at the same time? What if you want to connect an USB external storage device to your Raspberry Pi? No worries. Just purchase a self-powered USB hub, which gets its power by being plugged into an electrical outlet.

### Bus-Powered USB Hubs

Make sure you do not get a bus-powered USB hub. A bus-powered USB hub draws the power it needs from the computer it is connected to. Therefore, it would try to draw power from your Raspberry Pi.

Typically, a self-powered USB hub can supply up to 500mA to each device connected to it. It has a USB A cable that allows you to connect it to your Raspberry Pi via a USB port. Thus, you can turn one of the Raspberry Pi's USB ports into many!

## Using a Network Cable or Wi-Fi Adapter

Having your Raspberry Pi connected to the Internet and/or your local network is very handy. The Raspberry Pi comes with an RJ45 port for a wired Ethernet connection. Depending on how your local network is configured, connecting to the network may be as simple as plugging an Ethernet patch cable into the Raspberry Pi and into the back of your router. In this case, all you need to purchase is an Ethernet patch cable with two RJ45 connectors.

You can also set up your Raspberry Pi to connect via a wireless network. In this case, you need a USB wireless network adapter. You can get very small ones that are not too expensive. The downside of this method is that you need to use one of your Pi's USB ports. Also, wireless network configuration is not always very simple. But with a wireless setup, you have much more flexibility.

# Deciding How to Purchase Peripherals

Now that you have seen what the Raspberry Pi needs in the way of peripherals, you can decide which ones will be best for you. You can either buy the Raspberry Pi with its necessary peripherals in a prepackaged kit or purchase the Raspberry Pi and its necessary peripherals separately.

If you decide to purchase a prepackaged kit, keep in mind the following points:

► You will spend more money on this option than if you buy the Raspberry Pi and peripherals separately.

► Kits vary, so be sure to buy a kit that has the peripherals you want or be prepared to buy any that don't come with the kit.

► Many kits have the operating system preloaded on the SD card. If you get such a kit, you can skip downloading the operating system and loading it onto your card—and also skip the next section.

# Getting Your Raspberry Pi Working

Once you have made your purchase decisions and received your Raspberry Pi and its necessary peripherals, you can begin to really have some fun. The first time your Raspberry Pi boots up and you realize what a powerful little machine you now own, you'll really be amazed. The following sections describe what you need to do to prepare your Pi for booting.

## Do Your Research

As with many other things in life, if you plan ahead and do your research, getting your Raspberry Pi up and running should go smoothly and quickly. This up-front time and effort are very worthwhile. And many excellent resources can help. For example, the book *Hacking Raspberry Pi* will really help you have a pleasant Pi experience. Books like this one help you get your Raspberry Pi working and troubleshoot problems.

Also, there are many sources on the Internet that can assist you in your Raspberry Pi research. One of the best comes from the Raspberry Pi Foundation. It maintains a website (www.raspberrypi.org) filled with wonderful tidbits of information, including Frequently Asked Questions (FAQs), help forums, and a Quick Start Guide. At this site, you can also find software downloads and the latest news concerning the Raspberry Pi Foundation and the Pi itself. You should start your Raspberry Pi investigation at this resource.

# Choosing the Operating System

Once you have completed your initial research, the next step is to choose and download an operating system. The Raspberry Pi Foundation's website, www.raspberrypi.org, offers a choice of several operating systems.

DID YOU KNOW

## Preloaded SD Card

If you purchased a Raspberry Pi prepackaged kit, it probably contains an SD card with the operating system preloaded. If this is the case, you can skip ahead to the section "Plugging In the Peripherals."

This book is based on the Raspbian operating system, which is recommended for learning Python, and for those new to the Raspberry Pi. To download the operating system, open an Internet browser, such as Mozilla Firefox, and go to www.raspberrypi.org/downloads/, as shown in Figure 1.6.

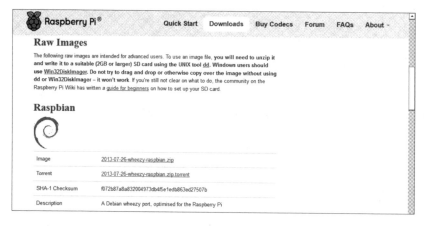

**FIGURE 1.6**
The Raspberry Pi operating system downloads page.

# Downloading the Operating System

You need an SD card reader on the system where the operating system file will be downloaded to. If you have different computers, such as a Windows machine and a Linux machine, available to you, choose the machine you feel the most comfortable using.

After choosing a machine, download the operating system from www.raspberrypi.org/downloads/. Click either the Raspbian file listed after Image that ends in ".zip" or the Raspbian file listed after Torrent that ends in ".torrent." These two file are the same, but the ".torrent" file typically downloads faster. How long it takes to download this file depends, of course, on the speed of your Internet connection.

BY THE WAY

**Need More Help?**

If you are feeling overwhelmed by this process, go to elinux.org/RPi_Easy_SD_Card_Setup, which offers a lot of advice to get you through the process of downloading the operating system and putting it on your SD card. And don't forget that you can buy a preloaded SD card. See elinux.org/RPi_Easy_SD_Card_Setup under the "Safe/Easy Way" section for a list of companies who sell these cards.

# Moving the Operating System to an SD Card

Once you have downloaded the operating system to your local computer, the next step is to move the operating system from your computer to the SD card. You will not be able to just copy the file to the SD card. Instead, you need to use an image writer program or utility to move it.

If you have experience with this, then go ahead and use the image writer program of your choice to move the operating system file to your SD card. If you have limited experience using an image writer or feel uncomfortable with the process, don't worry. Appendix A of this book describes the steps involved in transferring the Raspbian operating system file to your SD card properly.

# Plugging In the Peripherals

Now that you have your Raspberry Pi, all the necessary peripherals, and the SD card loaded with the Raspbian operating system, you can reap the rewards of your preparations. Go through the following steps, to make sure everything is working correctly:

**1.** Put the SD card into the card reader port on the Raspberry Pi, as shown in Figure 1.7.

**2.** Plug in the power cord to the Raspberry Pi. Do not attach the power cord to a power source yet.

**FIGURE 1.7**
A properly seated SD card.

BY THE WAY

## The Missing On/Off Button

The Raspberry Pi does not have an on/off switch. Therefore, when you plug it into the power source, it automatically boots.

3. Plug your USB keyboard into a USB port on the Raspberry Pi.

4. If using HDMI, plug in the HDMI cable to the Raspberry Pi's HDMI port. With your monitor or TV powered off, plug the other end of the cable into it. Turn on the monitor or TV. If you are using a TV, you may have to tune it to use the HDMI port, so do so now.

   If you are using another display output connection other than HDMI, such as video composite or DVI, then hook it up to the Raspberry Pi and your monitor or TV in a manner similar to the above.

5. You are now ready for the initial test drive (exciting isn't it?!). Sit down in front of your monitor or TV and plug the power cord into a power source.

If nothing happens, go directly to the "Troubleshooting Your Raspberry Pi" section, later in this hour.

If a lot of words go flying by on the screen and you see a menu similar to the one shown in the listing below, congratulations! Your Raspberry Pi boots!

```
Raspi-config
            info                  Information about this tool
            expand_rootfs         Expand root partition to fill SD card
            overscan              Change overscan
            configure_keyboard    Set keyboard layout
            change_pass           Change password for 'pi' user
            change_locale         Set locale
            change_timezone       Set timezone
            memory_split          Change memory split
            overclock             Configure overclocking
            ssh                   Enable or disable ssh server
            boot_behaviour        Start desktop on boot?
            update                Try to upgrade raspi-config


        <Select>                      <Finish>
```

Press the Tab key until you reach the <Finish> menu item and then press the Enter key. The command line appears, and it looks like this:

```
pi@raspberrypi ~ $
```

Pat yourself on the back. All your hard work has paid off, and you have your Raspberry Pi up and running.

Type sudo poweroff at the command prompt and press the Enter key to shut down and power off your Raspberry Pi.

## DID YOU KNOW

### Where Did the Menu Go?

Don't worry if you do not see the Raspberry Pi configuration menu the next time you boot your Pi. It is configured to show up only the first time you boot. However, in Hour 2, "Understanding the Raspbian Linux Distribution," you will learn how to call it up any time you please.

Whether your Raspberry Pi boots or not, be sure to read through the next section, and then you can safely proceed to Hour 2.

# Troubleshooting Your Raspberry Pi

The following sections discuss the most common areas to check when you are having problems getting a Raspberry Pi to boot.

## Check Your Peripheral Cords

One of your peripheral cords may not be fully seated in its port. Being "fully seated" means the connector on the cable is all the way plugged into its port. A cord that's not fully seated can cause a peripheral to work some of the time or not at all. To check your peripheral cords, follow these steps:

1.  Unplug the power source to your Raspberry Pi.

2.  Turn off your monitor or TV.

3.  For each cable connector hooked to your Raspberry Pi, unplug it and then plug it back in to the connector. Take time to make sure the connector is fully seated into the port.

4.  For each cable connector hooked to another device from your Raspberry Pi, unplug it and then plug it back into the device. Take time to make sure the connector is fully seated in the port.

5.  Turn on your monitor or TV.

6.  Plug the power source back into your Raspberry Pi.

## Check Your SD Card

If you Pi doesn't boot, you might not have used a SD card that works. To ensure that you have a usable SD card, go to elinux.org/RPi_SD_cards and double-check that a Raspberry Pi can use the SD card you have.

DID YOU KNOW

### Using LED Lights for Troubleshooting

The Raspberry Pi has no BIOS in it. Thus, it can only boot off the SD card when it receives power. There are LED lights on the Raspberry Pi that may help you troubleshoot your booting problem. If you see the red LED light on, but the green LED light is not lit and nothing is showing on the display, then you have either a bad SD card or a bad operating system image on the card. For more LED troubleshooting tips, see elinux.org/R-Pi_Troubleshooting#Normal_LED_status.

## Check Your Operating System Image

If you are using a verified SD card but the Raspberry Pi is still not booting, you may have a bad operating system image on the SD card. The image may have been damaged during the download, or it could have been damaged when you moved it to the SD card. You can verify the image by using the tips at elinux.org/RPi_Easy_SD_Card_Setup.

## Check Your Peripherals

If you've checked everything listed so far, make sure all your peripherals are verified to work with the Raspberry Pi. You can find this information at elinux.org/RPi_VerifiedPeripherals.

# Summary

In this hour, you learned what the Raspberry Pi is and why it exists, how to purchase one, and the peripherals you need to get it up and running. You read about the available operating systems for the Raspberry Pi and how to obtain a copy of Raspbian. You also learned how to get your Raspberry Pi up and running so you can proceed with learning Python programming. The hour concludes with some troubleshooting tips to consult if you have problems getting your Pi up and running.

In the next hour, you will be learning about the Raspbian operating system, as well as how to navigate through its interface to the Raspberry Pi.

# Q&A

**Q.** This book has only one hour on setting up my Raspberry Pi. Where can I get more help?

**A.** You can get additional help from the following sources:

> The Raspberry Pi Foundation and its forums, at www.raspberrypi.org/phpBB3/.

> The Embedded Linux Wiki for Raspberry Pi, at elinux.org/RPi_Hub.

> The book *Hacking Raspberry Pi*.

**Q.** What version of Python does this book cover?

**A.** Python v3. That topic is covered in Hour 3, "Setting Up a Programming Environment."

**Q.** Does this book contain a recipe for Raspberry Pie?

**A.** No, there was not enough room for all the possible recipe variations. However, open up your favorite web browser and type into your search engine "raspberry pie recipes." You will find a lot of links to recipes.

# Workshop

## Quiz

1. Python is easy to learn but has very little power so can't be used for complicated programs. True or false?

2. The Raspberry Pi can use different operating systems. Which one is recommended for those who are new to the Raspberry Pi?

3. The Raspberry Pi's on/off switch is hard to see on the circuit board and it is located near the:

   a. SoC

   b. RJ45 jack

   c. Power port

## Answers

1. False. Python is an extremely powerful programming language and is not considered wimpy in any way.

2. The Raspbian operating system is recommended for those starting out with a Raspberry Pi.

3. This is a trick question! The Raspberry Pi does not have an on/off switch. To turn on the Raspberry Pi, you must plug it into a power source. To turn off the Raspberry Pi, you must unplug it from the power source.

# Understanding the Raspbian Linux Distribution

---

**What You'll Learn in This Hour:**

- ▶ What is Linux?
- ▶ How to use the Raspbian command line
- ▶ The Raspbian graphical user interface

This hour, you will learn about Raspbian, the operating system that runs on your Raspberry Pi and supports the Python programming environment. By the end of this hour, you should know how to navigate to and from the Raspbian graphical user interface (GUI), what software comes preinstalled, and some basic shell commands.

## Learning About Linux

Linux is the third most popular desktop operating system in the world, after Microsoft Windows and Apple OS X. Therefore, the general public tends to be unaware of the Linux operating system. However, Linux is an incredibly robust and flexible operating system that can run on everything from large supercomputers all the way down to small embedded devices.

DID YOU KNOW

### Devices That Use Linux

You might be surprised to learn that the Kindle ebook reader runs on Linux. The IBM Watson supercomputer, which appeared on the television game show *Jeopardy!* in 2011, also runs on Linux.

---

Raspberry Pi's Raspbian operating system is a distribution of Linux. To understand the concept of a Linux distribution, think of an automobile. Cars have different features, such as body type, body color, automatic or manual windows, heated or regular seats, and so on. Different cars have different features. However, all cars have an engine. The "engine" in the Raspberry Pi's operating system is Linux. The various specific "features" are in the Raspbian distribution.

The Raspbian distribution is based on a Linux distribution called Debian. Debian, which was created in 1993, is a well-respected and stable distribution. It is the basis for many other popular Linux distributions, such as Ubuntu.

BY THE WAY

### Raspbian Software Packages

You can install and use more than 35,000 software packages on your Raspberry Pi, and many of them are free! You can find a small list of the packages at the Raspberry Pi store: store.raspberrypi.com.

Because it is based on Debian, Raspbian has the stability of and many of the same benefits as Debian. This means your little Raspberry Pi uses a very powerful operating system. Raspbian and Pi provide you with applications such as word processing, powerful 3D Python game graphic programs, and more.

You can find documentation and help for the Raspbian Linux distribution at www.raspbian.org. In addition, because Raspbian is based on the Debian Linux distribution, there is a lot of other documentation available. Most of the Debian documentation applies to Raspbian as well. The following are a few excellent references for Debian:

▶ *The Debian Administrator's Handbook*, which is available from debian-handbook.info

▶ *The Debian User Guide*, which you can easily access via the Raspbian GUI

▶ The Debian Project's website, www.debian.org/doc/, which offers documentation as well as helpful user forums

# Interacting with the Raspbian Command Line

When you first booted your Raspberry Pi, you did not have to provide a username and password. However, after the initial boot, on all subsequent boots, you see a Raspbian login screen. Listing 2.1 shows how you log in to your Raspberry Pi. By default, you enter the username `pi` and the password `raspberry`. Notice when the password is typed in, nothing appears on the screen. This is normal.

**LISTING 2.1**  Logging In to the Raspberry Pi

```
Debian GNU/Linux 7.0 raspberrypi tty1
raspberrypi login: pi
Password:
Last login: Wed Apr 17 20:34:50 UTC 2013 on tty1
Linux raspberrypi 3.6.11+ #371 PREEMPT Thu Feb 7 16:31:35 GMT 2013 armv6l
```

```
The programs included with the Debian GNU/Linux system are free software;
the exact distribution terms for each program are described in the
individual files in /usr/share/doc/*/copyright.

Debian GNU/Linux comes with ABSOLUTELY NO WARRANTY, to the extent
permitted by applicable law.
pi@raspberrypi ~ $
```

Once you have successfully completed the login process, you see the rest of the information shown in Listing 2.1. The Raspbian prompt looks like this:

```
pi@raspberrypi ~ $
```

This is also called the Linux *command line*. At the command line, you can enter commands to perform various tasks. In order for a command to work, you must type in the command in the proper case and then press the Enter key.

DID YOU KNOW

### What Is the Linux Shell?

When you enter commands at the command-line prompt, you are using a special utility called a *Linux shell*. The Linux shell is an interactive utility that allows you to run programs, manage files, control processes, and so on. There are several variations of the Linux shell utility. Raspbian uses the dash shell by default.

Listing 2.2 shows how you enter the whoami command. The whoami command displays the name of the user who entered the command. In this case, you can see that the user pi entered the command.

### LISTING 2.2   Entering a Command at the Command Line

```
pi@raspberrypi ~ $ whoami
pi
pi@raspberrypi ~ $
```

You can do a lot of work at the Linux command line. Table 2.1 lists some commands that will help you as you start to learn Python programming.

### TABLE 2.1   A Few Basic Command-Line Commands

| Command | Description |
|---------|-------------|
| cd | Changes your current location to a new location in the directory structure. |
| cat | Displays the contents of a file. |

| Command | Description |
|---------|-------------|
| mkdir | Makes a new directory in the directory structure. |
| ls | Displays a list of files in your current directory. |
| pwd | Shows where you are currently located in the directory structure (that is, the present working directory). |

In the following Try It Yourself, you will start to use some commands so you can begin to understand them better.

▼ TRY IT YOURSELF

### Log In and Issue Commands at the Command Line

In this section, you will try out a few commands at the Raspbian command line. As you'll see in the following steps, contrary to popular belief, using the command line is not hard at all:

1. Power-up your Raspberry Pi. You will see a lot of startup messages scroll by the screen. These are informational, and it is a good habit to view the messages as they scroll by. Don't worry if you don't know what they mean. Over time, you will learn.

2. At the raspberrypi login: prompt, type pi and press the Enter key. You should now see a Password: prompt.

3. At the Password: prompt, type raspbian and press the Enter key. If you are successful, you see the pi@raspberrypi ~ $ prompt. If you are not successful, you get the message "Login incorrect," and you see the raspberry pi login: prompt again.

BY THE WAY

### Blank Passwords

If you have never logged into a Linux command line, you may be surprised by the fact that nothing is displayed when you type in a password. Normally, in a graphical user interface environment, you see a large dot or asterisk displayed for each character you type into the password field. However, the Linux command line displays nothing as you type a password.

4. At the pi@raspberrypi ~ $ prompt, type the command whoami and press the Enter key. You should see the word pi displayed and then, on the next line down, another pi@ raspberrypi ~ $ prompt displayed.

5. Now at the prompt, type the command calendar and press the Enter key. You should see some interesting facts concerning today's date and the next few days.

BY THE WAY

## Exploring Files and Directories

In the next few steps, you will explore files and directories. It is important that you learn how to do this so you will know how and where to store the Python programs you create in this book.

6. Type the command `ls` and press the Enter key. You should see a list of files and subdirectories that are located in your current location in the directory structure. This is called your *present working directory*.

7. Type the command `pwd` and press the Enter key. This shows you the actual name of your present working directory. If you are logged into the `pi` user account, the displayed present working directory is `/home/pi`.

8. Type `mkdir py3prog` and press the Enter key to create a subdirectory called `py3prog`. You will use this directory to store all your Python programs and work in progress.

9. To see if you created the subdirectory, type the command `ls` and press the Enter key. Along with the list of files and subdirectories you saw in step 6, you should now see the `py3prog` subdirectory.

10. To make your present working directory the newly created `py3prog` subdirectory, type `cd py3prog` and press the Enter key.

11. Make sure you are in the correct directory by typing the command `pwd` and pressing the Enter key. You should see the directory name `/home/pi/py3prog` displayed.

12. Now go back to the `pi` user home directory by simply typing `cd` and pressing the Enter key. Make sure you made it to the home directory by typing the `pwd` command and pressing the Enter key. You should now see the directory name `/home/pi` displayed because you are back to the home directory.

BY THE WAY

## Commands to Manage

Now you will try out a few commands that will help you manage your Raspberry Pi.

13. (Warning: This next command will not work, and it is not supposed to work!) Type the command `reboot` and press the Enter key. You should get the message `reboot: must be superuser.`, as shown in Listing 2.3.

**LISTING 2.3** Attempting a Reboot Without `sudo`

```
pi@raspberrypi ~ $ reboot
reboot: must be superuser.
pi@raspberrypi ~
```

BY THE WAY

### Getting to Know `sudo`

You cannot run some commands unless you have special privileges. For example, the root user, also called the superuser, is an account that was originally set up in Linux as an all-powerful user login. Its primary purpose was to allow someone to properly administer the system. In some ways, the root account is similar to the administrator account in Microsoft Windows.

Due to security concerns, it is best to avoid logging into the root user account. On Raspbian, you are not even allowed to log into the root user account!

So how do you run commands for which you need root privileges, such as installing software or rebooting your Pi? The `sudo` command helps you here. `sudo` stands for "superuser do." User accounts that are allowed to use `sudo` can perform administrative duties. The `pi` user account on your Raspberry Pi is, by default, granted access to the `sudo` command. Therefore, if you are logged into the `pi` account, you can put the `sudo` command in front of any command that needs superuser privileges.

14. Type the command `sudo reboot` and press the Enter key. Your Raspberry Pi should now reboot.

15. At the `raspberrypi login:` prompt, type `pi` and press the Enter key. You should now see the `Password:` prompt.

16. At the `Password:` prompt, type `raspberry` and press the Enter key. If you are successful, you see the `pi@raspberrypi ~ $` prompt. If you are not successful, you get the message `Login incorrect` and see the `raspberry pi login:` prompt again.

17. To change the password for the `pi` account from the default to something new, type in the command `sudo raspi-config` and press the Enter key. You should see the text-based menu you saw when you first booted your Raspberry Pi:

```
Raspi-config
        info                    Information about this tool
        expand_rootfs           Expand root partition to fill SD card
        overscan                Change overscan
        configure_keyboard      Set keyboard layout
        change_pass             Change password for 'pi' user
        change_locale           Set locale
        change_timezone         Set timezone
```

```
    memory_split          Change memory split
    overclock             Configure overclocking
    ssh                   Enable or disable ssh server
    boot_behaviour        Start desktop on boot?
    update                Try to upgrade raspi-config
        <Select>                  <Finish>
```

18. Press the down-arrow key four times, until you reach the `change_pass` menu option. Press the Enter key.

19. You should now be at a screen that states, "You will now be asked to enter a new password for the pi user." Press the Enter key.

20. When you see the `Enter new UNIX password:` at the bottom left of your display screen, enter a new password for the `pi` account and press Enter. (Make it at least eight characters long and a combination of letters and numbers.) Again, as you type in the new password, you do not see it onscreen.

21. When you see the `Retype new UNIX password:` prompt at the bottom left of your screen, again type in your new password for the `pi` account at the prompt and press the Enter key. If you typed in the password correctly, you will get a screen that says `Password changed successfully`. In this case, press the Enter key to continue.

22. If you did not type in the password correctly, you get the message `There was an error running do_change_pass`. In this case, repeat steps 18–21 until you succeed.

23. Back at the main Raspbian configuration (`raspi-config`) menu screen, press the Tab key to highlight the `<Finish>` selection and press the Enter key to leave the menu.

24. In the lower-left corner of the display screen, you should now see that you are back to the Raspbian prompt. At the Raspbian prompt, type `sudo poweroff` and press the Enter key to log out of your Raspberry Pi and gracefully power it down.

Well done! You now know several Linux command-line commands. You can log in, move to subdirectories, list files that are in those subdirectories, and even do some management work, such as change your `pi` account password and reboot your system.

# Interacting with the Raspbian GUI

When you boot your Raspberry Pi and log in, by default you go to the Linux command line. But Raspbian also offers a graphical user interface (GUI).

To reach the GUI, you enter the command `startx` at the command-line prompt and press the Enter key. The Lightweight X11 Desktop Environment (LXDE) starts up, providing you with the GUI shown in Figure 2.1.

**FIGURE 2.1**
The Raspbian LXDE GUI.

BY THE WAY

## Linux Desktop Environments

One of the wonderful things about Linux is that you can change your desktop environment. You are not stuck with just one! Each desktop environment provides a unique way to graphically interact with the computer.

These are some of the most popular desktops:

▶ **KDE**—A graphical desktop that is similar to the Microsoft Windows environment

▶ **Xfce**—A lightweight but fully functional graphical desktop

▶ **GNOME**—A historically popular desktop that is the default desktop on many Linux distributions

▶ **LXDE**—A lightweight yet powerful graphical desktop that is specifically designed for smaller computers

Raspbian uses LXDE by default. This book's graphical interface descriptions are based on the LXDE desktop environment.

# The LXDE Graphical Interface

On the LXDE graphical interface, you see several icons. And no, that big Raspberry in the middle is not a desktop icon; it is a background image. The desktop icons are briefly described in Table 2.1.

**TABLE 2.1** Desktop Icons

| Icon | Description |
| --- | --- |
| Midori | A web browser |
| Scratch | A programming language development environment targeted for kids |
| Pi Store | A shortcut to the Raspberry Pi store |
| WiFi Configuration | A wizard that steps through the setup of a wireless network connection |
| Debian Reference | A Debian user guide provided with Raspbian |
| IDLE3 | The Python v3 development environment |
| LXTerminal | A window that provides an interface to the command line |
| OCR Resources | A shortcut to the OCR teaching resources for teachers using Raspberry Pi in the classroom |
| IDLE | The Python v2 development environment |
| Python Games | Games that demonstrate the use of the Python programming language |

The icons on the desktop behave as you would expect them to behave. You can double-click each one with your mouse to open it. You can also right-click an icon to open a drop-down menu and then select Open to open that program.

The LXTerminal icon provides a portal to the command-line interface. You can double-click the LXTerminal icon to start the program. Once the window is open, you can type in exactly the same commands as at the command-line prompt. For example, Figure 2.2 shows what happens when you type the whoami command in the LXTerminal. You can see that LXTerminal allows you to stay in the GUI and yet, enter command-line commands.

DID YOU KNOW

**I Just Want My GUI**

Because you can reach the command line via the LXTerminal program, you might want to have your Raspberry Pi boot straight into the GUI. To set this up, follow these steps:

1. At the command prompt, type sudo raspi-config and press the Enter key.
2. In the text-based menu, press the down-arrow key until you reach the boot_behavior menu option and press the Enter key.

3. When you see a new window and the question `Should we boot straight to the desktop?` press Tab until you reach the `<Yes>` option and then press Enter.

4. At the configuration menu, press Tab until you get to the `<Finish>` option and then press the Enter key.

5. When a new window opens and asks `Would you like to reboot now?`, press Tab until you reach the `<Yes>` option and press Enter. The Raspberry Pi reboots and takes you to the LXDE GUI. You are not required to provide a login name or password.

If you change your mind and want to log in to the command line after the Pi boots again, you can run the LXTerminal program and type in `sudo raspi-config` to change your boot behavior configuration option.

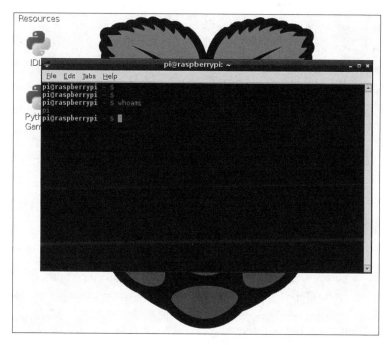

**FIGURE 2.2**
The LXTerminal command-line interface.

## The LXPanel

The bottom panel of the LXDE graphical interface, shown in Figure 2.3, includes several noteworthy items. The bottom panel for the LXDE is called the LXPanel. Its behavior and the various icons on it can be modified through a pop-up menu that you open by right-clicking anywhere on the panel.

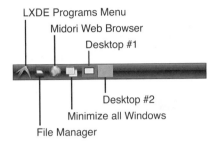

**FIGURE 2.3**
The icons on the left side of the LXDE LXPanel.

The first icon all the way to the left on the LXPanel is the LXDE Programs Menu icon. The unusual icon for this menu is a symbol for a fast flying bird. (The LXDE environment is quick and so lightweight it can fly. Get it?)

When you click this LXDE Programs Menu icon, you get several menu categories and options (see Table 2.2).

**TABLE 2.2   The LXDE Menu**

| Menu Category | Description |
| --- | --- |
| Accessories | Various useful programs such as a calculator and the LXTerminal program |
| Education | Programs such as Scratch and Squeak that are focused on helping kids with programming |
| Graphics | Programs such as Xpdf, a PDF reader |
| Internet | Various web browsers, including Midori |
| Other | An eclectic collection of packages from the other menu categories, as well programs that are not in other menu categories, such as Aptitude Package Manager |
| Programming | Programing development environments, such as IDLE |
| System Tools | Programs to help you manage your system, such as Task Manager |
| Preferences | Programs that allow you to modify the graphical interface environment, such as Customize Look and Feel |
| Run | A program that allows you to run a single command-line command, such as `sudo poweroff` |
| Logout | The LXDE Logout Manager window |

The next LXPanel icon is the File Manager. The File Manager window, shown in Figure 2.4, is similar to the Microsoft Windows File Manager in that it allows you to graphically navigate through your files and folders.

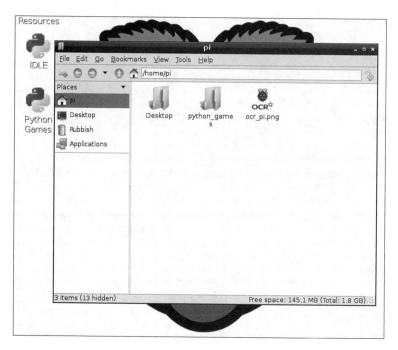

**FIGURE 2.4**
The LXDE File Manager.

After the File Manager icon is an icon to start the Midori web browser. Then comes the Minimize All Windows icon, which is a handy feature! If you have several windows open on your desktop, you can minimize them all to the LXPanel by simply clicking this icon.

The last two icons on the left side of the LXPanel, a light gray rectangle and a light blue rectangle, are the Virtual Desktop icons. The first one is called Desktop 1, and the second one is called Desktop 2. It is like having two virtual monitors. For example, you can open program windows on Desktop 1 and then click the Desktop 2 icon and see no open program windows—just an LXDE desktop. You switch back to Desktop 1 by clicking its icon, and there are your open program windows. Thus, you can have different program windows running on the different desktops and jump back and forth between them. This is a really nice feature!

On the far right of the LXPanel, as shown in Figure 2.5, is the Logout Manager icon. This icon opens the LXDE Logout Manager window, which allows you to log out of the GUI session.

CPU Usage Monitor ——————— Screen Lock

13:05 ——— Logout Manager

Digital Clock

**FIGURE 2.5**
The icons on the right side of the LXDE LXPanel.

BY THE WAY

## The Changing LXDE Logout Manager

Whether you have your Raspberry Pi set up to boot by default into the command line or into the GUI will affect the appearance of the LXDE Logout Manager.

If you have your Pi set up to boot to the command line, the LXDE Logout Manager will have only two buttons: Logout and Cancel. The Logout button will dump you—gracefully, of course—out of the LXDE graphical interface and into the command line. The Cancel button just cancels the logout request and takes you back to the GUI.

If you have your Pi set up to boot to the GUI automatically, then the LXDE Logout Manager has two additional buttons: Shutdown and Reboot. As you would expect, the Shutdown button shuts down the Raspberry Pi, and the Reboot button reboots it.

The next icon to the left of the Logout button on the LXPanel is ScreenLock. ScreenLock allows you to immediately lock your display screen. After you do this, the only way to get back into the system is to enter a username and password. However, by default, Raspbian does not come with a screen saver application installed, so the ScreenLock feature of LXDE does not work. Don't worry, though: You will be installing a screen saver in the Try It Yourself section of this hour.

After ScreenLock is the Digital Clock icon, which displays what your Raspberry Pi thinks is the current time. If you hover the mouse over it, the current date is displayed. You can right-click the Digital Clock icon to see the current month's calendar. You can right-click it again to hide the current month's calendar.

The rectangle to the left of the Digital Clock icon is the CPU Usage Monitor icon. It gives you a nice graphical display of how busy your Raspberry Pi currently is in terms of running programs, opening windows, and so on. If a window is sluggish to open in the GUI, glance over at this graph. You may see that your Pi is very busy!

▼ TRY IT YOURSELF

### Explore the LXDE Graphical Interface

Now that you have reviewed the various icons on the LXDE graphical interface and the features of the LXPanel, it's time to play with the GUI yourself. In the following steps, you will get a chance to try out items in both the command line and the LXDE GUI, as well as fix some potential problems and irritations.

1. If you have not already done so, connect your Raspberry Pi to your network.

WATCH OUT!

### Wired Versus Wi-Fi

In the next few steps, you will update your Raspbian Linux distribution software. When you do this, it is best to use a wired network connection. A Wi-Fi connection can be a little fussy and cause you a great deal of unnecessary work due to software bugs. The safest way to proceed is to connect to a wired network, update your software, and then attempt to connect to your Wi-Fi.

2. Power up your Raspberry Pi.

3. At the `raspberrypi login:` prompt, type `pi` and press the Enter key. You should now see a `Password:` prompt.

4. At the `Password:` prompt, type `raspberry` or the password you created in the last "Try it Yourself" section, and press the Enter key. You should see the `pi@raspberry~$` prompt.

WATCH OUT!

### Did You Change Your Password?

Earlier this hour, you may have changed your password from `raspberry` to something else. If you are following along with the Try It Yourself steps, be sure to enter that password in step 4.

5. At the `pi@raspberry~$` prompt, type `startx` and press the Enter key to start Raspbian's LXDE graphical interface.

6. Once you are in the LXDE graphical interface, click the LXTerminal icon to open a command-line interface. You should see the familiar `pi@rasbperry~$` prompt displayed in the LXTerminal window.

7. Click the LXTerminal window with your mouse to select it. Type `whoami` and press the Enter key. You should see the response `pi` displayed along with another prompt, just as you saw when you were typing in commands at the command line.

8. To get your Raspbian Linux distribution software up to date, in the same LXTerminal window, type the command `sudo apt-get dist-upgrade` and press the Enter key. You should see several messages concerning the software update, and then the question `Do you want to continue [Y/n]?`

9. Type `Y` and press the Enter key. If your software was already up to date, you will get a message similar to "0 upgraded, 0 newly installed..." However, if your software was terribly out of date this may take several minutes! The software update will continue on its merry way, until the software is all updated.

WATCH OUT!

**Problems Fetching Archives**

If your software update ends quickly and you get a message similar to `E: Unable to fetch some archives...`, then your Raspberry Pi is not connected to the network properly or is unable to reach the Internet. In order for the update to work correctly, you must be able to access the Internet from your Raspberry Pi.

10. Now that your system is up to date, you will be adding an extra package to your Raspberry Pi. In order for the ScreenLock on the LXPanel to work correctly, you need a screen saver software package installed. In the LXTerminal window, type `sudo apt-get install xscreensaver` and press the Enter key.

11. You should see several messages concerning the software update and then the question `Do you want to continue [Y/n]?` Type `Y` and then press the Enter key. When you get the prompt back, your screen saver has been installed.

12. Leave the LXTerminal window open for now and click on the LXDE Programs Menu icon on the far left of the LXPanel to open the menu.

13. Hover over Preferences in the LXDE menu to open the submenu and then click on Screensaver. The Screensaver Preferences window appears, as shown in Figure 2.6.

BY THE WAY

**Sluggish Windows**

Don't be surprised if the windows in the GUI open a little slowly. Your Raspberry Pi is working with all its might to get them opened quickly. If a window seems slow, look at the CPU Monitor Graph on the near right side of the LXPanel to see if your Pi is busy processing.

14. If you see another window that says `The XScreenSaver daemon doesn't seem to be running on display "o:" Launch it now?`, click the OK button on that window.

**15.** On the Screensaver Preferences window, make sure the Display Modes tab is selected, as shown in Figure 2.6.

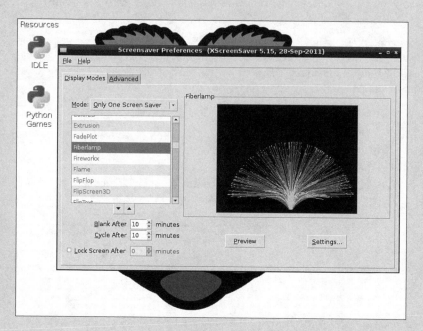

**FIGURE 2.6**
The LXDE Screensaver Preferences window.

**16.** Click the Modes drop-down and select Only One Screen Saver.

**17.** Still in the Screensaver Preferences window, under the Modes section, scroll through the different screen savers until you find Fiberlamp and click it to select it.

**18.** Now click the Preview button in the Screensaver Preferences window and wait a few seconds. You should see the screen saver in action.

**19.** Click anywhere on the screen saver window to return to the LXDE graphical interface.

**20.** Now close the Screensaver Preferences window by clicking the white X in the right-hand corner of the window and give it a few seconds to close.

**21.** Test your screen lock by clicking the ScreenLock icon on the LXPanel. In a few seconds, the screen saver should appear.

**22.** Click anywhere on the screen saver window. This time you don't return to the LXDE graphical interface. Instead, a new window pops up, stating `Please enter your password`.

23. Type in your password and press the Enter key.

WATCH OUT!

**Did You Change Your Password?**

Earlier this hour, you may have changed your password from `raspberry` to something else. If are following along with the Try It Yourself steps in this chapter, remember that you changed the password from `raspberry` to something else earlier in this chapter! Be sure to enter that password here in this step 23.

24. When the LXDE graphical interface appears again, click the LXTerminal window to select it.

25. In the LXTerminal window, type `exit` and press Enter to close the window.

Good work! Now you know how to use the LXDE graphical interface to change various items to your liking.

# Summary

In this hour, you read about the Raspbian Linux distribution. You can now enter commands at the Linux command line as well as navigate through the LXDE GUI. You know about various Debian and Raspbian documentation resources, and you know how to update the software packages on your Raspberry Pi. Now that you have looked around your Pi, in Hour 3, "Setting Up a Programming Environment," you will learn how to set up and explore the Python programming environment.

# Q&A

Q. I don't like entering commands at the Linux command line. Do I have to do this?

A. Nope. The LXDE GUI can handle a lot of the commands you enter at the Linux command line. However, if you know how to use both the command line and the GUI, you will have the most flexibility and troubleshooting capabilities.

Q. Can I install a different graphical interface besides LXDE?

A. Yes, you can! Some Raspberry Pi users prefer the Xfce desktop. See http://www.raspbian.org/RaspbianForums for help on obtaining a new interface.

**Q.** Does this book focus on Python programming in the command line or the GUI?

**A.** The book primarily focuses on teaching you Python programming using the GUI. (You can breathe a sigh of relief now.)

# Workshop

## Quiz

**1.** Raspbian is based on the Debian distribution, with Linux at its core. True or false?

**2.** Which command entered at the Linux command line will reboot your Raspberry Pi?

   **a.** `reboot`

   **b.** `restart`

   **c.** `sudo reboot`

**3.** Which graphical interface desktop environment comes with Raspbian by default?

## Answers

**1.** True. Raspbian is based on the Debian Linux distribution.

**2.** To reboot your Raspberry Pi from the command line, enter the command `sudo reboot`.

**3.** The Lightweight X11 Desktop Environment (LXDE) graphical interface comes with Raspbian by default.

# Setting Up a Programming Environment

**What You'll Learn in This Hour:**

▶ Why learn Python?

▶ How to check your Python environment

▶ The Python interactive shell

▶ Using a Python development environment

▶ How to create and run a Python script

This hour, you will explore the Python programming environment. You will learn about the various tools that can help as you learn how to program in Python. By the end of this hour, you will be familiar with the Python interactive shell and a Python development environment, and you will have written your first line of Python code!

# Exploring Python

You would not be reading this book if you were not interested in learning Python! The Python programming language is an extremely popular language. It is one of the most used programming languages. Python can be used on a wide variety of platforms, such as Windows, Linux-based systems, and Apple OS X. One of its best features is that it's free!

More good news: The Python programming language has easy-to-understand syntax. *Syntax* refers to the Python commands, their proper order in a Python statement, and additional characters, such as a quotation mark ("), needed to make a Python statement work properly. Python's syntax makes it easy for a beginner to start programming quickly. Despite its ease of use, Python contains a lot of rich and powerful features that make it useful for advanced programmers.

## A Little Python History

The Python programming language was invented in the early 1990s by Guido van Rossum. The name Python was based on the popular television show *Monty Python's Flying Circus*.

Through the years, the Python programming language has become extremely popular. It also has gone through some changes.

## Python v3 Versus Python v2

Python recently went from version 2 to version 3. Here are a few of the major differences between the two versions:

▶ Python v3 is based upon Unicode and provides a more predictable handling of it. Unicode is the way a computer encodes, represents, and handles individual characters. Python v2 is based on ASCII, which can handle only English characters. Unicode can handle English characters and non-English characters.

▶ Python v3 is a smaller language than Python v2. A favorite saying of Python developers is "Python fits in your brain." This sentiment is even more true of Python v3 than of Python v2, so it is even easier to learn Python quickly now.

▶ Several changes were made to Python v3 in order to improve its longevity as a programming language. Therefore, the time you spend learning it now will provide you benefits long into the future.

Many systems support both Python v2 and Python v3, including Raspbian. Python v2 is provided for backward compatibility purposes. In other words, you can run Python v2 programs on Raspbian. However, to move you in the right direction, this book focuses on Python v3.

# Checking Your Python Environment

The Raspbian Linux distribution comes with Python v3 and the necessary tools loaded by default. The following Python items are preloaded:

▶ A Python interpreter

▶ An interactive Python shell

▶ A Python development environment

▶ Text editors

Even though all you need should be preloaded, it makes sense to double-check all the tools. These checks take only a few minutes of your time.

## Checking the Python Interpreter and Interactive Shell

To check the Python interpreter and interactive shell versions on your system, open the LXTerminal in the LXDE GUI. Type `python3 -V` and press Enter, as shown in Listing 3.1.

**LISTING 3.1**   Checking the Python Version

```
pi@raspberrypi ~ $ python3 -V
Python 3.2.3
pi@raspberrypi ~ $
```

If you get the message command not found, then for some reason, the Python v3 interpreter is not installed. Go to the "Installing Python and Tools" section in this hour to remedy this situation.

# Checking the Python Development Environment

To see if a Python development environment has been installed, open the LXDE graphical interface and look for the IDLE 3 icon on the desktop, as shown in Figure 3.1.

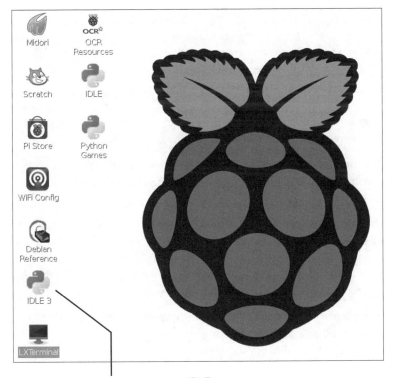

Python v3 development environment, IDLE

**FIGURE 3.1**
The Python v3 development environment, IDLE.

If you do not see the IDLE 3 icon on the desktop, check in the LXDE menu by clicking on the LXDE Programs Menu icon and hovering your mouse pointer over the Programming menu option. You should see the IDLE 3 icon there. If you don't, go to the "Installing Python and Tools" section of this hour to remedy this situation.

## Checking for a Text Editor

Finally, you should ensure that there is a text editor, called nano, installed. You will learn later in this hour exactly what a text editor is. Open the LXTerminal in the LXDE graphical interface. To see if the nano text editor is installed, type nano  -V and press Enter (see Listing 3.2).

**LISTING 3.2**   Checking the nano Version

```
pi@raspberrypi ~ $ nano -V
 GNU nano version 2.2.6 (compiled 16:52:03, Mar 30 2012)
 (C) 1999, 2000, 2001, 2002, 2003, 2004, 2005, 2006, 2007,
2008, 2009 Free Software Foundation, Inc.
 Email: nano@nano-editor.org      Web: http://www.nano-editor.org/
 Compiled options: --disable-wrapping-as-root --enable-color --enable-extra
 --enable-multibuffer --enable-nanorc --enable-utf8
pi@raspberrypi ~ $
```

If you get the message command not found, then for some reason, the nano text editor is not installed. Go to the "Installing Python and Tools" section in this hour to remedy this situation.

Hopefully, you have found nothing missing from your Python environment and all the tools you need are loaded onto your Raspberry Pi. If nothing is missing, you can skip the next section and go straight to "Checking the Keyboard."

# Installing Python and Tools

If you found anything missing from your Python environment, it's not a big problem. In this section, you will get everything you need installed very quickly by following these steps:

**1.** If your Raspberry Pi has a wired connection to the Internet ensure that it is connected to the Internet and boot up your Pi.

**2.** Start the LXDE GUI if it is not started automatically. If your Internet connection is wireless, ensure that it is working.

**3.** Open the LXTerminal. At the command prompt, type sudo apt-get install python3 idle3 nano and press Enter.

BY THE WAY

**But I Don't Need All Those Programs!**

Don't worry if the installation command in step 3 includes items that you already have installed. It will simply get an already installed tool updated, if needed.

You should see several messages concerning the software installs/updates and then the question Do you want to continue [Y/n]? Type Y and then press Enter. When the installs are complete, you see a prompt. You should now go back through the "Checking Your Python Development Environment" section to make sure all is well with your Python environment.

# Checking the Keyboard

If you live and work in the United Kingdom, then most likely you can skip this section. For those of you who live elsewhere, it is highly likely that your keyboard setup is not quite correct.

You probably have been using your keyboard with no problems so far. However, try a little test: Press the @ key on the keyboard. Do you see a double quote (") instead of the @ symbol? If you do, you need to work through this section to get your keyboard set up correctly.

If you have a typical U.S. keyboard, follow these steps to get your keyboard working properly for programming in Python:

1. If it is not already on, boot up your Raspberry Pi and open the LXDE GUI.

2. Double-click the LXTerminal icon to open the LXTerminal window.

3. Type sudo raspi-config and press Enter.

4. In the Raspbi-config window, press the down-arrow key until you get to configure_keyboard and then press Enter. It may take several seconds for the next window to open, so be patient!

5. When the next window says Please select the model of the keyboard for this machine., press Enter to accept the default selection.

6. When the next window says Please select the layout matching the keyboard for this machine, press the up-arrow key to scroll up the menu until you get to English (US). Press Enter.

WATCH OUT!

**The Wrong Keyboard**

If you are using a special keyboard, such as a Dvorak keyboard, the English (US) selection will not work for your keyboard, and you will end up having keys on the keyboard not producing the correct letters. This could prevent you from logging back in to your Raspberry Pi!

If you have a special keyboard, scroll through the selections in this window and pick the one that best matches your needs. If something goes wrong and your keyboard acts funny, don't worry. You can go back to Hour 1, "Setting Up the Raspberry Pi," and put a fresh copy of the Raspbian operating system image onto your SD card to get back to "normal" keyboard operations.

---

7. On the next three screens listed, modify the selections or press Enter to accept the defaults:

   ▶ `Key to function as AltGr` screen

   ▶ `Compost Key` screen

   ▶ `Use Control Alt Backspace` screen

8. In the `Raspbi-config` window, press Tab until you reach the `<Finish>` selection and then press Enter.

9. Because the keyboard changes will not take effect until you reboot your system, type `sudo reboot` in the LXTerminal window and press Enter.

10. After your Raspberry Pi reboots, test your keyboard. See if pressing the @ key now produces the symbol @ and pressing the " key produces a double quote (").

Remember that you can fix any disasters here by going back to Hour 1 and putting a fresh copy of the Raspbian operating system image onto your SD card. Doing so will get you back to "normal" keyboard operations.

# Learning About the Python Interpreter

Python is an interpreted programming language, instead of a compiled one. A compiled programming language has all its program's language statements (commands) turned into binary code at once, before it can be executed (run). With an interpreted programming language, each of its programming statements, one at a time, is checked for syntax errors, translated into binary code, and then executed.

You can learn about a variety of Python statements and concepts by using different tools that fall into three primary categories:

▶ **Interactive shell**—The interactive shell allows you to enter a single Python statement and have it immediately checked for errors and interpreted.

▶ **Development environment shell**—This tool provides many features to assist in the development of Python programs. As with the interactive shell, each Python statement is interpreted as it is entered. However, you can also develop entire Python programs, called scripts.

▶ **Text editors**—A text editor is a program that allows you to create and modify regular text files. A text editor does not format the text for display on a printed page, like a word processor does. Python statements are not interpreted as they are entered in a text editor. This tool only helps you to quickly create a Python script.

BY THE WAY

### Running Python Scripts

After a Python script file is created, it is run either using a command at the command line or via the development environment shell. And even though a script has multiple Python statements in it, each statement is still interpreted one at a time, as it is encountered in the file.

Now that you have had a short introduction to the various Python tools, you can start exploring them in more depth. Learning to use these tools will help you as you learn Python programming.

# Learning About the Python Interactive Shell

The Python interactive shell is primarily used to try out Python statements and check syntax. To enter the interactive Python shell, type in the command `python3` at the command line and press Enter.

BY THE WAY

### The Python v2 Interactive Shell

If you want to try out old Python v2 statements, you can still access the Python v2 interactive shell on Raspbian. Just type in the command `python2` and press Enter.

Figure 3.2 shows the interactive shell. Notice that the top bar of the window displays the Python interpreter's version number. After a little helpful information, the prompt is shown as three greater-than signs, >>>.

At this point, you can simply enter a Python statement and press Enter to have the shell interpret it. The Python interpreter checks the statement's syntax. If the syntax is correct, the statement is translated into binary code and executed.

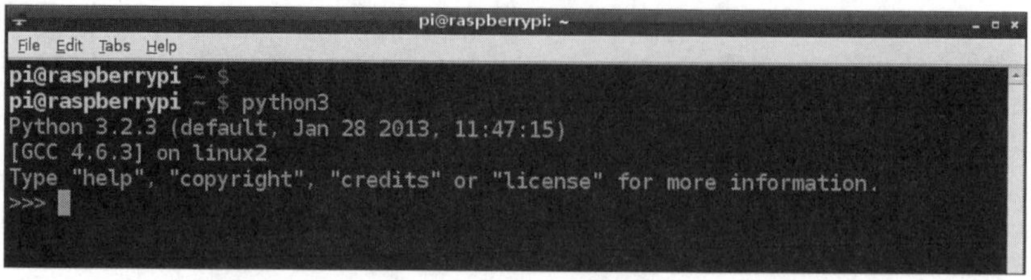

**FIGURE 3.2**
The Python interactive shell.

BY THE WAY

## GUI or Command Line?

The Python interactive shell examples in this hour are shown using the LXTerminal in the LXDE GUI. However, you can use the Python interactive shell at the command line, too.

Figure 3.3 shows a Python statement involving the print function: print ("I love my Raspberry Pi!"). The Python interactive shell interprets, converts, and executes the command and then prints I love my Raspberry Pi! to the screen.

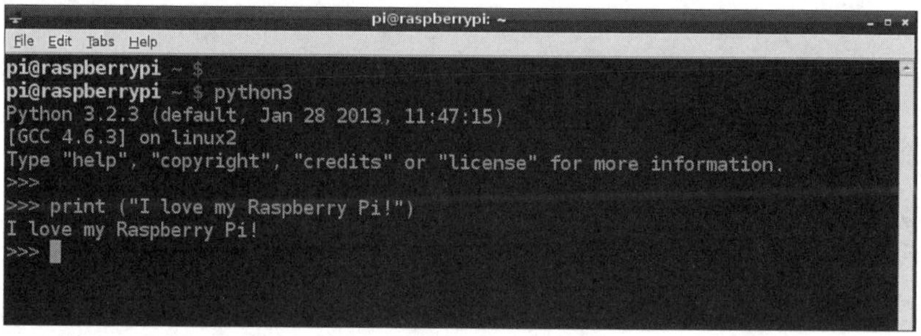

**FIGURE 3.3**
The print function in the Python interactive shell.

To get help using the interactive shell or Python statements, you can type help() and press Enter. Figure 3.4 shows the interactive shell's help utility.

You can type in a Python keyword, module, or topic to get help on it. To exit help on a particular keyword, module, or topic, press the Q key. To exit the interactive shell's help utility, press the Ctrl key and hold it. Then press and release the D key. This combination is written Ctrl+D. Instead, you can type quit and press Enter.

When you are done using the Python interactive shell, simply type exit () and press Enter. Python takes you out of the interactive shell and puts you back to the command line.

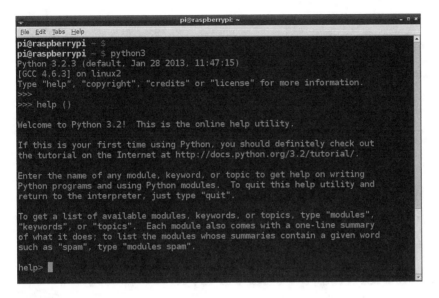

**FIGURE 3.4**
Interactive shell help.

TRY IT YOURSELF ▼

## Explore the Python Interactive Shell

This is your chance to try out the interactive shell yourself! Follow these steps to enter a `print` function statement into the Python interactive shell and then exit from it:

1. If you have not already done so, power up your Raspberry Pi and log in to the system.

2. If you do not have the LXDE GUI started automatically at boot, start it now by typing `startx` and pressing Enter.

3. Open the LXTerminal by double-clicking the LXTerminal icon.

4. At the command-line prompt, type `python3` and press Enter. You are now in the Python interactive shell.

WATCH OUT!

### Take Time

Before you press that Enter key, take time to review your Python statements. It is very easy to leave out a quotation mark or use the wrong case on a command (for example, `Print` instead of `print`). Getting into the habit of reviewing your commands before you press Enter will save you lots of frustration and time later on.

**5.** At the `>>>` prompt, type `print ("This is my first Python statement!")` and press Enter. You should see the shell display `This is my first Python statement!` Pretty cool! You have taken your first small step toward lots of great Python programming.

WATCH OUT!

**My Keyboard Doesn't Work!**

When you press the " (double quote) key, if you get an @ symbol instead, then your keyboard isn't correctly configured. Go back to the "Checking the Keyboard" section earlier in this hour.

**6.** To exit the Python interactive shell, type `exit ()` and press Enter.

# Learning About the Python Development Environment Shell

A development environment shell is a single tool for creating, running, testing, and modifying Python scripts. Often development environments color code key syntax for easier identification of various statement features. This color coding helps with a script's testing, modification, and debugging. Another nice feature is automatic code completion. As you type Python syntax, the development environment provides screen tips to help you complete your code.

In addition to these features, a development environment shell can provide syntax checking so that you can find any incorrect Python syntax without having to run the entire Python script. To maintain consistent indentation within a script, environment tools often provide automatic indentation.

Finally, debugging tools within the environment allow you to step through a Python script to uncover logic errors. What doesn't a development environment shell do? Well, it can't write a Python script for you, but it can help you accomplish that task.

IDLE is the default Python development environment shell installed on Raspbian, and it is the environment that this book focuses on. There are dozens of other Python development environment tools, including the following:

▶ **jEdit**—www.jedit.org

▶ **Komodo IDE**—www.activestate.com/komodo-ide/

▶ **SPE IDE**—pythonide.blogspot.com

# The IDLE Development Environment Shell

IDLE stands for Interactive DeveLopment Environment. This development environment provides a built-in text editor and many features that assist in the creation and testing of Python scripts.

To start up IDLE in the LXDE graphical interface, you just double-click the IDLE 3 icon on the desktop. You can also find it under the LXDE Programs Menu icon. Figure 3.5 shows the IDLE shell for Python v3.

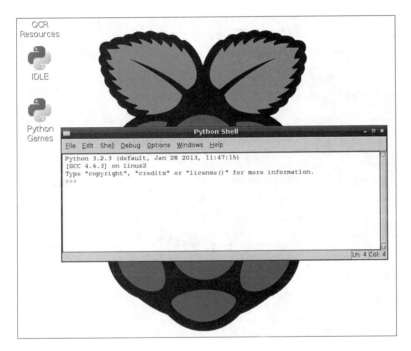

**FIGURE 3.5**
Python v3 IDLE in interactive mode.

The IDLE window's title bar says Python Shell. Notice that this window uses exactly the same verbiage as the Python interactive shell. This is because the IDLE environment uses the Python interactive shell for this development mode, which is called *interactive mode*.

DID YOU KNOW

**IDLE Everywhere**

One of the great things about learning IDLE is that this development environment is not just available on Linux. It is also available on Windows and OS X.

Interactive mode has many features that help in the creation and testing of Python scripts. There are lots of features in IDLE. The following are a few of the most important ones to help you get started in Python programming:

▶ **Menu-driven options and their matching control keys**—For example, to open a new IDLE window, you can click the File menu option and then select New Window from the drop-down menu. To use the control keys to open a new IDLE window, instead of using the menu, you can press the Ctrl+N key sequence.

▶ **Basic text editor**—To type a Python script, you can open up a new window from the main interactive IDLE window to get access to a very basic text editor. The text editor allows you to take such actions as cut and paste text using menu-driven selections or control keys.

▶ **Code completion**—As you type in Python statements, helpful hints appear on the screen, making recommendations on how to finish the syntax you've started.

▶ **Syntax checking**—When you enter a command and press Enter, the Python interpreter checks the syntax of your statement and reports any problems immediately. This is much better than finding out about syntax errors after an entire script is written.

▶ **Color coding**—The IDLE shell color codes syntax as you type it to help you follow the logic of your Python statements. Table 3.1 shows the color codes it uses.

**TABLE 3.1**  IDLE Color Codes

| Color | Python Item |
| --- | --- |
| Red | Comments |
| Orange | Python keywords |
| Green | String literals |
| Blue | Defined names (functions, classes) |
| Violet | Built-in functions |

▶ **Indentation support**—Python requires the use of indentation for some of its constructs. The IDLE shell recognizes these required indentations and automatically provides them for you. (For more information on indentation, see Hour 6, "Controlling Your Program.")

▶ **Debugger features**—The term *debugging* refers to removing incorrect syntax or logic from a program. With IDLE, the Python interpreter's syntax checking typically finds syntax errors. You can uncover logic problems by using the IDLE Debugger, which allows you to step through a program without adding additional Python statements.

▶ **Help**—Because everyone needs a little help, IDLE provides a nice help facility. You can access the help facility by selecting the Help menu option on the menu bar of the IDLE window and clicking IDLE Help in the drop-down menu.

Of course, trying out IDLE's features yourself will help you better learn to use the IDLE tool. The following Try It Yourself gives you an opportunity.

**TRY IT YOURSELF ▼**

## Explore the Python IDLE Tool

In the following steps, you'll try out a couple of the IDLE tool's features. Don't be overwhelmed by all the bells and whistles of this tool. Follow these steps to test the basic features and take a look around the environment:

1. If you have not already done so, power up your Raspberry Pi and log in to the system.

2. If you do not have the LXDE GUI started automatically at boot, start it now by typing `startx` and pressing Enter.

3. Open IDLE for Python v3 by double-clicking the IDLE 3 icon or clicking the LXDE Programs Menu icon; hovering over the Programming menu option; and clicking the IDLE 3 menu option. You are now in the main IDLE interactive mode window.

BY THE WAY

## IDLE 3 Not IDLE

You may notice an IDLE icon and an IDLE menu option alongside the IDLE 3 options. These selections offer the IDLE shell for Python v2. Be sure to select IDLE 3 to stay on course with this hour.

4. In the IDLE 3 window, at the `>>>` prompt, type `print` and then pause and look at the screen. You should notice that the print command has been colored violet. This is because the `print` statement is considered a built-in function in Python. (You will be learning more about the various built-in functions in the coming hours.) The color is provided to help you recognize the syntax of your Python statement and assist in the logic of your scripts. Look back to Table 3.1 for a reminder of the various IDLE color codes.

5. Press the space bar and type `("This is my first Python"` and then pause again and look at the screen. You should notice that the text `This is my first Python` is colored green because Python considers it a string literal. (You will be learning more about string literals, too, in the coming hours. For now, just notice the color.)

6. Instead of correctly finishing your Python statement, just press the Enter key. (You are deliberately trying to generate a syntax error to see how IDLE handles syntactical

problems.) You should get the message `Syntax Error: EOL error while scan-
ning string literal`. This is because you did not correctly close the `print` function.
(Well, actually, you were just following directions.)

7. In the IDLE 3 window, type `print` ( and then pause. You should see a screen tip appear
   in your window, similar to the one shown in Figure 3.6. IDLE attempts to help you by giving
   guidance via screen tips.

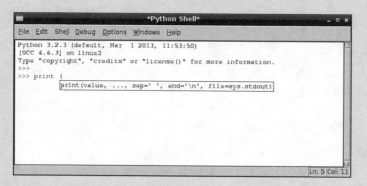

**FIGURE 3.6**
An IDLE script tip.

8. Finish the Python statement by typing `"This is my first Python statement
   in IDLE")`. Look at your Python statement and make sure it reads `print ("This
   is my first Python statement in IDLE")`. If you do not have it correct, then
   modify it by using the left- and right-arrow keys and the Delete key. When you are sure it
   is correct, press Enter. You should see output similar to what is displayed in Figure 3.7.
   Congratulations! You just correctly entered your first Python statement in IDLE.

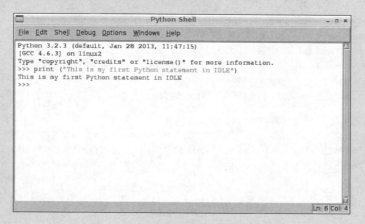

**FIGURE 3.7**
Output from a Python statement in IDLE.

**9.** Finally, exit the IDLE shell by pressing the key combination Ctrl+Q. The IDLE interactive mode window should close.

DID YOU KNOW

**Exiting IDLE**

You can leave the IDLE shell a couple different ways. As you did in step 9, you can use the key combination Ctrl+Q to exit. Also, you can use the menu options in IDLE to leave: To do so, click the File menu and then select Exit. The third way is to enter the Python statement `exit ()`. When you do this, you get a pop-up window titled Kill? that says `The program is still running! Do you want to kill it?` and then you can press the OK button. This last option is a little violent, but it will get you out of IDLE and back to the LXDE GUI.

Now that you've played with IDLE a bit, its basic features should be more useful to you. As your experience with Python grows, you might want to try out some of the IDLE power-user features as well.

BY THE WAY

**More IDLE, Please**

The official Python website maintains an IDLE document that is worth exploring for more info on using IDLE. You can find it at docs.python.org/3/library/idle.html.

# Creating Python Scripts

Instead of typing in each Python statement every time you need to run a program, you can create whole files of Python statements and then run them. These whole files of Python statements are called Python *scripts*.

You can run Python scripts from either the Python interactive shell or from IDLE. Listing 3.3 shows a file called `sample.py` that contains two Python statements.

**LISTING 3.3**  The `sample.py` Python Script

```
pi@raspberrypi ~ $ cat py3prog/sample.py
print ("Here is a sample python script.")
print ("Here is the second line of the sample script.")
pi@raspberrypi ~ $
```

BY THE WAY

### Where Is My samply.py?

You will not find this script, py3prog/sample.py, on your Rapsberry Pi. It was created for this book. Later in this chapter, you will be learning how to create your very own Python scripts.

# Running a Python Script in the Interactive Shell

To run the `sample.py` script in the Python interactive shell, at the command line, type `python3 py3prog/sample.py` and press Enter. Listing 3.4 shows the results you should get. As you can see, the shell runs the two Python statements without any problems.

**LISTING 3.4**  The Execution of `sample.py`

```
pi@raspberrypi ~ $ python3 py3prog/sample.py
Here is a sample python script.
Here is the second line of the sample script.
pi@raspberrypi ~ $
```

BY THE WAY

### Script Storage Location

It is a good idea to store your Python scripts in a standard location. This book uses the subdirectory `/home/pi/py3prog`.

# Running a Python Script in IDLE

To run the sample.py script in IDLE, start IDLE and in the main interactive mode window, either press the key combination Ctrl+O or select the File menu and then Open. The Open window appears. Navigate to the location of the Python script. In this case, `sample.py` is located in `/home/pi/py3prog`, as shown in Figure 3.8. Click the script to select it and then click the Open button.

When you click the Open button, another IDLE window opens, showing the Python script and its directory location and name in the window's title bar (see Figure 3.9).

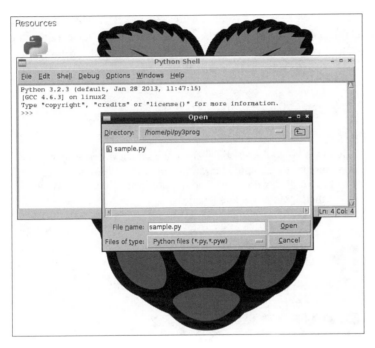

**FIGURE 3.8**
Opening a Python script in IDLE.

**FIGURE 3.9**
A Python script opened in IDLE.

Now, to run the Python script, in the Python script's window, press the F5 key, or click Run in the menu bar and then Run Module. The control switches to the originally opened IDLE window (the IDLE interactive mode window), and the results of the Python script are displayed, as shown in Figure 3.10.

WATCH OUT!

### Where Is My Script Output?

When you first start using IDLE, you might be confused about where the output of the Python script you are running is displayed. Just remember that output is always displayed in IDLE's main interactive mode window. This window has the words "Python Shell" on the title bar. This is true whether you are running a script or entering Python statements one by one.

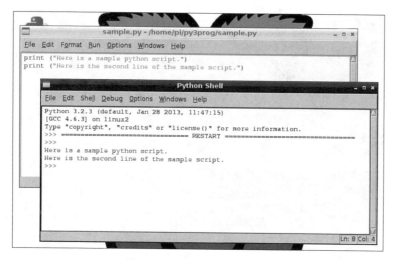

**FIGURE 3.10**
A Python script executed in IDLE.

Now that you have seen two methods for running a Python script, it's time to look at how to create a Python script. You have two methods to choose from here as well.

## Using IDLE to Create a Python Script

Creating a Python script in IDLE is easy. You simply open the IDLE text editor window from the IDLE interactive mode window by clicking Ctrl+N or by clicking the File menu and selecting New Window. You see a new window open, with the word "Untitled" on the title bar. You are now in the basic IDLE text editor. In this mode, when you type in your Python statements, they will not be interpreted, and no output will be displayed.

In the basic IDLE text editor, type in the Python statements to create your script. When you are all done, you save the statements to a file.

DID YOU KNOW

## Editing in IDLE

You are not limited to only using the arrow keys and the Delete key for editing text files. Take a look at all the options available in the Edit menu. You can undo an edit, find words, copy and paste, and so on. The text editor in IDLE may be a basic text editor, but it does offer you a lot of help.

To save the Python script to a file, start by pressing Ctrl+S or by clicking on the File menu and selecting Save. A Save As window appears, as shown in Figure 3.11. Navigate to the directory where you want the file to be stored. Type in the name of the file and click the Save button.

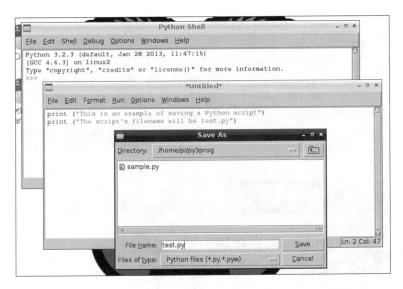

**FIGURE 3.11**
Saving a Python script from IDLE text editor.

DID YOU KNOW

## The "py" in Python Scripts

Notice in Figure 3.11 that the file has a `.py` on the end of it. This file extension identifies files as Python scripts. Thus, all your Python programs should be named something like filename`.py`.

## Using a Text Editor to Create a Python Script

There are other text editors available to you besides the one in IDLE. Two of them are available by default on Raspbian. One is Leaf Pad, which is geared toward school-age children. The other is nano.

The nano text editor is small and lightweight, so it is perfect for the Raspberry Pi. Compared to other more complicated text editors, nano is fairly easy to use. Its biggest advantage over the text editor in IDLE is that nano can be used in both the GUI and at the command line!

To start the nano text editor at the command line, you just type the command nano and press Enter. To start the nano text editor in the GUI, you click on the LXDE Programs Menu icon on the far left of the LXPanel, hover over the Other menu so its submenu is displayed, and then click the Nano menu selection.

Note that the nano text editor does not perform any syntax checking while you type in Python statements. It also does not do any color coding while you type statements. And it does not perform any auto-indentation. nano doesn't give you any handholding when you're creating and editing Python scripts.

Figure 3.12 shows the nano text editor in the GUI. Notice that in the GUI, the LXTerminal is opened, and the nano editor is being used within it. The title bar of the nano editor program window is the line where the left side starts with "GNU nano" and the nano editor version number. In the middle of the title bar are either the words "New Buffer" if you are creating a new file or the name of the file you are editing.

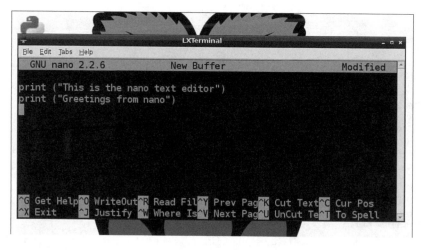

**FIGURE 3.12**
The nano text editor.

The nano editor's middle panel is the editing area. This is where you can add Python statements or make changes to existing ones.

BY THE WAY

**Messages and Questions**

Right above the bottom two lines of the nano editor window is a special messages/questions area. This area is usually blank. However, if nano has a special message or a question, such as `File Name to Write:`, this is where it shows up.

The bottom two lines of the nano editor window show the most commonly used keyboard command sequences. These keyboard sequences are actual nano text editor commands. This window uses the ^ symbol to indicate the Ctrl key. Therefore, the command ^G means use Ctrl+G. Table 3.2 lists some of the basic nano commands.

**TABLE 3.2**   A Few Basic nano Commands

| Key Combination | Command Performed |
|---|---|
| Ctrl+G | Opens the nano help information |
| Ctrl+X | Exits the currently open window |
| Ctrl+O | Saves the current contents to a file |
| Ctrl+F | Opens a file in the nano editor |

If you want to learn more about the nano text editor, you can press Ctrl+G and read through nano's help information. Another great source is the nano editor homepage, at www.nano-editor.org.

# Knowing Which Tool to Use and When

Now that you have looked at the text editor, the Python interactive shell, and IDLE, you might be trying to remember where you run Python scripts or which tool you use to test Python statements. Tables 3.3 through 3.5 help answer those questions and give you a reference point to keep it straight as you work through the next several hours.

**TABLE 3.3**   Testing Python Statements

| Tool | Command or Icon to Start the Tool |
|---|---|
| Python interactive shell | `python3` |
| Python development environment—interactive mode | IDLE 3 |

**TABLE 3.4**   Creating Python Scripts

| Tool | Command or Icon to Start the Tool |
|---|---|
| nano text editor | `nano` |
| Python development environment—text editor mode | IDLE 3, Ctrl+N |

**TABLE 3.5**   Running Python Scripts

| Tool | Command or Icon to Start the Tool |
|---|---|
| Python interactive shell | `python3` filename.py |
| Python development environment—editor mode and interactive mode | IDLE 3, Ctrl+O or F5 |

You should refer to these tables whenever you are not sure which tool to use when. A tool is no help unless you know when to use it!

# Summary

In this hour, you learned about the history of Python, how to ensure that the proper Python tools are installed, and how to make sure your keyboard is set up correctly. You took a first look at how to use tools to test Python statements, and you learned how to create and run Python scripts.

Up to this point in the book, you have been setting up and learning about the Python development environment. Now that hard work is about to pay off. In Hour 4, "Understanding Python Basics," you will be typing in some real Python statements.

# Q&A

**Q.** Do I have to use the nano text editor?

**A.** No. You can use the basic text editor within IDLE rather than nano. You can also try Leaf Pad or install another text editor, such as gedit. If you are really into pain and suffering, you can even install and use the vi/vim text editor.

**Q.** Can I use a word processor to create Python scripts?

**A.** Yes, you can! However, you must save the files you create as plain text files.

**Q.** Do I have to use IDLE?

**A.** No. However, it would be wise to use and learn at least one development environment tool. As you learn the concepts in this book, the scripts and code segments will be fairly small. But when you start writing scripts for yourself, they may get rather large! This is where knowing a development environment will be very helpful.

# Workshop

## Quiz

1. When you save a Python script, the file extension should be `.python`. True or false?

2. Where did the Python programming language get its name?

   **a.** *Monty Python's Flying Circus*

   **b.** The python snake

   **c.** Mount Python in Greece

3. In IDLE interactive mode, what color indicates a string literal?

## Answers

1. False. The file extension for Python scripts is `.py`.

2. The Python programming language got its name from *Monty Python's Flying Circus*.

3. In IDLE interactive mode, the color green indicates a string literal.

# PART II

# Python Fundamentals

# HOUR 4
# Understanding Python Basics

**What You'll Learn in This Hour:**

▶ How to produce output from a script
▶ Making a script readable
▶ How to use variables
▶ Assigning value to variables
▶ Types of data
▶ How to put information into a script

In this hour, you will get a chance to learn some Python basics, such as using the `print` function to display output. You will read about using variables and how to assign them values, and you will gain an understanding of their data types. By the end of the hour, you will know how to get data into a script by using the `input` function, and you will be writing your first Python script!

## Producing Python Script Output

Understanding how to produce output from a Python script is a good starting point for those who are new to the Python programming language. You can get instant feedback on your Python statements from the Python interactive interpreter and gently experiment with proper syntax. The `print` function, which you met in Hour 3, "Setting Up a Programming Environment," is a good place to focus your attention.

## Exploring the `print` Function

A *function* is a group of python statements that are put together as a unit to perform a specific task. You can simply enter a single Python statement to perform a task for you.

BY THE WAY

### The "New" print Function

In Python v2, print is not a function. It became a function when Python v3 was created.

The print function's task is to output items. The "items" to output are correctly called an *argument*. The basic syntax of the print function is as follows:

```
print (argument)
```

DID YOU KNOW

### Standard Library of Functions

The print function is called a *built-in* function because it is part of the Python standard library of functions. You don't need to do anything special to get this function. It is provided for your use when you install Python.

The *argument* portion of the print function can be characters, such as ABC or 123. It can also be values stored in variables. You will learn about variables later in this hour.

## Using Characters as print Function Arguments

To display characters (also called *string literals*) using the print function, you need to enclose the characters in either a set of single quotes or double quotes. Listing 4.1 shows using a pair of single quotes to enclose characters (a sentence) so it can be used as a print function argument.

**LISTING 4.1**   Using a Pair of Single Quotes to Enclose Characters

```
>>> print ('This is an example of using single quotes.')
This is an example of using single quotes.
>>>
```

Listing 4.2 shows the use of double quotes with the print function. You can see that the output that results from both Listing 4.1 and Listing 4.2 does not contain the quotation marks, only the characters.

**LISTING 4.2**   Using a Pair of Double Quotes to Enclose Characters

```
>>> print ("This is an example of using double quotes.")
This is an example of using double quotes.
>>>
```

BY THE WAY

### Choose One Type of Quotes and Stick with It

If you like to use single quotation marks to enclose string literals in a `print` function argument, then consistently use them. If you prefer double quotation marks, then consistently use them. Even though Python doesn't care, it is considered poor form to use single quotes on one `print` function argument and then double quotes on the next. This makes the code hard for humans to read.

Sometimes you need to output a string of characters that contain a single quote to show possession or a contraction. In such a case, you use double quotes around the `print` function argument, as shown in Listing 4.3.

**LISTING 4.3**    Protecting a Single Quote with Double Quotes

```
>>> print ("This example protects the output's single quote.")
This example protects the output's single quote.
>>>
```

At other times, you need to output a string of characters that contain double quotes, such as for a quotation. Listing 4.4 shows an example of protecting a quote, using single quotes in the argument.

**LISTING 4.4**    Protecting a Double Quote with Single Quotes

```
>>> print ('I said, "I need to protect my quotation!" and did so.')
I said, "I need to protect my quotation!" and did so.
>>>
```

DID YOU KNOW

### Protecting Single Quotes with Single Quotes

You can also embed single quotes within single quote marks and double quotes within double quote marks. However, when you do, you need to use something called an "escape sequence," which is covered later in this hour.

## Formatting Output with the `print` Function

You can perform various output formatting features by using the `print` function. For example, you can insert a single blank line by using the `print` function with no arguments, like this:

```
print ()
```

The screen in Figure 4.1 shows a short Python script that inserts a blank line between two other lines of output.

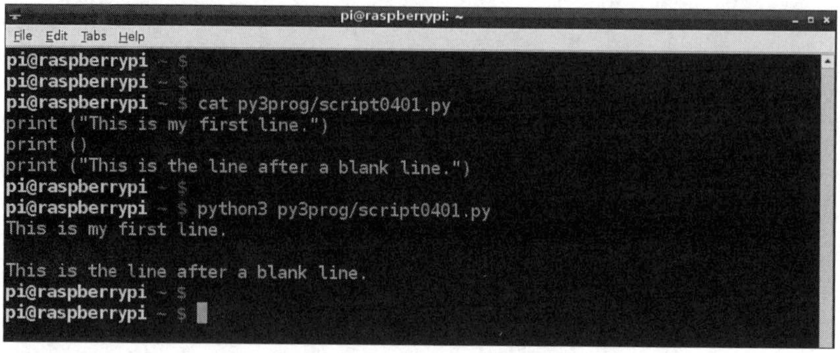

**FIGURE 4.1**
Adding a blank line in script output.

Another way to format output using the `print` function is via triple quotes. Triple quotes are simply three sets of double quotes.

Listing 4.5 shows how you can use triple quotes to embed a linefeed character by pressing the Enter key. When the output is displayed, each embedded linefeed character causes the next sentence to appear on the next line. Thus, linefeed moves your output to the next new line. Notice that you cannot see the linefeed character embedded on each line in the code; you can only see its effect in the output.

**LISTING 4.5**   Using Triple Quotes

```
>>> print ("""This is line one.
... This is line two.
... This is line three.""")
This is line one.
This is line two.
This is line three.
>>>
```

BY THE WAY

### But I Prefer Single Quotes

Triple quotes don't have to be three sets of double quotes. You can use three sets of single quotes instead to get the same result!

By using triple quotes, you can also protect single and double quotes that you need to be displayed in the output. Listing 4.6 shows the use of triple quotes to protect both single and double quotes in the same character string.

**LISTING 4.6   Using Triple Quotes to Protect Single and Double Quotes**

```
>>> print ("""Raz said, "I didn't know about triple quotes!" and laughed.""")
Raz said, "I didn't know about triple quotes!" and laughed.
>>>
```

# Controlling Output with Escape Sequences

An *escape sequence* is a series of characters that allow a Python statement to "escape" from normal behavior. The new behavior can be the addition of special formatting for the output or the protection of characters typically used in syntax. Escape sequences all begin with the backslash (\\) character.

An example of using an escape sequence to add special formatting for output is the \\n escape sequence. The \\n escape sequence forces any characters listed after it onto the next line of displayed output. This is called a *newline*, and the formatting character it inserts is a linefeed. Listing 4.7 shows an example of using the \\n escape sequence to insert a linefeed. Notice that it causes the output to be formatted exactly as it was Listing 4.5, with triple quotes.

**LISTING 4.7   Using an Escape Sequence to Add a Linefeed**

```
>>> print ("This is line one.\nThis is line two.\nAnd this is line three.")
This is line one.
This is line two.
And this is line three.
>>>
```

Typically, the print function puts a linefeed only at the end of displayed output. However, the print function in Listing 4.7 is forced to "escape" its normal formatting behavior because of the addition of the \\n escape sequence.

DID YOU KNOW

## Quotes and Escape Sequences

Escape sequences work whether you use single quotes, double quotes, or triple quotes to surround your print function argument.

You can also use escape sequences to protect various characters used in syntax. Listing 4.8 shows the backslash (\) character used to protect a single quote so that it will not be used in the `print` function's syntax. Instead, the quote is displayed in the output.

**LISTING 4.8**    Using an Escape Sequence to Protect Quotes

```
>>> print ('Use backslash, so the single quote isn\'t noticed.')
Use backslash, so the single quote isn't noticed.
>>>
```

You can use many different escape sequences in your Python scripts. Table 4.1 shows a few of the available sequences.

**TABLE 4.1**    A Few Python Escape Sequences

| Escape Sequence | Description |
| --- | --- |
| \' | Displays a single quote in output. |
| \" | Displays a double quote in output. |
| \\ | Displays a single backslash in output. |
| \a | Produces a "bell" sound with output. |
| \f | Inserts a formfeed into the output. |
| \n | Inserts a linefeed into the output. |
| \t | Inserts a horizontal tab into the output. |
| \u#### | Displays the Unicode character denoted by the character's four hexadecimal digits (####). |

Notice in Table 4.1 that not only can you insert formatting into your output, you can produce sound as well! Another interesting escape sequence involves displaying Unicode characters in your output.

# Now for Something Fun!

Thanks to the Unicode escape sequence, you can print all kinds of characters in your output. You learned a little about Unicode in Hour 3. You can display Unicode characters by using the \u escape sequence. Each Unicode character is represented by a hexadecimal number. You can find these hexadecimal numbers at www.unicode.org/charts. There are lots of Unicode characters!

The hexadecimal number for the pi (∏) symbol is 03c0. To display this symbol using the Unicode escape sequence, you must precede the number with \u in your `print` function argument. Listing 4.9 displays the pi symbol to output.

**LISTING 4.9**   Using a Unicode Escape Sequence

```
>>> print ("I love my Raspberry \u03c0!")
I love my Raspberry π!
>>>
```

## Create Output with the `print` Function

This hour you have been reading about creating and formatting output by using the `print` function. Now it is your turn to try out this versatile Python tool. Follow these steps:

1. If you have not already done so, power up your Raspberry Pi and log in to the system.

2. If you do not have the LXDE GUI started automatically at boot, start it now by typing `startx` and pressing Enter.

3. Open the LXTerminal by double-clicking the LXTerminal icon.

4. At the command-line prompt, type `python3` and press Enter. You are taken to the Python interactive shell, where you can type Python statements and see immediate results.

5. At the Python interactive shell prompt (`>>>`), type `print ('I learned about the print function.')` and press Enter.

6. At the prompt, type `print ('I learned about single quotes.')` and press Enter.

7. At the prompt, type `print ("Double quotes can also be used.")` and press Enter.

BY THE WAY

### Multiple Lines with Triple Double Quotes

In steps 8 through 10, you will not be completing the `print` function on one line. Instead, you will be using triple double quotes to allow multiple lines to be entered and displayed.

8. At the prompt, type `print ("""I learned about things like...` and press Enter.

9. Type `triple quotes`, and press Enter.

10. Type `and displaying text on multiple lines.""")` and press Enter. Notice that the Python interactive shell did not output the Python print statement's argument until you had fully completed it with the closing parenthesis.

11. At the prompt, type `print ('Single quotes protect "double quotes" in output.')` and press Enter.

**12.** At the prompt, type `print ("Double quotes protect 'single quotes' in output.")` and press Enter.

**13.** At the prompt, type `print ("A backslash protects \"double quotes\" in output.")` and press Enter.

**14.** At the prompt, type `print ('A backslash protects \'single quotes\' in output.')` and press Enter. Using the backslash to protect either single or double quotes will allow you to maintain your chosen method of consistently using single (or double quotes) around your print function argument.

**15.** At the prompt, type print `("The backslash character \\ is an escape character.")` and press Enter.

**16.** At the prompt, type `print ("Use escape sequences to \n insert a linefeed.")` and press Enter. Notice how part of the sentence, "Use escape sequences to," is on one line and the end of the sentence "insert a linefeed." is on another line. This is due to your insertion of the escape sequence \n in the middle of the sentence.

**17.** At the prompt, type `print ("Use escape sequences to \t\t insert two tabs or")` and press Enter.

**18.** At the prompt, type `print ("insert a check mark: \u2714")` and press Enter.

You can do a lot with the print function to display and format output! In fact, you could spend this entire hour just playing with output formatting. However, there are additional important Python basics you need to learn, such as formatting scripts for readability.

# Formatting Scripts for Readability

Just as the development environment, IDLE, will help you as your Python scripts get larger, a few minor practices will also be helpful to you. Learn these tips early on, so they become habits as your Python skills grow (and as the length of your scripts grow!).

## Long Print Lines

Occasionally you will have to display a very long line of output using the `print` function. It may be a paragraph of instructions you have to provide to your script user. The problem with long output lines is that they make your script code hard to read and the logic behind the script harder to follow. Python is supposed to "fit in your brain." The habit of breaking up long output lines will help you meet that goal. There are a couple of ways you can accomplish this.

## A Script User?

You may be one of those people who have never heard the term "user" in association with computers. A *user* is a person who is using the computer or running the script. Sometimes the term "end user" is used instead. You should always keep the "user" in mind when you write your scripts, even if the "user" is just you!

The first way to break up a long output line of characters, is to use something called string concatenation. *String concatenation* takes two or more strings of text and "glues" them together, so they become one string of text. The "glue" in this method is the plus (+) symbol. However, to get this to work properly, you also need to use the backslash (\) to escape out of the normal print function behavior of putting a linefeed at the end of a string of characters. Thus, the two items you need are +\, as shown in Listing 4.10.

**LISTING 4.10**    String Concatenation for Long Text Lines

```
>>> print ("This is a really long line of text " +\
... "that I need to display!")
This is a really long line of text that I need to display!
>>>
```

As you can see in Listing 4.10, the two strings are concatenated and displayed as one string in the output. However, there is an even simpler and cleaner method of accomplishing this!

You can forgo the +\ and simply keep each character string in its own sets of quotation marks. The characters strings will be automatically concatenated by the print function! The print function handles this perfectly and it is a lot cleaner looking. This method is demonstrated in Listing 4.11.

**LISTING 4.11**    Combining for Long Text Lines

```
>>> print ("This is a really long line of text "
... "that I need to display!")
This is a really long line of text that I need to display!
>>>
```

It is always a good rule to keep your Python syntax simple to provide better readability of the scripts. However, sometimes you need to use complex syntax. This is where comments will help you. No, not comments spoken aloud, like "I think this syntax is complicated!" We're talking about comments that are embedded in your Python script.

## Creating Comments

In scripts, *comments* are notes from the Python script author. A comment's purpose is to provide understanding of the script's syntax and logic. The Python interpreter ignores any comments. However, comments are invaluable to humans who need to modify or debug scripts.

To add a comment to a script, you precede it with the pound or hash symbol (#). The Python interpreter ignores anything that follows the hash symbol.

For example, when you write a Python script, it is a good idea to insert comments that include your name, when you wrote the script, and the script's purpose. Figure 4.2 shows an example. Some script writers believe in putting these type of comments at the top of their scripts, while others put them at the bottom. At the very least, if you include a comment with your name as the author in your script, when the script is shared with others, you will get credit for its writing.

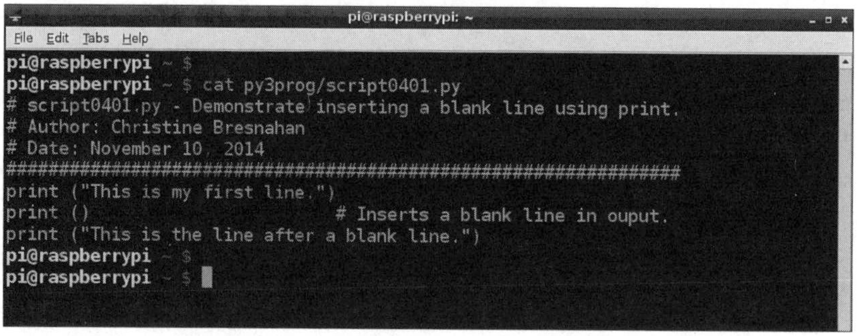

**FIGURE 4.2**
Comments in a Python script.

You can also provide clarity by breaking up sections of your scripts using long lines of the # symbol. Figure 4.2 shows a long line of hash symbols used to separate the comment section from the main body of the script.

Finally, you can put comments at the end of a Python statement. Notice in Figure 4.2 that the `print ()` statement is followed by the comment # `Inserts a blank line in output`. A comment placed at the end of a statement is called an *end comment*, and it provides clarity about that particular line of code.

Those few simple tips will really help you improve the readability of your code. Putting these tips into practice will save you lots of time as you write and modify Python scripts.

# Understanding Python Variables

A *variable* is a name that stores a value for later use in a script. A variable is like a coffee cup. A coffee cup typically holds coffee, of course! But a coffee cup can also hold tea, water, milk, rocks, gravel, sand...you get the picture. Think of a variable as a "holder of objects" that you can look at and use in your Python scripts.

BY THE WAY

### An Object Reference

Python really doesn't have variables! Instead, they are "object references." However, for now, just think of them as variables.

When you name your coffee cup...err, variable...you need to be aware that Python variable names are case sensitive. For example, the variables named `CoffeeCup` and `coffeecup` are two different variables. There are other rules associated with creating Python variable names:

▶ You cannot use a Python keyword as a variable name.

▶ The first character of a variable name cannot be a number.

▶ There are no spaces allowed in a variable name.

# Python Keywords

The list of Python keywords changes every so often. Therefore, it is a good idea to take a look at the current list of keywords before you start creating variable names. To look at the keywords, you need to use a function that is part of the standard library. However, this function is not built -in, like the `print` function is built -in. You have this function on your Raspbian system, but before you can use it, you need to `import` the function into Python. The function's name is `keyword`. Listing 4.12 shows you how to import into Python and determine keywords.

### LISTING 4.12   Determining Python Keywords

```
>>> import keyword
>>> print (keyword.kwlist)
 ['False', 'None', 'True', 'and', 'as',
'assert', 'break', 'class', 'continue',
'def', 'del', 'elif', 'else', 'except',
'finally', 'for', 'from', 'global', 'if',
'import', 'in', 'is', 'lambda', 'nonlocal',
'not', 'or', 'pass', 'raise', 'return',
'try', 'while', 'with', 'yield']
>>>
```

In Listing 4.12, the command `import keyword` brings the `keyword` function into the Python interpreter so it can be used. Then the statement `print (keyword.kwlist)` uses the `keyword` and `print` functions to display the current list of Python keywords. These keywords cannot be used as Python variable names.

## Creating Python Variable Names

For the first character in your Python variable name, you must not use a number. The first character in the variable name can be any of the following:

▶ A letter a through z

▶ A letter A through Z

▶ The underscore character (_)

After the first character in a variable name, the other characters can be any of the following:

▶ The numbers 0 through 9

▶ The letters a through z

▶ The letters A through Z

▶ The underscore character (_)

DID YOU KNOW

### Using Underscore for Spaces

Because you cannot use spaces in a variable's name, it is a good idea to use underscores in their place, to make your variable names readable. For example, instead of creating a variable name like `coffeecup`, use the variable name `coffee_cup`.

After you determine a name for a variable, you still cannot use it. A variable must have a value assigned to it before it can be used in a Python script.

## Assigning Value to Python Variables

Assigning a value to a Python variable is fairly straightforward. You put the variable name first, then an equal sign (=), and finish up with the value you are assigning to the variable. This is the syntax:

```
variable = value
```

Listing 4.13 creates the variable `coffee_cup` and assigns a value to it.

**LISTING 4.13**   Assigning a Value to a Python Variable

```
>>> coffee_cup = 'coffee'
>>> print (coffee_cup)
coffee
>>>
```

As you see in Listing 4.13, the `print` function can output the value of the variable without any quotation marks around it. You can take output a step further by putting a string and a variable together as two `print` function arguments. The `print` function knows they are two separate arguments because they are separated by a comma (,), as shown in Listing 4.14.

**LISTING 4.14**   Displaying Text and a Variable

```
>>> print ("My coffee cup is full of", coffee_cup)
My coffee cup is full of coffee
>>>
```

# Formatting Variable and String Output

Using variables adds additional formatting issues. For example, the `print` function automatically inserts a space whenever it encounters a comma in a statement. This is why you do not need to add a space at the end of `My coffee cup is full of`, as shown in Listing 4.14. There may be times, however, when you want something else besides a space to separate a string of characters from a variable in the output. In such a case, you can use a separator in your statement. Listing 4.15 uses the `sep` separator to place an asterisk (*) in the output instead of a space.

**LISTING 4.15**   Using Separators in Output

```
>>> coffee_cup = 'coffee'
>>> print ("I love my", coffee_cup, "!", sep='*')
I love my*coffee*!
>>>
```

Notice you can also put variables in between various strings in your `print` statements. In Listing 4.15, four arguments are given to the `print` function:

▶ The string `"I love my"`

▶ The variable `coffee_cup`

► The string `"!"`

► The separator designation `'*'`

The variable `coffee_cup` is between two strings. Thus, you get two asterisks (*), one between each argument to the `print` function. Mixing strings and variables in the `print` function gives you a lot of flexibility in your script's output.

## Avoiding Unassigned Variables

You cannot use a variable until you have assigned a value to it. A variable is created when it is assigned a value and not before. Listing 4.16 shows an example of this.

**LISTING 4.16**   Behavior of an Unassigned Variable

```
>>> print (glass)
Traceback (most recent call last):
File "<stdin>", line 1, in <module> Name
Error: name 'glass' is not defined
>>>
>>> glass = 'water'
>>> print (glass)
water
>>>
```

## Assigning Long String Values to Variables

If you need to assign a long string value to a variable, you can break it up onto multiple lines by using a couple methods. Earlier in the hour, in the "Formatting Scripts for Readability" section, you looked at using the `print` function with multiple lines of outputted text. The concept here is very similar.

The first method involves using string concatenation (+) to put the strings together and an escape character (\) to keep a linefeed from being inserted. You can see in Listing 4.17 that two long lines of text were concatenated together in the assignment of the variable `long_string`.

**LISTING 4.17**   Concatenating Text in Variable Assignment

```
>>> long_string = "This is a really long line of text" +\
... " that I need to display!"
>>> print (long_string)
This is a really long line of text that I need to display!
>>>
```

Another method is to use parentheses to enclose your variable's value. Listing 4.18 eliminates the +\ and uses parentheses on either side of the entire long string in order to make it into one long string of characters.

**LISTING 4.18**  Combining Text in Variable Assignment

```
>>> long_string = ("This is a really long line of text"
... " that I need to display!")
>>> print (long_string)
This is a really long line of text that I need to display!
>>>
```

The method used in Listing 4.18 is a much cleaner method. It also helps improve the readability of the script.

BY THE WAY

**Assigning Short Strings to Variables**

You can use parentheses for assigning short strings to variables, too! This is especially useful if it helps you improve the readability of your Python script.

# More Variable Assignments

The value of a variable does not have to only be a string of characters; it can also be a number. In Listing 4.19, the number of cups consumed of a particular beverage are assigned to the variable cups_consumed.

**LISTING 4.19**  Assigning a Numeric Value to a Variable

```
>>> coffee_cup = 'coffee'
>>> cups_consumed = 3
>>> print ("I had", cups_consumed, "cups of",
... coffee_cup, "today!")
I had 3 cups of coffee today!
>>>
```

You can also assign the result of an expression to a variable. The equation 3+1 is completed in Listing 4.20, and then the value 4 is assigned to the variable cups_consumed.

**LISTING 4.20**   Assigning an Expression Result to a Variable

```
>>> coffee_cup = 'coffee'
>>> cups_consumed = 3 + 1
>>> print ("I had", cups_consumed, "cups of",
... coffee_cup, "today!")
I had 4 cups of coffee today!
>>>
```

You will learn more about performing mathematical operations in Python scripts in Hour 5, "Using Arithmetic in Your Programs."

## Reassigning Values to a Variable

After you assign a value to a variable, the variable is not stuck with that value. It can be reassigned. Variables are called *variables* because their values can be varied. (Say that three times fast.)

In Listing 4.21, the variable `coffee_cup` has its value changed from `coffee` to `tea`. To reassign a value, you simply enter the assignment syntax with a new value at the end of it.

**LISTING 4.21**   Reassigning a Variable

```
>>> coffee_cup = 'coffee'
>>> print ("My cup is full of", coffee_cup)
My cup is full of coffee
>>> coffee_cup = 'tea'
>>> print ("My cup is full of", coffee_cup)
My cup is full of tea
>>>
```

DID YOU KNOW

### Variable Name Case

Python script writers tend to use all lowercase letters in the names of variable whose values might change, such as `coffee_cup`. For variable names that are never reassigned values, all uppercase letters are used (for example, `PI = 3.14159`). The unchanging variables are called *symbolic constants*.

# Learning About Python Data Types

When a variable is created by an assignment such as *variable = value*, Python determines and assigns a data type to the variable. A *data type* defines how the variable is stored and the rules governing how the data can be manipulated. Python uses the value assigned to the variable to determine its type.

So far, this hour has focused on strings of characters. When the Python statement `coffee_cup = 'tea'` was entered, Python saw the characters in quotation marks and determined the variable `coffee_cup` to be a *string literal* data type, or `str`. Table 4.2 lists a few of the basic data types Python assigns to variables.

**TABLE 4.2**  Python Basic Data Types

| Data Type | Description |
| --- | --- |
| float | Floating-point number |
| int | Integer |
| long | Long integer |
| str | Character string or string literal |

You can determine what data type Python has assigned to a variable by using the `type` function. In Listing 4.22, you can see that the variables have been assigned two different data types.

**LISTING 4.22**  Assigned Data Types for Variables

```
>>> coffee_cup = 'coffee'
>>> type (coffee_cup)
<class 'str'>
>>> cups_consumed = 3
>>> type (cups_consumed)
<class 'int'>
>>>
```

Python assigned the data type `str` to the variable `coffee_cup` because it saw a string of characters between quotation marks. However, for the `cups_consumed` variable, Python saw a whole number, and thus it assigned it the integer data type, `int`.

DID YOU KNOW

**The `print` Function and Data Types**

The `print` function assigns to its arguments the string literal data type, `str`. It does this for anything that is given as an argument, such as quoted characters, numbers, variables values, and so on. Thus, you can mix data types in your `print` function argument. The `print` function will just convert everything to a string literal data type and spit it out to the display.

Making a small change in the `cups_consumed` variable assignment statement causes Python to change its data type. In Listing 4.23, the number assigned to `cups_consumed` is reassigned from 3 to 3.5. This causes Python to reassign the data type to `cups_consumed` from `int` to `float`.

**LISTING 4.23**   Changed Data Types for Variables

```
>>> cups_consumed = 3
>>> type (cups_consumed)
<class 'int'>
>>> cups_consumed = 3.5
>>> type (cups_consumed)
<class 'float'>
>>>
```

You can see that Python does a lot of the "dirty work" for you. This is one of the many reasons Python is so popular.

# Allowing Python Script Input

There will be times that you need a script user to provide data into your script from the keyboard. In order to accomplish this task, Python provides the `input` function. The `input` function is a built-in function and has the following syntax:

*variable = input (user prompt)*

In Listing 4.24, the variable `cups_consumed` is assigned the value returned by the `input` function. The script user is prompted to provide this information. The prompt provided to the user is designated in the `input` function as an argument. The script user inputs an answer and presses the Enter key. This action causes the `input` function to assign the answer 3 as a value to the variable `cups_consumed`.

**LISTING 4.24**   Variable Assignment via Script Input

```
>>> cups_consumed = input("How many cups did you drink? ")
How many cups did you drink? 3
>>> print ("You drank", cups_consumed, "cups!")
You drank 3 cups!
>>>
```

For the user prompt, you can enclose the prompt's string characters in either single or double quotes. The prompt is shown enclosed in double quotes in Listing 4.24's `input` function.

BY THE WAY

### Be Nice to Your Script User

Be nice to the user of your script, even if it is just yourself. It is no fun typing in an answer that is "squished" up against the prompt. Add a space at the end of each prompt to give the end user a little breathing room for prompt answers. Notice in the input function in Listing 4.24 that there is a space added between the question mark (?) and the enclosing double quotes.

The input function treats all input as strings. This is different from how Python handles other variable assignments. Remember that if cups_consumed = 3 were in your Python script, it would be assigned the data type integer, int. When using the input function, as shown in Listing 4.25, the data type is set to string, str.

**LISTING 4.25**  Data Type Assignments via Input

```
>>> cups_consumed = 3
>>> type (cups_consumed)
<class 'int'>
>>> cups_consumed = input("How many cups did you drink? ")
How many cups did you drink? 3
>>> type (cups_consumed)
<class 'str'>
>>>
```

To convert variables which are input from the keyboard, from strings, you can use the int function. The int function will convert a number from a string data type to an integer data type. You can use the float function to convert a number from a string to a floating-point data type. Listing 4.26 shows how to convert the variable cups_consumed to an integer data type.

**LISTING 4.26**  Data Type Conversion via the int Function

```
>>> cups_consumed = input ("How many cups did you drink? ")
How many cups did you drink? 3
>>> type (cups_consumed)
<class 'str'>
>>> cups_consumed = int(cups_consumed)
>>> type (cups_consumed)
<class 'int'>
>>>
```

You can get really tricky here and use a nested function. *Nested functions* are functions within functions. The general format, is as follows:

```
variable = functionA(functionB(user_prompt))
```

Listing 4.27 uses this method to properly change the input data type from a string to an integer.

**LISTING 4.27**   Using Nested Functions with `input`

```
>>> cups_consumed = int(input("How many cups did you drink? "))
How many cups did you drink? 3
>>> type (cups_consumed)
<class 'int'>
>>>
```

Using nested functions makes a Python script more concise. However, the trade-off is that the script is a little harder to read.

▼ TRY IT YOURSELF

**Explore Python Input and Output with Variables**

You are now going to explore Python input and output using variables. In the following steps, you will write a script to play with, instead of using the interactive Python shell:

1. If you have not already done so, power up your Raspberry Pi and log in to the system.

2. If you do not have the LXDE GUI started automatically at boot, start it now by typing `startx` and pressing Enter.

3. Open the LXTerminal by double-clicking the LXTerminal icon.

4. At the command-line prompt, type `nano py3prog/script0402.py` and press Enter. The command puts you into the nano text editor and creates the file `py3prog/script0402.py`.

5. Type the following code into the nano editor window, pressing Enter at the end of each line:

```
# script0402.py - My first real Python script.
# Written by <your name here>
# Date: <today's date>
#
########### Define Variables ##########
#
amount = 4              #Number of vessels.
vessels = 'glasses'    #Type of vessels used.
liquid = 'water'       #What is contained in the vessels.
location = 'on the table' #Location of vessels.
#
########### Output Variable Description ################
#
```

```
print ()
#
print ("The variables are as follows:")
#
print ("name: amount", "data type:", type (amount), "value:", amount)
#
print ("name: vessels", "data type:", type (vessels), "value:", vessels)
#
print ("name: liquid", "data type:", type (liquid), "value:", liquid)
#
print ("name: location", "data type:", type (location), "value:", location)
print ()
#
########### Output Sentence Using Variables #############
#
print ("There are", amount, vessels, "full of", liquid, location, end='.\n')
print ()
```

BY THE WAY

## Be Careful!

Be sure to take your time here and avoid making typographical errors. Double-check and make sure you have entered the code into the nano text editor window as shown above. You can make corrections by using the Delete key and the up- and down-arrow keys.

6. Write out the information you just typed in the text editor to the script by pressing Ctrl+O. The script file name will show along with the prompt `File name to write`. Press Enter to write out the contents to the `script0402.py` script.

7. Exit the nano text editor by pressing Ctrl+X.

8. Type `python3 py3prog/script0402.py` and press Enter to run the script. If you encounter any errors, note them so you can fix them in the next step. You should see output like the output shown in Figure 4.3. The output is okay, but it's a little sloppy. You can clean it up in the next step.

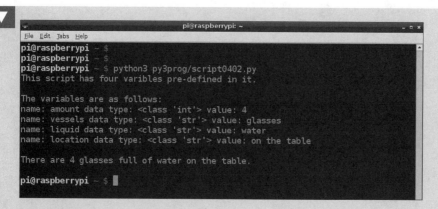

**FIGURE 4.3**
Output for the Python script `script0402.py`.

9. At the command-line prompt, type `nano py3prog/script0402.py` and press Enter. The command puts you into the nano text editor, where you can modify the `script0402.py` script.

10. Go to the `Output Variable Description` portion of the script and add a separator to the end of each line. The lines of code to be changed and how they should look when you are done are shown here:

```
print ("name: amount", "data type:", type (amount), "value:", amount, sep='\t')
#
print ("name: vessels", "data type:", type (vessels), "value:", vessels,
➥sep='\t')
#
print ("name: liquid", "data type:", type (liquid), "value:", liquid, sep='\t')
#
print ("name: location", "data type:", type (location), "value:",
➥location,sep='\t')
```

11. Write out the modified script by pressing Ctrl+O. Press Enter to write out the contents to the `script0402.py` script.

12. Exit the nano text editor by pressing Ctrl+X.

13. Type `python3 py3prog/script0402.py` and press Enter to run the script. You should see output like the output shown in Figure 4.4. Much neater!

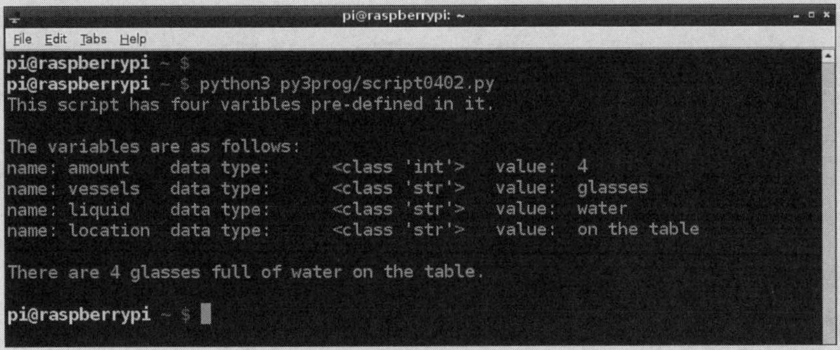

**FIGURE 4.4**
The `script0402.py` output, properly tabbed.

**14.** To try adding some input into your script, at the command-line prompt, type `nano`
   `py3prog/script0402.py` and press Enter.

**15.** Go to the bottom of the script and add the additional Python statements shown here:

```
#
################## Get Input ####################
#
print ()
print ("Now you may change the variables' values.")
print ()
#
amount=int(input("How many vessels are there? "))
print ()
#
vessels = input("What type of vessels are being used? ")
print ()
#
liquid = input("What type of liquid is in the vessel? ")
print ()
#
location=input("Where are the vessels located? ")
print ()
#
################ Display New Input to Output ##########
#
print ("So you believe there are", amount, vessels, "of", liquid, location,
➥end='. \n')
print ()
#
################## End of Script ####################
```

**16.** Write out the modified script by pressing Ctrl+O. Press Enter to write out the contents to the `script0402.py` script.

**17.** Exit the nano text editor by pressing Ctrl+X.

**18.** Type `python3 py3prog/script0402.py` and press Enter to run the script. Answer the prompts any way you want. (You are supposed to be having fun here!) Figure 4.5 shows what your output should look like.

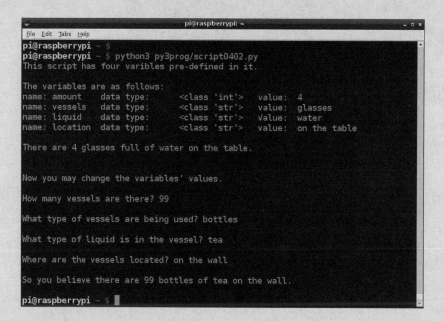

**FIGURE 4.5**
The complete `script0402.py` output.

Run this script as many times as you want. Experiment with the various types of answers you put in and see what the results are. Also try making some minor modifications to the script and see what happens. Experimenting and playing with your Python script will enhance your learning.

# Summary

In this hour, you got a wonderful overview of Python basics. You learned about output and formatting output from Python; creating legal variable names and assigning values to variables; and various data types and when they are assigned by Python. You explored how Python can handle input from the keyboard and how to convert the data types of the variables receiving that input. Finally, you got to play with your first Python script. In Hour 5, your Python exploration will continue as you delve into mathematical algorithms with Python.

# Q&A

**Q.** Can I do any other kind of output formatting besides what I learned about in this chapter?

**A.** Yes, you can also use the `format` function, which is covered in Hour 5.

**Q.** Which is better to use with a `print` function, double quotes or single quotes?

**A.** Neither one is better than the other. Which one you use is a personal preference. However, whichever one you choose, it's best to consistently stick with it.

**Q.** Bottles of tea on the wall?!

**A.** This is a family-friendly book. Feel free to modify your answers to `script0402.py` to your liking.

# Workshop

## Quiz

1. The `print` function is part of the Python standard library and is considered a built-in function. True or false?

2. When is a variable created and assigned a data type?

3. Which of the following is a valid Python data type?

   **a.** `int`

   **b.** `input`

   **c.** `print`

## Answers

1. True. The `print` function is a built-in function of the standard library. There is no need to import it.

2. A variable is created and assigned a data type when it is assigned a value. The value and data type for a variable can be changed with a reassignment.

3. `int` is a Python data type. `input` and `print` are both built-in Python *functions*.

# Using Arithmetic in Your Programs

**What You'll Learn in This Hour:**

▶ Using basic math
▶ Working with fractions
▶ Working with complex numbers
▶ Using the `math` module in Python scripts
▶ Using other Python math libraries

Just about every Python script that you write requires some type of mathematical operation. Whether you need to increment a counter or calculate the Fourier transform of a signal, you need to know how to incorporate mathematical operators and functions in your Python code. This hour walks through all the basics you need to know to work with numbers and perform calculations in your Raspberry Pi Python scripts.

# Working with Math Operators

Python supports all the basic math calculations that you'd expect from a programming language. This section walks you through the basics of how to use math operators in your Python scripts.

## Python Math Operators

To get a feel for how Python handles numbers, you can open an IDLE window and experiment with some simple math calculations right at the command line. You can use the IDLE command prompt as a calculator, entering any type of mathematical equation for it to evaluate and return an answer.

Here's an example of some basic math calculations performed in IDLE:

```
>>> 1 + 1
2
>>> 5 - 2
3
>>> 2 * 5
10
>>> 15 / 3
5.0
>>>
```

As you can see, Python supports all the basic math operators that you learned in school (using an asterisk for multiplication and a forward slash for division).

BY THE WAY

### Division Data Types

Notice that when we performed the division, Python automatically converted the output to a floating-point data type, even though the inputs were both integers. This is a new feature in Python v3.2!

Besides the basic math operators that you were taught in elementary school, Python also supports some other types of mathematical operators. Table 5.1 shows all the Python mathematical operators available for you to use in your scripts.

**TABLE 5.1**   Python Math Operators

| Operator | Description |
| --- | --- |
| + | Addition |
| – | Subtraction |
| * | Multiplication |
| / | Division |
| % | Modulus |
| // | Floor division |
| ** | Exponentiation |
| & | Binary AND |
| \| | Binary OR |
| ^ | Binary XOR |
| ~ | Binary ones complement |
| << | Binary left shift |

| Operator | Description |
|----------|-------------|
| >> | Binary right shift |
| and | Logical AND |
| or | Logical OR |
| not | Logical NOT |

The floor division operator (//) returns the integer portion of a division result (what we use to call the "goes into" part, back in long division classes). The modulus operator returns the remainder of the division (what we use to call the "left over" part).

You'll notice from the table that there are two types of AND and OR operators. There's a subtle difference between the binary and logical operators. The binary operators are used in what's called *bitwise calculations*. You use bitwise calculations to perform binary math using binary values.

If you're using binary operators, you'll probably want to also specify your values in binary notation. To do that, just use the 0b symbol in front of the number, like this:

```
>>> a = 0b01100101
>>> b = 0b01010101
>>> c = a & b
>>> bin(c)
'0b1000101'
>>>
```

To display the value of the c variable in binary notation, you just use the bin() function.

The logical operators work with Boolean True and False logic values. These are most often used in if-then comparisons (see Hour 6, "Controlling Your Program"). Here's an example of how they work:

```
>>> a = 101
>>>b = 85
>>> if ((a > 100) and (b < 100)): print("It worked!")

It worked!
>>> if ((a > 100) and (b > 100)): print("It worked!")

>>>
```

After you enter the if statement, IDLE produces a blank line, waiting for you to complete the statement. Just press the Enter key to finish the statement. In these examples, we compared two conditions using the logical and operator. When both conditions are `True`, Python runs the `print()` statement. If either one is `False`, Python skips the `print()` statement.

## Order of Operations

As you might expect, Python follows all the standard rules of mathematical calculations, including the order of operations. In the following example, Python performs the multiplication first and then the addition operation:

```
>>> 2 + 5 * 5
27
>>>
```

And just like in math, Python allows you to change the order of operations by using parentheses:

```
>>> (2 + 5) * 5
35
>>>
```

You can nest parentheses as deeply as you need to in your calculations. Just be careful to make sure that you match up all the opening and closing parentheses in pairs. If you don't, as in the following example, Python will continue to wait for the missing parenthesis:

```
>>> ((2 + 5) * 5
```

When you press the Enter key, IDLE returns a blank line instead of displaying the result. It's waiting for you to close out the missing parenthesis. To close out the command, just supply the missing parenthesis on the blank line:

```
)
35
>>>
```

IDLE completes the calculation and displays the result.

## Using Variables in Math Calculations

Probably the most useful feature of using math in Python is the ability to use variables inside your equations. The variables can contain values of any numeric data type in Python math calculation. The following example shows that if you mix data types in your calculations, Python will stick with the floating-point data type for the result:

```
>>> test1 = 5
>>> test1 * 2.0
10.0
>>>
```

Be careful to assign a value to a variable before using it in a calculation, or Python will complain, as in this example:

```
>>> test10 * 5
Traceback (most recent call last):
  File "<pyshell#0>", line 1, in <module>
    test10 * 5
NameError: name 'test10' is not defined
>>>
```

Not only can you use variables within a calculation, you can also assign the result of a calculation to a variable. Python automatically sets the variable to the data type required to hold the calculation result:

```
>>> test1 = 2 + 5
>>> result = test1 * 5
>>> print(result)
35
>>>
```

The `result` variable contains the result of the calculation, which you can then display by using the `print()` function. As you can see in the example, you can use both variables with numbers anywhere in the calculations.

The ability to assign numbers and calculation results to variables is crucial to using math in Python scripts. Listing 5.1 shows the `script0501.py` script, which performs a simple math calculation and then displays the result.

### LISTING 5.1   The `script0501.py` Script

```
#!/usr/bin/python
test1 = 2 + 5
result = test1 * 5
print(result)
```

When you run the `script0501.py` script, all you should see is the output from the `print()` function:

```
$ python3 script0501.py
35
$
```

All the math calculations performed in the script are hidden from view!

## Floating-Point Accuracy

As you've been playing around with your calculations, you may have seen some odd behavior with some of the floating-point calculations. Here's an example of what we mean:

```
>>> 5.2 * 9
46.800000000000004
>>>
```

The result of 5.2 multiplied by 9 should be 46.8, but the result displayed in IDLE has a stray value added to the actual result.

This is caused by the way the underlying CPU handles floating-point arithmetic. Because the floating-point data type converts the numbers into a special format, the calculations are somewhat inaccurate.

You can't get around this problem in your calculations, but you can use some Python tricks to help make things more presentable when displaying the results.

## Displaying Numbers

One way to solve the floating-point accuracy issue is to display only the pertinent part of the result. You can use the `print()` function to reformat the numbers that it displays.

By default, the `print()` function displays the actual result that's calculated by Python:

```
>>> result = 5.2 * 9
>>> print(result)
46.800000000000004
>>>
```

However, you can use some Python tricks to help out with formatting the output.

Despite what the output looks like, the `print()` function actually produces a string object for the output (in this case, the string just happens to look like the number result). Since the output is a string object, you can use the `format()` function on the output (see Hour 10, "Working with Strings"), which allows you to define how Python displays the string object.

The `format()` function allows you to separate the variable from the output text in the string object by using the {} placeholder symbol:

```
>>> print ("The result is {}".format(result))
The result is 46.800000000000004
>>>
```

That hasn't helped yet, but now you can use the special features of the {} placeholder to help you reformat the output.

You just define an output template in the {} placeholder, and Python will use it to format the number output. For example, to restrict the output to two decimal places, you use the template {0:.2f} to produce this output:

```
>>> print("The result is {0:.2f}".format(result))
The result is 46.80
>>>
```

Now this is much better! The first number in the template defines what position in the number to start to display. The second number (the .2) defines the number of decimal places to include in the output. The f in the template tells Python that the number is a floating-point format.

## Operator Shortcuts

Python provides a few shortcuts for mathematical operations. If you're performing an operation on a variable and plan on storing the result in the same variable, you don't have to use the long format:

```
>>> test = 0
>>> test = test + 5
>>> print(test)
5
>>>
```

Instead, you can use an *augmented assignment*:

```
>>> test = 0
>>> test += 5
>>> print(test)
5
>>>
```

This feature works for addition, subtraction, multiplication, division, modulus, floor division, and exponentiation.

# Calculating with Fractions

Python supports some other cool math features that you don't often run across in other programming languages. One of those features is the ability to work directly with fractions. This section walks through how to work with fractions in your Python scripts.

## The Fraction Object

The fractions Python module defines a special object called Fraction. The Fraction object holds the numerator and the denominator of a fraction as separate properties of the object.

To use a `Fraction` object, you need to import it from the `fractions` module. After you import the object class, you can create an instance of a `Fraction` object, like this:

```
>>> from fractions import Fraction
>>> test1 = Fraction(1, 3)
>>> print(test1)
1/3
>>>
```

The first parameter of the `Fraction()` method is the fraction numerator, and the second parameter is the denominator of the fraction value. Now you can perform any type of fraction operation on the variable, like this:

```
>>> result = test1 * 4
>>> print(result)
4/3
>>>
```

Starting in Python v3.3, the `Fraction()` constructor can also convert floating-point values into a `Fraction` object, as in the following example:

```
>>> test2 = Fraction(5.5)
>>> print(test2)
11/2
>>>
```

Now that you know how to create fractions, the next step is to start using them in your calculations!

## Fraction Operations

After you create a `Fraction` object, you can use any type of mathematical calculation on the object with other `Fraction` objects, as in this example:

```
>>> test1 = Fraction(1, 3)
>>> test2 = Fraction(4, 3)
>>> result = test1 * test2
>>> print(result)
4/9
>>>
```

Python will also work out common denominator problems with your fractions, like this:

```
>>> test1 = Fraction(1,3)
>>> test2 = Fraction(1,2)
>>> result = test1 + test2
>>> print(result)
5/6
>>>
```

Now you can perform calculations with fractions just as easily as with decimal numbers. That can make life a lot easier if you're working in an environment that uses fractions!

# Using Complex Number Math

For the scientific and engineering communities, Python also supports complex numbers, as well as complex number calculations. A complex number is represented by a combination of a real number and an imaginary number.

BY THE WAY

**Imaginary Numbers**

By definition, an imaginary number is the square root of –1, which theoretically doesn't exist, thus the term *imaginary*.

A complex number is represented by the real number, a plus sign, and the complex number, followed by a *j*. For example, in the complex number 1 + 2j, 1 is the real component, and 2 is the imaginary component.

Trying to work with calculations that use complex numbers is, well, complex! The combination of the real and imaginary parts of the complex number causes the calculations to behave somewhat differently from the math operations you're probably use to seeing. This section walks through how to handle complex numbers in your Python scripts.

## Creating Complex Numbers

You define complex numbers by using the `complex()` function, which is built into the core Python language. As you can see in this example, to create the complex number, you just specify the real component value as the first parameter and then the imaginary component value as the second parameter:

```
>>> test = complex(1, 3)
>>> print(test)
(1+3j)
>>>
```

When you need to display the complex number value, Python displays it using the *j* format, making it easier to view.

## Complex Number Operations

After you define a complex number, you can use it in any type of mathematical calculation, like this:

```
>>> result = test * 2
>>> print(result)
(2+6j)
>>>
```

And as you would expect, you can perform complex number calculations by using other complex numbers, as in the following example:

```
>>> test1 = complex(1, 2)
>>> test2 = complex(2, 3)
>>> result = test1 + test2
>>> print(result)
(3+5j)
>>> result2 = test1 * test2
>>> print(result2)
(-4+7j)
>>>
```

Complex math is not for the faint of heart. If you have to work with complex numbers, though, at least you have a friend in Python!

# Getting Fancy with the `math` Module

For more advanced math support, you can use the methods found in the Python `math` module. It provides some additional mathematical methods commonly found in advanced calculations for trigonometry, statistics, and number theory.

Fortunately, the Raspbian distribution installs the Python `math` module by default in the Python installation, so you don't have to install it as a separate package. However, you do have to use the `import` statement to import the module into your Python script to be able to use the methods:

```
>>> import math
>>> math.factorial(5)
120
>>>
```

If there's just one function you need to use from the `math` module, but you use it lots of times in your script, you can import just that function using the `from` statement:

```
>>> from math import factorial
>>> factorial(7)
5040
>>>
```

The math module provides lots of mathematical functions for you to use. The following sections provide a rundown of what it provides.

## Number Theory Functions

Number theory functions provide handy features such as absolute values, factorials, and determining whether a value is a number and, if it is, what type of number. Table 5.2 lists the number theory functions that you'll find in the math module.

**TABLE 5.2**   Python Number Functions

| Function | Description |
|----------|-------------|
| ceil($x$) | Returns the smallest integer value greater than $x$. |
| copysign($x, y$) | Copies the sign of $y$ of $x$. |
| fabs($x$) | Returns the absolute value of $x$. |
| factorial($x$) | Returns the factorial of $x$. |
| floor($x$) | Returns the largest integer value smaller than $x$. |
| fmod($x, y$) | Returns the modulus of $x$ and $y$. |
| frexp($x$) | Returns a mantissa and exponent of $x$ as a pair. |
| fsum(*iterable*) | Returns the sum of the values stored in a list. |
| isfinite($x$) | Returns TRUE if $x$ is not infinity and is a number. |
| isinf($x$) | Returns TRUE if $x$ is a positive or negative infinity. |
| isnan($x$) | Returns TRUE if $x$ is not a number. |
| ldexp($x, i$) | Returns the value of $x$ * (2 ** $i$). |
| modf($x$) | Returns the fraction and integer parts of $x$. |
| trunc($x$) | Returns the integer part of $x$ as an integer value. |

Most of these functions are pretty self-explanatory if you're working with math. The fsum() function may need a little more explanation, though. It sums the values in a series, but you must specify the series as either a Python list or tuple (see Hour 8, "Using Lists and Tuples"). Here's an example:

```
>>> math.fsum([1, 2, 3])
6.0
>>>
```

You just put the numbers you need to sum in the list and plug that into the `fsum()` function.

## Power and Logarithmic Functions

If you work with logarithms and exponents, the Python `math` module has some functions for you. Table 5.3 shows the logarithmic functions available.

**TABLE 5.3**   Python Logarithmic Functions

| Function | Description |
|---|---|
| `exp(x)` | Returns the value of `e**x`. |
| `expm1(x)` | Returns the value of `e**(x-1)`. |
| `log(x, [base])` | Returns the natural log of `x` or the log of `x` with base of `base`. |
| `log1p(x)` | Returns the natural log of `1+x`(`base e`). |
| `log2(x)` | Returns the base-2 log of `x`. |
| `log10(x)` | Returns the base-10 log of `x`. |
| `pow(x, y)` | Returns `x**y`. |
| `sqrt(x)` | Returns the square root of `x`. |

The `pow()` function performs the same function as the standard `**` math symbol. It's mostly included in Python for completeness, as the `pow()` function is used in many other programming languages.

## Trigonometric Functions

If trigonometry is your thing, you'll be glad to know there are plenty of trig functions in the `math` module as well. Table 5.4 shows what's available.

**TABLE 5.4**   Python Trigonometric Functions

| Function | Description |
|---|---|
| `acos(x)` | Returns the arc cosine of `x` in radians. |
| `asin(x)` | Returns the arc sine of `x` in radians. |
| `atan(x)` | Returns the arc tangent of `x` in radians. |

| Function | Description |
|---|---|
| atan2(y, x) | Returns the arc tangent of (y/x) in radians. |
| cos(x) | Returns the cosine of x in radians. |
| hypot(x, y) | Returns sqrt(x*x + y*y) to find the hypotenuse. |
| sin(x) | Returns the sine of x in radians. |
| tan(x) | Returns the tangent of x in radians. |
| degrees(x) | Converts x in radians to degrees. |
| radians(x) | Converts x in degrees to radians. |

Notice that the trigonometric functions require you to specify the parameter in radians. If you're working with degrees, don't forget to convert first, like this:

```
>>> angle = 90
>>> radangle = math.radians(angle)
>>> anglesine = math.sin(radangle)
>>> print(anglesine)
1.0
>>>
```

Now you're all set to start working on your triangle calculations!

## Hyperbolic Functions

Somewhat related to the trigonometric functions are hyperbolic functions. Whereas trigonometric functions are derived from circular calculations, hyperbolic functions are derived from a hyperbola calculation. Table 5.5 shows the hyperbolic functions that the math module supports.

**TABLE 5.5**   Python Hyperbolic Functions

| Function | Description |
|---|---|
| acosh(x) | Returns the inverse hyperbolic cosine of x. |
| asinh(x) | Returns the inverse hyperbolic sine of x. |
| atanh(x) | Returns the inverse hyperbolic tangent of x. |
| cosh(x) | Returns the hyperbolic cosine of x. |
| sinh(x) | Returns the hyperbolic sine of x. |
| tanh(x) | Returns the hyperbolic tangent of x. |

Just as with the trigonometric functions, you must specify the hyperbolic function parameters in radians.

## Statistical Math Functions

The `math` module includes a few statistical math functions for good measure, as shown in Table 5.6.

**TABLE 5.6**  Python Statistical Math Functions

| Function | Description |
|---|---|
| `erf(x)` | Returns the statistical error function of $x$. |
| `erfc(x)` | Returns the complementary error function of $x$. |
| `gamma(x)` | Returns the gamma function of $x$. |
| `lgamma(x)` | Returns the natural log of the absolute value of the gamma function of $x$. |

The error function is a core computation used in statistical analysis equations. You need it to compute the normal cumulative distribution and the statistical Q-function. You need the complementary (or inverse) error function to calculate the normal quartile of a statistical series.

# Using the `NumPy` Math Libraries

Besides the host of functions available in the standard `math` module, many engineers, scientists, and statisticians who use fancy mathematical calculations have created and shared their own extended Python `math` modules.

One of the core Python libraries for advanced mathematical computing is `NumPy`. The `NumPy` module provides methods for multidimensional array manipulations, which are required for many advanced scientific and statistical calculations. It consists of the following:

▶ A multidimensional array object class

▶ Methods for array manipulation

The `NumPy` multidimensional array objects are somewhat different from standard Python lists or tuples in that you can easily use them in mathematical calculations that require arrays. Python handles the array objects differently from lists and tuples. The following section walks through how to use the `NumPy` features.

## `NumPy` Data Types

The `NumPy` module provides five core data types that you can use to store data in arrays:

▶ `bool`—Booleans

▶ `int`—Integers

▶ `uint`—Unsigned integers

▶ `float`—Floating-point numbers

▶ `complex`—Complex numbers

Within those five core data types, you can also specify a bit size at the end of the data type name, such as `int8`, `float64`, or `complex128`. If you don't specify the bit size, Python will assume the bit size based on the CPU platform (such as 32-bit or 64-bit).

To use the NumPy module functions in your programs, just import the numpy module, however, you may have to be patient, as it can take some time to load all of the library methods!

## Creating NumPy Arrays

The Raspberry Pi Python v3 installation already includes the NumPy module, so you can write your advanced array manipulations right out of the box.

There are several different ways to create arrays in NumPy. One way is to create an array from an existing Python list or tuple, as in this example:

```
>>> import numpy
>>> a = numpy.array(([1, 2, 3], [0, 2, 4], [3, 2, 1]))
>>> print(a)
[[1 2 3]
 [0 2 4]
 [3 2 1]]
>>>
```

This example creates a 3-by-3 array using three Python lists.

If you don't define a data type, Python assumes the data type for the data. If you need to change the data type of the array values, you can specify it as a second parameter to the `array()` function. For example, the following example causes the values to be stored in the floating-point data type:

```
>>> a = numpy.array(([1,2,3], [4,5,6]), dtype="float")
>>> print(a)
[[ 1.   2.   3.]
 [ 4.   5.   6.]]
>>>
```

You can also generate default arrays of either all zeros or all ones, like this:

```
>>> x = numpy.zeros((3,5))
>>> print(x)
[[ 0.   0.   0.   0.   0.]
 [ 0.   0.   0.   0.   0.]
```

```
 [ 0.   0.   0.   0.   0.]]
>>> y = numpy.ones((5,2))
>>> print(y)
[[ 1.   1.]
 [ 1.   1.]
 [ 1.   1.]
 [ 1.   1.]
 [ 1.   1.]]
>>>
```

You can also create an array of regularly incrementing values by using the `arrange()` function, as shown here:

```
>>> c = numpy.arange(10)
>>> print(c)
[0 1 2 3 4 5 6 7 8 9]
>>>
```

You can also specify the starting and ending values, as well as the increment value.

## Using NumPy Arrays

The beauty of NumPy lies in its ability to handle array math. These functions are somewhat of a pain using standard Python lists or tuples, as you have to manually loop through the list or tuple to add or multiply the individual array values. With NumPy, it's just a simple calculation, as shown here:

```
>>> a = numpy.array(([1, 2, 3], [4, 5, 6]))
>>> b = numpy.array(([7, 8, 9], [0, 1, 2]))
>>> result1 = a + b
>>> print(reuslt1)
[[ 8 10 12]
 [ 4  6  8]]
>>> result2 = a * b
>>> print(result2)
[[ 7 16 27]
 [ 0  5 12]]
>>>
```

Now working with arrays in Python is a breeze!

## Summary

Python supports a wide range of mathematical features for just about any type of calculations you need to perform in your scripts. You can perform standard math functions such as addition, subtraction, and division directly by using the standard Python math operators.

If you need to incorporate more advanced math functions in your calculations, you can import the `math` module into your script. The `math` module provides functions for number theory, trigonometry, and basic statistics.

Finally, you may at some point need to get into advanced scientific or statistical calculations. Python users have created some handy libraries for you. The most popular is the `NumPy` library, which contains the tools required to perform calculations using multidimensional arrays for linear algebra, advanced statistics, and signal processing.

In the next hour, we'll take a look at how to add control to your Python programs using the `if` family of control statements. That allows you to add dynamic features to your Python programs!

# Q&A

**Q. What data type should you use to store monetary values in Python?**

**A.** You should use the floating point data type so that the value can contain two decimal places for the cents value.

**Q. Does Python support the incrementor (`++`) and decrementor (`--`) operators?**

**A.** While the incrementor and decrementor operators are popular in other programming languages, currently Python doesn't provide support for those operators.

# Workshop

## Quiz

1. What math function should you use to find the square root of a number?

    a. `pow()`

    b. `sqrt()`

    c. `log2()`

    d. `sin()`

2. You must import the Python `math` library in order to use the Python trigonometric functions. True or false?

3. How do you create a fraction by using the `Fraction` class in Python?

## Answers

1. You should use the sqrt() function to find the square root of a value.

2. True. The Python standard library only supports standard math functions and features. You'll need to import the separate math library module to use any of the more advanced math features such as working with trigonometric functions.

3. The `Fraction` class uses the `Fraction()` method to create a fraction value. The format of the `Fraction()` method is `Fraction(numerator, denominator)`, which specifies the numerator and the denominator of the fraction as separate values.

HOUR 6

# Controlling Your Program

## What You'll Learn in This Hour:

- ► How to use if-then statements
- ► How to group multiple statements
- ► How to add else sections
- ► Stringing together if-then statements
- ► Testing conditions

In all the Python scripts discussed so far, Python processes each individual statement in the script in the order in which it appears. This works out fine for sequential operations, where you want all the operations to process in the proper order. However, this isn't how all programs operate.

Many programs require some sort of logic flow control between the statements in the script. This means that Python executes certain statements given one set of circumstances but has the ability to execute other statements given a different set of circumstances. A whole class of statements, called *structured commands*, allow Python to skip over or loop through statements based on conditions of variables or values.

There are quite a few structured commands available in Python, and we look at them individually. In this hour, we look at the if statement.

## Working with the if Statement

The most basic type of structured command is the if statement. The if statement in Python has the following basic format:

```
if (condition): statement
```

If you have ever used if statements in other programming languages, this format may seem somewhat odd because there's no "then" keyword in the statement.

Python uses the semicolon to act as the "then" keyword. Python evaluates the condition in the parentheses and then either executes the statement after the semicolon if the condition returns a `True` logic value or skips the statement after the semicolon if the condition returns a `False` logic value.

▼ TRY IT YOURSELF

## Using the `if` statement

Let's walk through a few examples to show using the if statement:

1. Open the Python3 IDLE interface on your graphical desktop (see Hour 3, "Setting Up a Programming Environment").

2. Set a value for a variable:

   ```
   >>> x = 50
   ```

3. Test the variable value using an `if` statement:

   ```
   >>> if (x == 50): print("The value is 50")

   The value is 50
   ```

4. Try another test using another condition:

   ```
   >>> if (x < 100): print("The value is less than 100")

   The value is less than 100
   ```

5. Try a test condition that should fail:

   ```
   >>> if (x > 100): print("The value is more than 100")

   >>>
   ```

With the `if` statement, each time you enter the statement and press the Enter key, the IDLE interface pauses on the next line to see if you're going to enter any more statements. You just press the Enter key again to close out the statement.

In the first example, the condition checks to see if the variable x is equal to 50. (We talk about the double equal sign in the "Comparison Conditions" section, later in this chapter.) Since it is, Python executes the `print()` statement on the line and prints the string.

Likewise, the second example checks whether the value stored in the x variable is less than 100. Since it is, Python again executes the `print()` statement to display the string.

However, in the third example, the value stored in the x variable is not greater than 100, so the condition returns a `False` logic value, causing Python to skip the `print()` statement after the semicolon.

# Grouping Multiple Statements

The basic `if` statement format allows you to process one statement based on the outcome of the condition. More often than not, though, you want to group multiple statements together, based on the outcome of the condition. This is another place where the Python `if` statement format deviates from other programming languages.

Many programming languages use either braces or a keyword to indicate the group of statements that the `if` statement controls. Instead of grouping statements together using either braces or a special keyword, Python uses indentation.

To group a bunch of statements together, you must place them each on separate lines in the script and indent them from the location of the `if` statement. Here's an example:

```
>>> if (x == 50):
        print("The x variable has been set")
        print("and the value is 50")


The x variable has been set
and the value is 50
>>>
```

When you enter the code to test this in IDLE, after you press the Enter key for the `if` statement, IDLE automatically indents the next line for you. All the statements that you enter after that are considered part of the "then" section of the statement and are controlled by the condition.

When you're done entering statements, you just press the Enter key on an empty line.

If you're working with `if` statements in a Python script, you have to remember to manually indent the "then" section statements. You indicate the lines outside the "if" section by not indenting them. Listing 6.1 shows an example of doing this.

**LISTING 6.1**  Using `if` Statements in Python Scripts

```
1: #!/usr/bin/python
2: x = 50
3: if (x == 50):
4:     print("The x variable has been set")
5:     print("and the value is 50")
6: print("This statement executes no matter what the value is")
```

Notice how lines 4 and 5 are indented from the location of the `if` statement on line 3. The `print()` statement on line 6 isn't indented, though. This means it's not part of the "then" section.

To test it, run the `script0601.py` program from the command line:

```
$ python3 script0601.py
The x variable has been set
And the value is 50
This statement executes no matter what the value is
$
```

Now if you change the code to set the value of x to `25`, you get the following output:

```
$ ./script0601.py
This statement executes no matter what the value is
$
```

Python skips the `print()` statements inside the "then" section but picks up with the next `print()` statement that's not indented.

## Adding Other Options with the `else` Statement

In the `if` statement, you only have one option of whether to run statements. If the condition returns a `False` logic value, Python just moves on to the next statement in the script. It would be nice to be able to execute an alternative set of statements when the condition is `False`. That's exactly what the `else` statement allows you to do.

The `else` statement provides another group of commands in the statement:

```
>>> x = 25
>>> if (x == 50):
     print("The value is 50")
else:
     print("The value is not 50")


The value is not 50
>>>
```

When you use the `else` statement with the `if` statement, you must be careful how you place the `else` statement. If you try to keep it indented, you get an error message from Python:

```
>>> if (x == 50):
     print("The value is 50")
     else:

SyntaxError: invalid syntax
>>>
```

The same applies when you're using the if and else statements in Python scripts. When you're creating your script code file, make sure you line up the else statement properly in the text. Listing 6.2 demonstrates this.

**LISTING 6.2**   Using the else statement in a Python script

```
#!/usr/bin/python
x = 25
if (x == 50):
    print("The value is 50")
else:
    print("The value is not 50")
```

The code shown in Listing 6.2 has the else statement at the same indentation level as the if statement. When you run the script0602.py script, only one of the print() statements will execute, like this:

```
$ python3 script0602.py
The value is not 50
$
```

The same applies when you have multiple statements in either the "if" or "else" sections, and when you have additional code after the if and else statements block. Listing 6.3 demonstrates a more complicated if/else statement.

**LISTING 6.3**   Multiple Statements in the if and else Sections

```
#!/usr/bin/python
x = 25
if (x == 50):
    print("The x variable has been set")
    print("and the value is 50")
else:
    print("The x variable has been set")
    print("And the value is not 50")
print("This ends the test")
```

You can control the output by adjusting the value you assign to the x variable. When you run the script as is, you get this output:

```
$ python3 script0603.py
The x variable has been set
And the value is not 50
This ends the test
$
```

If you change the value of x to 50, you get this output:

```
$ ./script0603.py
The x variable has been set
And the value is 50
This ends the test
$
```

Everything in the if/else statement blocks is based on the indentation of the statements, so be very careful when you construct the statement!

# Adding More Options Using the elif Statement

So far you've seen how to control a block of statements by using either the if statement or the if and else combination. That gives you quite a bit of flexibility in controlling how your scripts work. However, there's more!

Sometimes you need to compare a value against multiple ranges of conditions. One way to solve that is to string multiple if statements back-to-back, as shown in Listing 6.4.

**LISTING 6.4**   The script0604.py File

```
#!/usr/bin/python
x = 45
if (x > 100):
    print("The value of x is very large")
if (x > 50):
    print("The value of x is medium")
if (x > 25):
    print("The value of x is small")
if (x <= 25):
    print("The value of x is very small")
```

When you run the script0604.py script, Python executes only one of the print() statements, based on the value stored in the x variable:

```
$ python3 script0604.py
The value of x is small
$
```

This works, but it is a somewhat ugly way to solve the problem. Fortunately, there's an easier solution.

Python supports the `elif` statement, which lets you string together multiple `if` statements and end with a catch-all `else` statement. The basic format of the `elif` statement looks like this:

```
if (condition1):statement1
elif (condition2): statement2
else: statement3
```

When Python runs this code, it first checks the `condition1` result. If that returns a `True` value, Python runs `statement1` and then exists the `if/elif/else` statements.

If `condition1` evaluates to a `False` value, Python then checks the `condition2` result. If that returns a `True` value, Python runs `statement2` and then exits the `if/elif/else` statement.

If `condition2` evaluates to a `False` value, Python runs `statement3` and then exits the `if/elif/else` statement.

Listing 6.5 shows an example of how to use the `elif` statement in a program.

**LISTING 6.5**   Using the `elif` statement

```
x = 45
if (x > 100):
    print("The value of x is very large")
elif (x > 50):
    print("The value of x is medium")
elif (x > 25):
    print("The value of x is small")
else:
    print("The value of x is very small")
```

When you run the `script0605.py` code, only one `print()` statement runs, based on the value you set the `x` variable to. By default, you see this output:

```
$ python3 script0605.py
The value of x is small
$
```

You can see that you have complete control over just what code statements Python runs in the script!

# Comparing Values in Python

The operation of the `if` statement revolves around the comparisons you make. Python provides quite a variety of comparison operators that allow you to check all types of data. This section walks through the different types of comparisons that are available in your Python scripts.

## Numeric Comparisons

The most common type of comparisons have to do with comparing numeric values. Python provides a set of operators for performing numeric comparisons in your `if` statement conditions. Table 6.1 shows the numeric comparison operators that Python supports.

**TABLE 6.1**   Numeric Comparison Operators

| Operator | Description |
|---|---|
| == | Equal |
| != | Not equal |
| <> | Not equal |
| > | Greater than |
| >= | Greater than or equal |
| > | Less than |
| <= | Less than or equal |

The comparison operators return a logical `True` value if the comparison succeeds and a logical `False` value if the comparison fails. For example, the following statement:

```
if (x >= y): print("x is larger than y")
```

executes the `print()` statement only if the value of the x variable is greater than or equal to the value of the y variable.

WATCH OUT!

### The Equality Comparison Operator

Be careful with the equal comparison! If you accidentally use a single equal sign, it becomes an assignment statement and not a comparison. Python processes the assignment and then exits with a `True` value all the time. That's probably not what you want to do.

## String Comparisons

Unlike numeric comparisons, string comparisons can sometimes be a little tricky. While comparing two string values for equality is easy:

```
x = "end"
if (x == "end"): print("Sorry, that's the end of the game")
```

trying to use a greater-than or less-than comparison in strings can get confusing. When is one string value greater than another string value?

Python performs what's called a *lexicographical comparison* of string values. This method converts letters in the string to the ASCII numeric equivalent and then compares the numeric values.

Here's a test string comparison:

```
>>> a = "end"
>>> if (a < "goodbye"):
        print("end is less than goodbye")
elif (a > "goodbye"):
        print("end is greater than goodbye")

end is less than goodbye
>>>
```

Python compares the values "end" and "goodbye" and determines which one is "greater." Since the string value "end" would come before "goodbye" in a sort method, it is considered "less than" the "goodbye" string.

Now, try this example:

```
>>> a = "End"
>>> if (a < "goodbye"):
    print("End is less than goodbye")
elif (a > "goodbye"):
    print("End is greater than goodbye")


End is less than goodbye
>>>
```

Changing the capitalization of "End" still makes it less than "goodbye".

Next, compare the same word capitalized and in all lowercase letters:

```
>>> if (a == "end"):
    print("End is equal to end")
elif (a < "end"):
    print("End is less than end")
elif (a > "end"):
    print("End is greater than End")


End is less than end
>>>
```

The capitalized version of the string evaluates to be less than the lowercase version. This is an important feature to know when comparing string values in Python!

## List and Tuple Comparisons

Python allows you to compare objects that contain multiple values, such as lists and tuples. In this example, Python considers the list stored in the a variable less than the list stored in the b variable:

```
>>> a = [1, 2, 3]
>>> b = [4, 5, 6]
>>> if (a < b):
    print("a is less than b")
elif (a > b):
    print("a is greater than b")


a is less than b
>>>
```

In the previous example, both lists are the same length. To see how Python handles comparing two lists of different lengths, you can try this example:

```
>>> c = [7,8]
>>> if (a < c):
    print("a is less than c")
elif (a > c):
    print("a is greater than c")


a is less than c
>>>
```

Even though the c variable contains a shorter list, Python evaluates the a variable list to be less than the c variable list.

## Boolean Comparisons

Since Python evaluates the if statement condition for a logic value, testing Boolean values is pretty easy:

```
>>> x = True
>>> if (x): print("The value is True")
```

```
The value is True
>>> x = False
>>> if (x): print("The value is True")

>>>
```

Setting a variable value directly to a logical `True` or `False` value is pretty straightforward. However, you can also use Boolean comparisons to test other features of a variable.

If you set a variable to a value, Python also makes a Boolean comparison:

```
>>> a = 10
>>> if (a): print("The a variable has been set")

The a variable has been set
>>>
```

The same applies if you assign a string value to a variable:

```
>>> b = "this is a test"
>>> if (b): print("The variable has been set")

The variable has been set
>>>
```

However, if a variable contains a value of `0`, it evaluates to a `False` Boolean condition:

```
>>> c = 0
>>> if (c): print("The b variable has been set")

>>>
```

So be careful when evaluating variables for Boolean values!

## Evaluating Function Results

A feature related to Boolean comparisons is Python's ability to test the result of functions. When you run a function in Python, the function returns a *return code*. You can test the return code by using the `if` statement to determine whether the function succeeded or failed.

A good example of this is using the `isdigit()` method in Python. The `isdigit()` method checks whether the supplied value can be converted to a number, and it returns a `True` Boolean value if it can. You can use this to check whether a value provided to your script by a user is a number. Listing 6.6 shows an example of how to use it.

**LISTING 6.6**   Using Functions in Conditions

```
#!/usr/bin/python

name = input("Please enter your name: ")
age = input("Please enter your age: ")

if (age.isdigit()):
    print("Hello", name, ",your age is", age)
else:
    print("Sorry, the age you entered is not a number")
```

The `if` statement checks whether the `age` variable is a digit. If it is, the script displays the data. If not, it displays an error message.

Here's an example of running the program to test it out:

```
$ python3 script0606.py
Please enter your name: Rich
Please enter your age: test
Sorry, the age you entered is not a number
$ python script0606.py
Please enter your name: Rich
Please enter your age: 10
Hello Rich, your age is 10
$
```

The first test uses bad data for the age value, and the script catches that! The second test uses correct data, and that works just fine.

# Checking Complex Conditions

So far in this hour, all the examples have used just one comparison check within the condition. Python also allows you to group multiple comparisons together in a single `if` statement. This section show some tricks you can use to combine more than one condition check into a single `if` statement.

## Using Logic Operators

Python allows you to use the logic operators (see Hour 5, "Using Arithmetic in Your Programs") to group comparisons together. Since each individual condition check produces a Boolean result value, Python applies the logic operation to the condition results. The result of the logic operation determines the result of the `if` statement:

```
>>> a = 1
>>> b = 2
>>> if (a == 1) and (b == 2): print("Both conditions passed")

Both conditions passed
>>> if (a == 1) and (b == 1): print("Both conditions passed")

>>>
```

When you use the and logic operator, both of the conditions must return a True value for Python to process the "then" statement. If either one fails, Python skips the "then" code block.

You can also use the or logical operator to compound condition checks:

```
>>> if (a == 1) or (b == 1): print("At least one condition passed")

At least one condition passed
>>>
```

In this situation, if either condition passes, Python processes the "then" statement.

## Combining Condition Checks

You can combine condition checks into a single condition check without using logic operators. Take a look at this example:

```
>>> c = 3
>>> if a < b < c: print("they all passed")

they all passed
>>> if a < b > c: print("they all passed")

>>>
```

In this example, Python first checks whether the a variable value is less than the b variable value. Then it checks whether the b variable value is less than the c variable value. If both of those conditions pass, Python runs the statement. If either condition fails, Python skips the statement.

# Negating a Condition Check

There's one final if statement trick that Python programmers like to use. Sometimes when you're writing if-else statements, it comes in handy to reverse the order of the "then" and "else" code blocks.

This can be because one of the code blocks is longer than the other, so you want to list the shorter one first, or it may be because the script logic says it makes more sense to check for a negative condition.

You can negate the result of a condition check by using the logical not operator (see Hour 5):

```
>>> a = 1
>>> if not(a == 1): print("The 'a' variable is not equal to 1")

>>> if not(a == 2): print("The 'a' variable is not equal to 2")

The 'a' variable is not equal to 2
>>>
```

The not operator reverses the normal result from the equality comparison, so the opposite action occurs from what would have happened without the not operator.

BY THE WAY

### Negating Conditions

You may have noticed that you can negate a condition result by either using the not operand or by using the opposite numeric operand (such as != instead of ==). Both methods produce the same result in your Python script.

# Summary

This hour covers the basics of using the if structured command. The if statement allows you to set up one or more condition checks on the data you use in your Python script. This comes in handy when you need to program any type of logical comparisons in your Python scripts. The if statement by itself allows you to execute one or more statements based on the result of a comparison test. You can add the else statement to provide an alternative group of statements to execute if the comparison fails.

You can expand the comparisons by using one or more elif statements in the if statement. You can just continue stringing elif statements together to continue comparing additional values.

In the next hour, we'll take a look at some more advanced control statements you can use to make your scripts more dynamic. We'll discuss the Python statements that allow you to loop through sections of your code multiple times!

# Q&A

**Q.** Does Python support the `select` and `case` statements that are often found in other programming languages?

**A.** No, you have to use the `elif` statement to string together multiple `if` condition checks.

**Q.** Is there a limit on how many statements I can place in an `if` or `else` code block?

**A.** No, you can add as many statements as you need to control inside the `if` or `else` code blocks.

**Q.** Is there a limit on how many `elif` statements I can place in an `if` statement?

**A.** No, you can nest as many elif statements together as you need to check in your code. However, you may want to be careful, as the more elif statements the longer it will take for Python to evaluate the code values.

# Workshop

## Quiz

1. What comparison should you use to check whether the value stored in the z variable is greater than or equal to 10?

    **a.** `>`

    **b.** `<`

    **c.** `>=`

    **d.** `==`

2. How would you write the `if` statement to display a message only if the value stored in the z variable is between 10 and 20 (not including those values)?

3. How would you write the `if` statement to give a game player status messages if a guess falls within 5, 10, or 15 of the actual value?

## Answers

1. c. A common mistake in writing conditions is to forget that the greater and less-than symbols don't include the specified number!

2. You can nest the variable between to numbers using greater-than or less-than symbols:

```
if 10 < z < 20: print("This is the message")
```

3. You can use the elif statement to add additional checks for a range of values. It's important to remember to check smaller ranges first, as the larger ranges will include the smaller ranges:

```
if (z == answer): print("Correct, you guessed the answer!")
elif (z > answer - 5) or (z < answer + 5): print("You're within 5 of the
answer")
elif (z > answer - 10) or (z < answer + 10): print("You're within 10 of the
answer")
elif (z > answer - 15) or (z < answer + 15): print("You're within 15 of the
answer")
```

# HOUR 7
# Learning About Loops

---

**What You'll Learn in This Hour:**

▶ How to perform repetitive tasks

▶ How to use the `for` loop

▶ How to use the `while` loop

▶ How to use nested loops

In this hour, you will be learning about additional structured commands that help you reach your script's goals using Python. Specifically, the focus is on repetitive tasks and what constructs are needed to accomplish those tasks.

## Performing Repetitive Tasks

One of the great benefits of using a computer is that it doesn't get bored performing a task over and over again. Doing a task over and over again is called *repetition*.

A synonym for *repetition* is *iteration*. In the programming world, *iteration* is the process of performing a defined set of tasks repeatedly until either a desired result is achieved or the set of tasks has been performed a desired number of times.

When referring to a loop in Python, the term *iteration* is used. One time through a loop is called *one iteration*. Going through a loop multiple times is referred to as *iterating through the loop*. Now, just iterate through these last three paragraphs again and again, until you reach the desired result of understanding the iteration terms.

# Using the `for` Loop for Iteration

In Python, the `for` loop construct is called a "count-controlled" loop, because the loop's set of tasks will be performed a set number of times. If you want a set of tasks to be performed five times, you can use a `for` loop in Python to accomplish this task.

The syntax structure of the `for` loop in Python is as follows:

```
for variable in data_list:
    set_of_Python_statements
```

Notice in the `for` loop structure that there is no ending statement. In some programming or scripting languages, you see a "done" or "end" type of statement. In a `for` loop, the Python statements to be included are indented under the `for` construct. This is similar to the if-then statement structure.

BY THE WAY

## Indentation in Loops

Just as with the if-then statements you learned about in Hour 6, "Controlling Your Program," the Python statements have to be indented to be part of a loop. Remember that in IDLE, the development environment editor does this for you automatically. However, in a text editor, you need to remember to tab or space over yourself.

The operation of a `for` loop is as follows:

▶ The *variable* in the `for` construct is assigned the first value in the data list.

▶ The Python statement(s) in the loop is executed and has the option of using the assigned variable's value during execution.

▶ Upon completion of the loop's Python statement(s), the variable is reassigned the next value in the data list.

▶ The Python statement(s) in the loop is then executed and has the option of using the variable's reassigned value during execution.

▶ The `for` loop continues until all the values have been assigned to the variable and the Python statement(s) in the loop is executed during each assignment.

Reading about structure is not as helpful as diving into specific examples. The following sections will help you better understand `for` loops.

## Iterating Using Numeric Values in a List

You can have the `for` loop iterate through numbers by providing the numbers in a data list, as shown in Listing 7.1. The only Python statement in this loop is `print (the_number)`, which prints the current number being used from the data list.

**LISTING 7.1** A for Loop

```
>>> for the_number in [1, 2, 3, 4, 5]:
...     print (the_number)
...
1
2
3
4
5
>>>
```

Notice the format of the data list in the `for` loop construct in Listing 7.1. The numbers are contained within two square brackets, and the numbers are separated with commas. The variable `the_number` is assigned a number in the data list, starting with the first number (1). After the Python statement `print (the_number)` within the `for` loop is completed, the variable, the_ number, is then assigned to the next number in the data list. Figure 7.1 shows stepping through the `for` loop in this manner.

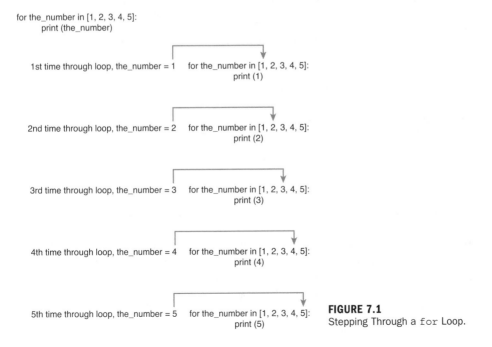

**FIGURE 7.1**
Stepping Through a for Loop.

The loop continues until `the_number` has been assigned to the last number in the data list (5) and the Python statement in the loop has been completed. Thus, all the numbers in the data list are used, one at a time, in an iteration of the loop.

# Watching for a Few "Gotchas"

You need to be careful about a couple potential problems with the `for` loop structure. The first "gotcha" is forgetting to put a colon at the end of your `for` loop's data list. Listing 7.2 shows the error message you get as a result of making this mistake.

**LISTING 7.2**   A Missing Colon on a `for` Loop

```
>>> for the_number in [1, 2, 3, 4, 5]
  File "<stdin>", line 1
    for the_number in [1, 2, 3, 4, 5]
                                     ^
SyntaxError: invalid syntax
>>>
```

BY THE WAY

## Python Interactive Shell Versus Text Editor

When you are testing loop structure in a Python interactive shell, you need to press the Enter key two times after the last Python statement in the loop. This alerts the interactive shell that you are ready for the loop to be interpreted and the results displayed. However, in a text editor, this extra press of the Enter key is not needed.

The next "gotcha" is not using commas to separate your numeric data list. In Listing 7.3, you can see that no error is generated, but this is probably not the result being sought.

**LISTING 7.3**   Missing Commas in a `for` Loop Data List

```
>>> for the_number in [12345]:
...     print (the_number)
...
12345
>>>
```

Don't forget to keep your indentation consistent. If you are using spaces for indenting, then continue to use exactly the same number of spaces for indentation for *each* Python statement in the loop. If you are using tabs for indenting, then continue to use exactly the same number of tabs for indentation. In Listing 7.4, you can see how Python complains when spaces are used for one indentation and tabs are used for the other.

**LISTING 7.4**  Inconsistent Indentation

```
>>> for the_number in [1, 2, 3, 4, 5]:
...      print ("Spaces used for indentation")
...      print ("Tab used for indentation")
  File "<stdin>", line 3
    print ("Tab used for indentation")
                                     ^
TabError: inconsistent use of tabs and spaces in indentation
>>>
```

This next item is not really a "gotcha" but a reminder to be aware that the numbers you use in a data list do not have to be in numeric order. Listing 7.5 shows an example of this.

**LISTING 7.5**  Non-Numeric Order of Numbered Lists

```
>>> for the_number in [1, 5, 15, 9]:
...      print (the_number)
...
1
5
15
9
>>>
```

As you can see, the data from the list is processed in the order in which it was placed in the data list. Python has no complaints in processing this list. It simply follows the order of the list.

BY THE WAY

**Spaces in a Data List**

You are not limited in terms of the number of spaces you can put between a comma and a number in the data list. The data list [1, 5,    15, 9] is legal in a for loop. However, that is poor form. It is best to put only one space between a comma and the next data item in a data list.

# Assigning Data Types from a List

Python behaves as you would expect it to with data types in for loops. In Listing 7.6, you can see that Python assigns the data type int (integer) to the variable the_number as it assigns each data list number to the variable.

**LISTING 7.6**    Data Types of Numbered Lists

```
>>> for the_number in [1, 5, 15, 9]:
...     print (the_number)
...     type (the_number)
...
1
<class 'int'>
5
<class 'int'>
15
<class 'int'>
9
<class 'int'>
>>>
```

Python also changes the data type, as needed, in the assignment (see Listing 7.7). For example, changing the integer 5 to a floating-point number 5.5 causes the data type to be changed as well.

**LISTING 7.7**    Changing Data Type

```
>>> for the_number in [1, 5.5, 15, 9]:
...     print (the_number)
...     type (the_number)
...
1
<class 'int'>
5.5
<class 'float'>
15
<class 'int'>
9
<class 'int'>
>>>
```

# Iterating Using Character Strings in a List

Besides iterating through numbers in a data list, you can also process through character strings in a for loop data list. In Listing 7.8, five words are used in the data list instead of numbers.

**LISTING 7.8**    Character Strings in a Data List

```
>>> for the_word in ['Alpha','Bravo','Charlie','Delta','Echo']:
...     print (the_word)
...
```

```
Alpha
Bravo
Charlie
Delta
Echo
>>>
```

The loop iterates through each word in the data list, just as it does through a list of numbers. Notice, however, that you need to have quotation marks around each word.

BY THE WAY

**Quotation Mark Choices**

You can use double quotation marks around each word in a data list rather than single quotes, if you prefer. You can even use single quotation marks on some words in the data list and double quotation marks on the rest of the words! However, such inconsistency is considered poor form. So pick one quotation mark style for your data list strings and stick with it.

# Iterating Using a Variable

Data lists are not limited to numbers and character strings alone. You can use variables in a for loop data list as well. In Listing 7.9, the variable top_number is assigned to the number 10.

**LISTING 7.9**   Variables in a Data List

```
>>> top_number=10
>>> for the_number in [1,2,3,4,top_number]:
...     print (the_number)
...
1
2
3
4
10
>>>
```

As you can see, the for loop construct has no problem handling this slight change. The loop evaluates the variable top_number as 10, and the iteration processes correctly for that number.

# Iterating Using the range Function

Instead of listing all the numbers individually in a data list, you can use the range function to create a contiguous number list for you. The range function really shines when it's used in loops.

BY THE WAY

### A Function or Not?

The range function is not really a function. It is actually a data type that represents an immutable sequence of numbers. However, for now, you can just think of it as a function.

To include the range function in a loop, you replace your numeric data list as shown in Listing 7.10. The single number between parentheses is called the *stop number*. In this example, the stop number is set to 5. Notice that the range of numbers starts at 0 and ends at 4.

**LISTING 7.10** Using the range Function in a for Loop

```
>>> for the_number in range (5):
...     print (the_number)
...
0
1
2
3
4
>>>
```

In Listing 7.10, using the range function causes a data list of numbers to be created: [0, 1, 2, 3, 4]. The range function, by default, starts at 0 and then produces a list of numbers all the way up to the stop number minus 1. Thus, with 5 listed as the stop number, the range function stops producing numbers at 5 minus 1, or 4.

DID YOU KNOW

### Integers Only

The range function can only accept integer numbers as arguments. No floating points or character strings are allowed.

You can alter the behavior of the range function by including a start number. The syntax looks like this:

range(*start, stop*)

and is demonstrated in Listing 7.11.

**LISTING 7.11**   Using a Stop Number in a `range` Function

```
>>> for the_number in range (1,5):
...       print (the_number)
...
1
2
3
4
>>>
```

Variables can be used in place of the numbers in the `range` function. Listing 7.12 shows how this works.

**LISTING 7.12**   Using Variables in a `range` Function

```
>>> start_number = 3
>>> stop_number = 6
>>> for the_number in range (start_number, stop_number):
...       print (the_number)
...
3
4
5
>>>
```

BY THE WAY

**Range of Expressions**

You can use a mathematical expression as your start or stop number. This is a slick trick to help add clarity to your Python statements. For example, if you want the numbers 1 to 5 to be used in the loop, you can use `range (1, 5+1)` as your `range` statement. At a glance, you will see the number where the `range` function stops.

To change the increment of the number list produced by the range function, you include *a step number* in your `range` arguments. By default, the `range` function increments the numbers in the list by 1. By adding a step number, using the format `range (start, stop, step)`, you can modify this increment. In Listing 7.13, the increment is changed from the default of 1 to 2.

**LISTING 7.13**   Using a Step Number in a `range` Function

```
>>> for the_number in range (2,9,2):
...       print (the_number)
...
```

```
2
4
6
8
>>>
```

You can use the `range` function to produce a list of numbers that goes "backward." You accomplish this by making your step number negative. Of course, you have to carefully think through your start and stop numbers, too. Listing 7.14 produces the same results as Listing 7.13, only backward. Notice the difference in the range arguments in Listing 7.14 and Listing 7.13.

**LISTING 7.14**    Stepping "Backward" with a `range` Function

```
>>> for the_number in range (8,1,-2):
...      print (the_number)
...
8
6
4
2
>>>
```

Now that you have a taste of the `for` loop, it's time to try out a practical `for` loop example for yourself.

▼ TRY IT YOURSELF

**Validate User Input with a `for` loop**

An important part of obtaining input from a script user is validating the input. This is called *input verification*. In the script you write in the following steps, you are going to allow the script user three attempts to get the input right. Also, you are going to get a chance to try something new, a `break`. Unfortunately, you don't get to pour a cup of tea with this kind of a break. Follow these steps:

1. If you have not already done so, power up your Raspberry Pi and log in to the system.

2. If you do not have the LXDE GUI started automatically at boot, start it now by typing `startx` and pressing Enter.

3. Open the LXTerminal by double-clicking the LXTerminal icon.

**4.** At the command-line prompt, type `nano py3prog/script0701.py` and press Enter. The command puts you into the nano text editor and creates the file `py3prog/script0701.py`.

**5.** Type the following code into the nano editor window, pressing Enter at the end of each line:

BY THE WAY

## Be Careful!

Be sure to take your time here and avoid making typographical errors. You can make corrections by using the Delete key and the up- and down-arrow keys.

```
# script0701.py - The Secret Word Validation.
# Written by <your name here>
# Date: <today's date>
#
########### Define Variables ###########
#
max_attempts = 3              #Number of allowed input attempts.
the_word = 'secret'           #The secret word.
#
############# Get Secret Word ###############
#
for attempt_number in range (1, max_attempts + 1):
    secret_word = input("What is the secret word?")
    if secret_word == the_word:
        print ()
        print ("Congratulations! You know the secret word!")
        print ()
        break    # Stops the scripts execution.
    else:
        print ()
        print ("That is not the secret word.")
        print ("You have", max_attempts - attempt_number, "attempts left.")
        print ()
```

WATCH OUT!

## Proper Indentation

Remember that when you use a text editor, you need to make sure you do the indentation properly. If you do not indent the `for` and the `if` code blocks properly, Python will give you an error message. Look back to the section "Watching for a Few 'Gotchas,'" earlier this hour, for help on this, if needed.

6. Write out the modified script you just typed in the text editor by pressing Ctrl+O. Press Enter to write out the contents to the `script0701.py` script.

7. Exit the nano text editor by pressing Ctrl+X.

8. Type `python3 py3prog/script0701.py` and press Enter to run the script. The first time you run the script, answer the question correctly by entering `secret`. You should see output similar to that shown in Figure 7.2.

**FIGURE 7.2**
The `script0701.py` output with the correct answer.

The script stops after you enter the correct answer because of the `break` statement. The `break` statement causes the loop to terminate. In other words, it lets you "break out" of the loop.

9. Type `python3 py3prog/script0701.py` again and press Enter to run the script. This time, answer the question incorrectly, answering anything except `secret`. You should see output similar to that shown in Figure 7.3.

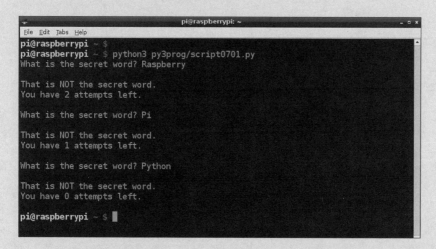

**FIGURE 7.3**
The `script0701.py` output with incorrect answers.

Input verification is an important tool. The little script you just created is a small example of what you can do with a `for` loop to verify user input. A `while` loop, as you'll see next, is another type of loop to use for input verification in Python scripts.

# Using the `while` Loop for Iteration

In Python, the `while` loop construct is called a "condition-controlled" loop because the loop's set of tasks are performed until a desired condition is met. Once the condition is met, the iterations stop. For example, you might want a loop's set tasks to be performed until a certain condition is no longer true. In such a case, you would use a `while` loop in Python.

The syntax structure of the `while` loop in Python is as follows:

```
while condition_test_statement:
    set_of_Python_statements
```

Just like the `for` loop, the `while` loop uses indentation to denote the Python statements associated with it (code block). `condition_test_statement` examines a condition, and if the condition is found to be true, Python statements within the loop's code block are executed. For each iteration, the condition is checked. If the condition is examined and found to be false, the iterations stop.

## Iterating Using Numeric Conditions

You can use a number or mathematical equation in a `while` loop's condition test statement. Listing 7.15, for example, shows a mathematical condition used in a `while` loop.

**LISTING 7.15**   A while Loop

```
1: >>> the_number = 1
2: >>> while the_number <= 5:
3: ...       print (the_number)
4: ...       the_number = the_number + 1
5: ...
6: 1
7: 2
8: 3
9: 4
10: 5
11: >>>
```

In Listing 7.15, the test statement in line 2 checks the variable `the_number`. As long as that variable remains less than or equal to 5, the subsequent Python statements will be executed. The last Python statement in the `while` loop on line 4 increases the variable's value by 1. Thus, when the_number is equal to 6, the while loop's test statement returns false. At that time, the iterations through the loop stop.

# Iterating Using String Conditions

Character strings can be part of the while loop's condition test statement. In Listing 7.16, the while test statement examines the variable the_name and sees if it is not equal to (!=) an empty string. The while loop continues to ask for names and build that list, as long as the script user does not just press the Enter key for a name.

**LISTING 7.16**   Test Condition Using Character Strings

```
1:>>> list_of_names = ""
2:>>> the_name = "Start"
3:>>> while the_name != "":
4:...     the_name = input("Enter name: ")
5:...     list_of_names = list_of_names + the_name
6:...
7:Enter name: Raz
8:Enter name:
9:>>>
```

A nice tweak on the while loop in Listing 7.16 is to replace the very long code in line 5 with a more efficient statement that uses an operator shortcut. You learned about operator shortcuts, also called *augmented assignment operators*, in Hour 5, "Using Arithmetic in Your Programs." With an operator shortcut, line 5 would now look like this:

```
list_of_names += the_name
```

Another nice tweak you can make is to include an optional else clause in the while loop. If you included these two changes, Listing 7.16 would become Listing 7.17.

**LISTING 7.17**   An else Clause in a while Loop

```
1: >>> list_of_names = ""
2: >>> the_name = "Start"
3: >>> while the_name != "":
4: ...      the_name = input("Enter name: ")
5: ...      list_of_names += the_name
6: ... else:
7: ...      print (list_of_names)
8: ...
9: Enter name: Raz
10: Enter name: Berry
11: Enter name: Pi
12: Enter name:
13: RazBerryPi
14: >>>
```

In Listing 7.17, the list_of_names variable is printed after the while loop terminates. However, you should know that an else clause is executed whenever the while loop's test statement returns false, which could be the very first time it is tested! Listing 7.18 shows a few changes made to the Python statements to demonstrate this potential problem.

**LISTING 7.18**   And else Clause Problem Due to a Pretest

```
1: >>> list_of_names = ""
2: >>> the_name = "Start"
3: >>> while the_name != "Start":
4: ...      the_name = input("Enter name: ")
5: ...      list_of_names += the_name
6: ... else:
7: ...      print (list_of_names)
8: ...
9:
10: >>>
```

The test statement in the while loop returns false before the loop's statements even iterates one time. However, the else section still executes, and thus a blank line prints on line 9! You can see that the else clause operates very differently in a while loop than it does in an if-then-else statement.

# **Using** while True

An infinite loop can be created using a while loop. An *infinite loop* is a loop that never ends. Adding a break statement to this type of a while loop makes it usable. Take a look at Listing 7.19. The while test statement has been modified on line 3 to while True:, and this causes the loop to be infinite. This means the while loop will iterate indefinitely. Thus, line 5 tests for an added sentinel value within the loop. If the Enter key is pressed without a name being typed first, the if statement returns a true value, and break is executed. Thus, the infinite loop stops.

**LISTING 7.19**    while True and break

```
1: >>> list_of_names = ""
2: >>> the_name = "Start"
3: >>> while True:
4: ...      the_name = input("Enter name: ")
5: ...      if the_name == "":
6: ...              break
7: ...      list_of_names += the_name
8: ... else:
9: ...      print (list_of_names)
10: ...
11: Enter name: Raz
12: Enter name: Berry
13: Enter name: Pi
14: Enter name:
15: >>>
```

Another item to notice in Listing 7.19 is that the else clause is not executed. This is because when you issue a break in a loop, any Python statements in the else clause are skipped. You simply "jump" right out of the while loop. To get the list of names printed out, you remove the else clause and move the print statement to an if statement before the break, as shown in Listing 7.20.

**LISTING 7.20**    An else Clause Fix

```
>>> list_of_names = ""
>>> the_name = "Start"
>>> while True:
...      the_name = input("Enter name: ")
...      if the_name == "":
...              print (list_of_names)
...              break
...      list_of_names += the_name
...
Enter name: Raz
Enter name: Berry
```

```
Enter name: Pi
Enter name:
RazBerryPi
>>>
```

## Use a while Loop to Enter Data

A loop is a useful tool for entering data. In the following steps, you will be creating a script to enter a fictitious club's member list by using a while loop. The script will ask up front for the number of member names you will be entering, and then the while loop will ask for the member's first, middle, and last names. Follow these steps:

1. If you have not already done so, power up your Raspberry Pi and log in to the system.

2. If you do not have the LXDE GUI started automatically at boot, start it now by typing startx and pressing Enter.

3. Open the IDLE window by double-clicking the IDLE 3 icon.

4. Open the IDLE text editor window by pressing Ctrl+N.

5. Type the code shown in Figure 7.4 into the IDLE text editor window, pressing Enter when you need to get to the next line.

**FIGURE 7.4**
The `script0702.py` script.

BY THE WAY

### Be Careful!

Be sure to take your time here and avoid making typographical errors. You can make corrections by using the Delete key and the up- and down-arrow keys.

Notice that no input verification code for entered member names is present. That will be added in the next section of this hour.

**6.** Test your new script in IDLE by pressing the F5 key and entering answers to the questions. Your results should look similar to the results in Figure 7.5.

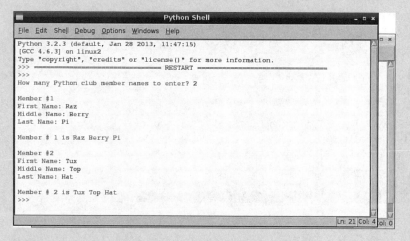

**FIGURE 7.5**
The `script0702.py` script output.

**7.** If you want to save this script, press Ctrl+S to open the Save As window, double-click the `py3prog` folder icon, type `script0702.py` in the File Name bar, and then click the Save button.

**8.** Exit the IDLE environment by pressing Ctrl+Q.

No input verification is done on the `while` loop that you entered in this section. In the next part of this hour, you will see how to clean up `script0702.py` by using a ***nested loop***.

# Creating Nested Loops

A *nested loop* is a loop statement that is inside a loop statement. For example, a `for` loop used within the code block of a `while` loop would be a nested loop. Listing 7.21 shows a script that uses nested loops. It has a `for` loop that contains three `while` loops in the `for` loop's code block. `script0703.py` is a slight improvement over the script you wrote in the last Try it Yourself section of this hour.

**LISTING 7.21**    A Nested Loop in `script0703.py`

```
1: pi@raspberrypi ~ $ cat py3prog/script0703.py
2: # script0703.py - Demonstration of a nested loop.
3: # Author: Blum and Bresnahan
4: # Date:  May 01
5: ##############################################################
6: #
7: # Find out how many club member names need to be entered
8: names_to_enter = int(input("How many Python club member names to enter? "))
9: #
10: for member_number in range (1, names_to_enter + 1):  #Loop to enter names
11:      print ()
12:      print ("Member #" + str(member_number))
13: #
14:      first_name = ""                    # Initialize first_name
15:      middle_name = ""                   # Initialize middle_name
16:      last_name = ""                     # Initialize last_name
17: #
18:      while first_name == "":            #Loop to keep out blanks
19:           first_name = input("First Name: ")     #Get first name
20:      while middle_name == "":           #Loop to keep out blanks
21:           middle_name = input("Middle Name: ")   #Get middle name
22:      while last_name == "":             #Loop to keep out blanks
23:           last_name = input("Last Name: ")       #Get last name
24: #
25:      print ()          # Display a member's full name
26:      print ("Member #", member_number, "is", first_name, middle_name, last_
name)
27:
28:
29:  pi@raspberrypi ~ $
```

The first improvement is that the main `while` loop has been replaced by a `for` loop, starting on line 10. Using a `for` loop eliminates the need to keep track of the number of names that have been entered along the way.

Nested within the `for` loop are three `while` loops on lines 18, 20, and 22 in Listing 7.21. These `while` loops improve the input verification. They ensure that a script user cannot accidentally leave a name blank.

In Listing 7.22, you can see the new script run and an example of the input verification improvement. When the script user accidently presses the Enter key instead of entering Raz's middle name on lines 6 and 7, the script loops back to the `input` statement and asks again.

**LISTING 7.22    Output of `script0703.py`**

```
1: pi@raspberrypi ~ $ python3  py3prog/script0703.py
2: How many Python club member names to enter? 1
3:
4: Member #1
5: First Name: Raz
6: Middle Name:
7: Middle Name:
8: Middle Name: Berry
9: Last Name: Pi
10:
11: Member # 1 is Raz Berry Pi
12: pi@raspberrypi ~ $
```

Listings 7.21 and 7.22 show a very simple example of a nested loop. Nested loops are often used in Python scripts for processing data tables, running image algorithms, manipulating games, and so on.

# Summary

In this hour, you got a little loopy. You learned how to create a `for` loop and a `while` loop. Also, you were introduced to concepts such as pretesting, sentinels, and input verification. Finally, you got to try out both a `for` loop and a `while` loop and look at a nested loop. In Hour 8, "Using Lists and Tuples," you will be moving on from structured commands and investigating lists and tuples.

# Q&A

**Q. Does Python v3 have the `xrange` function?**

**A.** Yes and no. In Python v2, the `xrange` function was available, along with the `range` function. In Python v3, the `range` function is the old `xrange` function, and the Python v2 `range` function is gone. The creators of Python made this change because `xrange` is more efficient in terms of memory usage than `range`. Unfortunately, to convert a Python v2 `xrange` to Python v3, you need to remove the `x` in front of the word `range`.

**Q.** Is it poor form to use a `break` statement in a loop?

**A.** This depends on who you ask. If you can avoid using a `break`, that is best. However, there are times when you cannot determine another method, so you have to use `break`. Most hard-core programmers consider using `break` to be poor form.

**Q.** I am running my Python script, and it is stuck in an infinite loop! What do I do?

**A.** You can stop the execution of a Python script by pressing Ctrl+C. If this doesn't work, try Ctrl+Z.

# Workshop

## Quiz

1. A `for` loop is a count-controlled loop, and a `while` loop is a condition-controlled loop. True or false?

2. What type of loop is pretested?

3. If you want to produce a list of numbers for a `for` loop, starting at `1` and going to `10`, with a step of `1`, which `range` statement should you use?

   **a.** `range (10)`

   **b.** `range (1, 10, 1)`

   **c.** `range (1, 11)`

## Answers

1. True. A `for` loop iterates a set number of times, and thus each iteration is counted. A while loop iterates until a certain condition is met and then it stops.

2. A `while` loop is a pretested loop, because the condition test statement is run before the statements in the loop's code block are executed.

3. Answer c is correct. The range (1, 11) produces the following list of numbers for the `for` loop to use: [1, 2, 3, 4, 5, 6, 7, 8, 9, 10]. Remember that the last number, the stop number, in the `range` function is not included in the output list.

# PART III

# Advanced Python

# Using Lists and Tuples

**What You'll Learn in This Hour:**

▶ Working with tuples and lists

▶ Using multidimensional lists

▶ Building lists with comprehensions

When you work with variables, sometimes it comes in handy to group data values together so that you can iterate through them later in your script. You can't easily do that with separate variables, but Python provide a solution for you.

Most programming languages use *array variables* to hold multiple data values but point to them using a single indexed variable name. Python is a little different in that it doesn't use array variables. Instead, it uses a couple other variable types, called *lists* and *tuples*. This hour examines how to use lists and tuples to store and manipulate data in Python scripts.

# Introducing Tuples

The tuple data type in Python allows you to store multiple data values that don't change. In programming-speak, these data values are said to be *immutable*.

After you create a tuple, you can either work with the tuple as a single object or reference each individual data value inside the tuple in your Python script code.

The following sections walk through how to create and use tuples in your scripts.

## Creating Tuples

There are four different ways to create a tuple value in Python:

▶ Create an empty tuple value by using parentheses, as in this example:

```
>>> tuple1 = ()
>>> print(tuple1)
()
>>>
```

▶ Add a comma after a value in an assignment, as in this example:

```
>>> tuple2 = 1,
>>> print(tuple2)
(1,)
>>>
```

▶ Separate multiple data values with commas in an assignment, as in this example:

```
>>> tuple3 = 1, 2, 3, 4
>>> print(tuple3)
(1, 2, 3, 4)
>>>
```

▶ Use the `tuple()` built-in function in Python and specify an iterable value (such as a list value, which we'll talk about later), as in this example:

```
>>> list1 = [1, 2, 3, 4]
>>> print(list1)
[1, 2, 3, 4]
>>> tuple4 = tuple(list1)
>>> print(tuple4)
(1, 2, 3, 4)
>>>
```

As you may have noticed in these examples, Python denotes the tuple by grouping the data values using parentheses.

You're not limited to storing numeric values in tuples. You can also store string values:

```
>>> tuple5 = "Sunday", "Monday", "Tuesday", "Wednesday", "Thursday",
"Friday", "Saturday"
>>> print(tuple5)
('Sunday', 'Monday', 'Tuesday', 'Wednesday', 'Thursday', 'Friday',
'Saturday')
>>>
```

You can use either single or double quotes to delineate the string values (see Hour 10, "Working with Strings").

WATCH OUT!

## Tuples Are Permanent

Once you create a tuple, you can't change the data values, nor can you add or delete data values.

## Accessing Data in Tuples

After you create a tuple, most likely you'll want to be able to access the data values you stored in it. To do that, you need to use an index.

An *index* points to an individual data value location within a tuple variable. You use the index value to retrieve a specific data value stored in the tuple from other Python statements in your scripts.

The index value 0 references the first data value you stored in the tuple. Starting at 0 can be confusing, so be careful when trying to reference the data values! Here's an example:

```
>>> tuple6 = (1, 2, 3, 4)
>>> print(tuple6[0])
1
>>>
```

To reference a specific index in the tuple variable, you just place square brackets around the index value and add it to the end of the tuple variable name.

If you try to reference an index value that doesn't exist, Python produces an error, like this:

```
>>> print(tuple6[5])
Traceback (most recent call last):
  File "<pyshell#33>", line 1, in <module>
    print(tuple6[5])
IndexError: tuple index out of range
>>>
```

## Accessing a Range of Values

Besides just retrieving a single data value from a tuple, Python also allows you to retrieve a subset of the data values. If you need to retrieve a sequential subset of data values from the tuple (called a *slice*), you just use the index format [i:j], where i is the starting index value, and j is the ending index value. Here's an example of doing that:

```
>>> tuple7 = tuple6[1:3]
>>> print(tuple7)
(2, 3)
>>>
```

### Starting and Ending Tuple Slices

Notice that the first value in the new tuple is the starting index value defined for the slice, but the ending value is the index value just before the ending index value defined for the slice. This can be somewhat confusing. To help remember this format, when determining a tuple slice, just use the equation i <= x < j, where x is the index values you want to retrieve.

Finally, there's one more format you can use for extracting data elements from a tuple, [i:j:k], where i is the starting index value, j is the ending index value, and k is a step amount to use to increment the index values in between the start and ending values. Here's an example of how this works:

```
>>> tuple8 = 1, 2, 3, 4, 5, 6, 7, 8, 9, 10
>>> tuple9 = tuple8[0:6:2]
>>> print(tuple9)
(1, 3, 5)
>>>
```

The tuple9 value consists of the data values contained in the tuple8 variable starting at index 0, until index 6, skipping every 2 index values. Thus, the resulting tuple consists of index values 0, 2, and 4, which creates the tuple (1, 3, 5).

# Working with Tuples

Since tuple values are immutable, there aren't any Python functions available to manipulate the data values contained in a tuple. However, there are some functions available to help you gain information about the data contained in a tuple.

## Checking Whether a Tuple Contains a Value

There are two comparison operations you can use with tuple variables to check whether a tuple contains a specific data value.

The in comparison operator returns a Boolean True value if the specified value is contained in the tuple data elements:

```
>>> if 7 in tuple8: print("It's there!")

It's there!
>>> if 12 in tuple8:
    print("It's there!")
else:
    print("It's not there!")

It's not there!
>>>
```

You can also add the `not` logical operator with the `in` comparison operator to reverse the result:

```
>>> if 7 not in tuple8:
        print("It's not there!")
else:
        print("It's there!")


It's there!
>>>
```

Sometimes adding the `not` logical operator to the comparison comes in handy, such as if you want to reverse the order of the "then" and "else" code blocks to place the shorter block first before the longer code block.

## Finding the Number of Values in a Tuple

Python includes the `len()` function to allow you to easily determine how many data values are in a tuple. Here's an example of its use:

```
>>> len(tuple8)
10
>>>
```

WATCH OUT!

### Referencing the Last Value in a Tuple

Be careful when you use the `len()` function with tuples. A common beginner's mistake is to think the value returned by `len()` is the index of the last data value in the tuple. Remember that the tuple index starts at `0`, so the ending tuple index value is one less than the value the `len()` function returns!

## Finding the Minimum and Maximum Values in a Tuple

Python provides the `min()` and `max()` functions to give an easy way to find the smallest (`min()`) and largest (`max()`) values in a tuple, as in this example:

```
>>> min(tuple8)
1
>>> max(tuple8)
10
>>>
```

The `min()` and `max()` functions can also work with tuples that store string values. Python determines the minimum and maximum values by using standard ASCII comparisons:

```
>>> min(tuple4)
'Friday'
>>> max(tuple4)
'Wednesday'
>>>
```

This is a quick way to find the range of values stored in a tuple!

### Concatenating Tuples

While you can't change the data elements contained within a tuple value, you can concatenate two or more tuple values to create a new tuple value:

```
>>> tuple10 = 1, 2, 3, 4
>>> tuple11 = 5, 6, 7, 8
>>> tuple12 = tuple10 + tuple11
>>> print(tuple12)
(1, 2, 3, 4, 5, 6, 7, 8)
>>>
```

This can be somewhat misleading if you're not familiar with tuples. The plus sign isn't used as the addition operator; with tuples, it's used as the concatenation operator. Notice that concatenating the two tuple values creates a new tuple value that contains all the data elements from the original two tuple values.

# Introducing Lists

Lists are similar to tuples, storing multiple data values referenced by a single list variable. However, lists are mutable, and you can change the data values as well as add or delete data values stored in the list. This adds a lot of versatility for your Python scripts!

The following sections show how to create lists, as well as how to extract the data you store in a list and work with the data.

## Creating a List

Very much like with tuples, there are four different ways to create a list variable:

▶ Create an empty list by using an empty pair of square brackets, as in this example:

```
>>> list1 = []
>>> print(list1)
[]
>>>
```

▶ Place square brackets around a comma-separated list of values, as in this example:

```
>>> list2 = [1, 2, 3, 4]
>>> print(list2)
[1, 2, 3, 4]
>>>
```

▶ Use the `list()` function to create a list from another iterable object, as in this example:

```
>>> tuple11 = 1, 2, 3, 4
>>> list3 = list(tuple11)
>>> print(list3)
[1, 2, 3, 4]
>>>
```

▶ Use a list comprehension.

The list comprehension method of creating lists is a more complicated process of generating a list from other data. We'll discuss how it works toward the end of this hour. Notice that with lists, Python uses square brackets around the data values, not parentheses as with tuples.

Just as with tuples, lists can contain any type of data, not just numbers, as in this example:

```
>>> list4 = ['Rich', 'Barbara', 'Katie Jane', 'Jessica']
>>> print(list4)
['Rich', 'Barbara', 'Katie Jane', 'Jessica']
>>>
```

## Extracting Data from a List

The examples in the previous section show how to extract all the data values from a list at the same time, by just referencing the list variable. You can retrieve individual data elements from list values by using index values, just as with tuple values. Here's an example:

```
>>> print(list2[0])
1
>>> print(list2[3])
4
>>>
```

You can also use a negative number for the list index. A negative index retrieves values starting from the end of the list:

```
>>> print(list2[-1])
4
>>>
```

Notice that when you use negative index values, the -1 value starts at the end of the list, since -0 is the same as 0.

Lists also support the slicing method of retrieving a subset of the data elements contained in the list value, as in the following example:

```
>>> list4 = list2[0:3]
>>> print(list4)
[1, 2, 3]
>>>
```

# Working with Lists

As mentioned earlier, the main difference between lists and tuples is that you can change the data elements contained in a list value. This means there are lots of things you can do with lists that you can't do with tuples! This section walks through the different operations you can perform with list values.

## Replacing List Values

The most basic operation you can perform with a list value is to replace an individual data value contained in the list. Doing this is as easy as using an assignment statement in your scripts, referencing the individual list data value by its index, and assigning it a new value. For example, this example replaces the second data value (referenced by index value 1) with the value 10:

```
>>> list1 = [1, 2, 3, 4]
>>> list1[1] = 10
>>> print(list1)
[1, 10, 3, 4]
>>>
```

When you print the list value, it now contains the value 10 as the second data value.

In a much trickier operation, it's possible to replace a subset of data values with another list or tuple value. You reference the subset by using the list slicing method, as shown here:

```
>>> list1 = [1, 2, 3, 4]
>>> tuple1 = 10, 11
>>> list1[1:3] = tuple1
>>> print(list1)
[1, 10, 11, 4]
>>>
```

Python replaces the data elements from index 1 up to index 3 with the data elements stored in the tuple1 value.

## Deleting List Values

You can remove data elements from within a list value by using the del statement, as shown here:

```
>>> print(list1)
[1, 10, 11, 4]
>>> del list1[1]
>>> print(list1)
[1, 11, 4]
>>>
```

You can also use slicing to remove a subset of data elements from the list, as in this example:

```
>>> list5 = [1, 2, 3, 4, 5, 6, 7, 8, 9, 10]
>>> del list5[3:6]
>>> print(list5)
[1, 2, 3, 7, 8, 9, 10]
>>>
```

The slicing method allows you to customize exactly which data elements to remove from the list.

## Popping List Values

Python provides a special function that can both retrieve a specific data element and remove it from the list value. The pop() function allows you to extract a value from anywhere in a list. For example, this example pops the fifth index value from the list6 list:

```
>>> list6 = [1, 2, 3, 4, 5, 6, 7, 8, 9, 10]
>>> list6.pop(5)
6
>>> print(list6)
[1, 2, 3, 4, 5, 7, 8, 9, 10]
>>>
```

When you pop a value from a list, the index values shift over to replace the popped index value.

If you don't specify an index value in the pop() function, it returns the last data value in the list, like this:

```
>>> list6.pop()
10
>>> print(list6)
[1, 2, 3, 4, 5, 7, 8, 9]
>>>
```

## Adding New Data Values

You can add new data values to an existing list by using the `append()` function, as shown here:

```
>>> list7 = [1.1, 2.2, 3.3]
>>> list7.append(4.4)
>>> print(list7)
[1.1, 2.2, 3.3, 4.4]
>>>
```

The `append()` function adds the new data value to the end of the existing list.

You can insert a new data value into a list at a specific index location by using the `insert()` function. The `insert()` function takes two parameters. The first parameter is the index value before which to place the new data value, and the second parameter is the value to insert. Thus, to insert a new data value at the front of the list, you use this:

```
>>> list7.insert(0, 0.0)
>>> print(list7)
[0.0, 1.1, 2.2, 3.3, 4.4]
>>>
```

To insert a data value in the middle of the list, you use this:

```
>>> list7.insert(3, 2.5)
>>> print(list7)
[0.0, 1.1, 2.2, 2.5, 3.3, 4.4]
>>>
```

The `insert()` statement inserts the value 2.5 before index 3 in the list, making it now the new index 3 value and pushing the other index locations down one position in the list.

You can use a combination of the `append()` and `pop()` functions to create a storage area commonly called a *stack* in your Python scripts. You push data values onto the stack and then retrieve them in the opposite order from which you pushed them (called last-in, first-out [LIFO]). To do this, you just use the `append()` function to add new data values to an empty list and then retrieve them by using the `pop()` function, without specifying the index. Listing 8.1 shows an example of doing this in a script.

**LISTING 8.1**   Using `append()` and `pop()` to work with a list

```
#!/usr/bin/python3

list1 = []

# push some data values into the list
list1.append(10.0)
list1.append(20.0)
```

```
list1.append(30.0)
print("The starting list is", list1)

# pop some values and see what happens
result1 = list1.pop()
print("The first item removed is", result1)
result2 = list1.pop()
print("The second item removed is", result2)

# add one more data value and see where it goes
list1.append(40.0)
print("The final version is", list1)
```

The `script0801.py` script creates an empty list by using the `list1` variable, and then it appends a few values into it. Then it retrieves a couple values by using the `pop()` function. When you run the `script0801.py` program, you should get this output:

```
pi@raspberrypi ~/scripts $ python3 script0801.py
The starting list is [10.0, 20.0, 30.0]
The first item removed is 30.0
The second item removed is 20.0
The final version is [10.0, 40.0]
pi@raspberrypi ~/scripts $
```

Using stacks is a common way to store values while performing calculations in long equations, as you can push values and operations into the stack and then pop them out in reverse order to process them.

## Concatenating Lists

You have to be a little careful about using the `append()` function with lists. If you try to append a list onto a list, you may not get what you were looking for, as shown here:

```
>>> list8 = [1, 2, 3]
>>> list9 = [4, 5, 6]
>>> list8.append(list9)
>>> print(list8)
[1, 2, 3, [4, 5, 6]]
>>>
```

When you use a list object with the `append()` function, Python appends the list as a single data value! Thus, the `list8[3]` value is now itself a list value in this example:

```
>>> print(list8[3])
[4, 5, 6]
>>>
```

If you wanted to concatenate the list8 and list9 lists, you need to use the extend() function, like this:

```
>>> list8 = [1, 2, 3]
>>> list9 = [4, 5, 6]
>>> list8.extend(list9)
>>> print(list8)
[1, 2, 3, 4, 5, 6]
>>>
```

Now the result is a list that contains the individual data values from the two lists. This also works using the addition sign, as with tuples:

```
>>> list8 = [1, 2, 3]
>>> list9 = [4, 5, 6]
>>> result = list8 + list9
>>> print(result)
[1, 2, 3, 4, 5, 6]
>>>
```

Again, the result is a single list of data values.

## Other List Functions

In addition to the list functions already discussed, Python includes a few other handy list functions by default. For example, you can count how many times a specific data value appears within a list by using the count() function, as shown here:

```
>>> list10 = [1, 5, 8, 1, 34, 75, 1, 23, 34, 100]
>>> list10.count(1)
3
>>> list10.count(34)
2
>>>
```

The 1 value occurs three times in the list, and the 34 value occurs twice in the list.

You can use the sort() function to sort the data values in a list, as in this example:

```
>>> list11 = ['oranges', 'apples', 'pears', 'bananas']
>>> list11.sort()
>>> print(list11)
['apples', 'bananas', 'oranges', 'pears']
>>>
```

### Sorting in Place

Notice that the `sort()` function replaces the original order of the data values with the sorted order in the list itself. This will change the index location of the individual data values, so be careful when referencing data values in the new list!

You can find the location of a data value within a list by using the `index()` function. The `index()` function returns the index location value of the first occurrence of the data value within the list:

```
>>> list11.index('bananas')
1
>>>
```

You can easily reverse the order of the data values stored in a list by using the `reverse()` function, like this:

```
>>> list12 = [1, 2, 3, 4, 5]
>>> list12.reverse()
>>> print(list12)
[5, 4, 3, 2, 1]
>>>
```

All the data values are still in the list, just in the opposite order from where they started.

# Using Multidimensional Lists to Store Data

Python supports the use of multidimensional lists—that is, lists that contain data values that themselves can be lists!

In a multidimensional list, more than one index value is associated with each specific data element contained in the multidimensional list. It can get somewhat complicated trying to keep track of your data in multidimensional lists, but these lists do come in handy!

You create a multidimensional list the same way you create normal lists, just with defining lists as the data values. Here's an example:

```
>>> list13 = [[1, 2, 3], [4, 5, 6], [7, 8, 9]]
>>> print(list13)
[[1, 2, 3], [4, 5, 6], [7, 8, 9]]
>>>
```

To reference an individual data value within a multidimensional list, you must specify the index value for the main list, as well as the index value for the data value list. You place square brackets around each index value and place them in order from the outermost list to the innermost list. Here are some examples:

```
>>> print(list13[0][0])
1
>>> print(list13[0][2])
3
>>> print(list13[2][1])
8
>>>
```

The first example retrieves the first data value contained in the first list. The last example retrieves the second data value contained in the third list. This demonstrates a two-dimensional list. You can continue this further by using list data values for the list data values within the list, creating a three-dimensional list! You can continue on even further, but anything more than three dimensions starts getting extremely complicated.

# Working with Lists and Tuples in Your Scripts

Lists and tuples are powerful tools to have at your disposal in your Python scripts. Once you load your data into a list or tuple, there are lots of Python functions you can use to extract information on the data. That can make having to perform mathematical calculations a lot easier!

The following sections show some of the most common data functions you can use with your lists and tuples.

## Iterating Through a List or Tuple

One of the most popular uses of lists and tuples is iterating through individual items using a loop. When you do this, you can grab each data value contained in the list or tuple individually and process the data.

To iterate through the data values, you need to use the `for` statement (discussed in Hour 7, "Learning About Loops"), like this:

```
>>> list14 = [1, 15, 46, 79, 123, 427]
>>> for x in list14:
        print("One value in the list is", x)


One value in the list is 1
One value in the list is 15
One value in the list is 46
```

```
One value in the list is 79
One value in the list is 123
One value in the list is 427
>>>
```

The x variable contains an individual data value from the list for each iteration of the for statement. The print statement displays the current value of x in each iteration.

## Sorting and Reversing Revisited

In the "Working with Lists" section earlier this hour, you saw how to sort and reverse the data values inside a list. There are also functions that allow you to sort or reverse the data values but return the result as a separate list, keeping the original list intact.

The sorted() function returns a sorted version of the list data values:

```
>>> list15 = ['oranges', 'apples', 'pears', 'bananas']
>>> result1 = sorted(list15)
>>> print(list15)
['oranges', 'apples', 'pears', 'bananas']
>>> print(result1)
['apples', 'bananas', 'oranges', 'pears']
>>>
```

The original list15 variable remains the same, and the result1 variable contains the sorted version of the list.

The reversed() function returns a reversed version of list data values, but it is a little tricky. Instead of returning a list, it returns an iterable object, which can be used in a for statement but cannot be directly accessed. Here's an example:

```
>>> list15 = ['oranges', 'apples', 'pears', 'bananas']
>>> result2 = reversed(list15)
>>> print(result2)
<list_reverseiterator object at 0x01559F70>
>>> for fruit in result2:
    print("My favorite fruit is", fruit)


My favorite fruit is bananas
My favorite fruit is pears
My favorite fruit is apples
My favorite fruit is oranges
>>>
```

If you try to print the `result2` variable, you get a message that it's a `reverseiterator` object and not printable. You can, however, use the `result2` variable in the `for` statement to iterate through the reversed values.

# Creating Lists by Using List Comprehensions

As mentioned earlier this hour, in the "Creating a List" section, there's a fourth way of creating lists: using a list comprehension. Using a *list comprehension* is a shortcut way to create a list by processing the data values contained in another list or tuple.

This is the basic format of a list comprehension statement:

```
[expression for variable in list]
```

The variable represents each data value contained in the list, as a normal `for` statement. A list comprehension applies the expression on each variable to create the new data values in the new list. Here's an example of how it works:

```
>>> list17 = [1, 2, 3, 4]
>>> list18 = [x*2 for x in list17]
>>> print(list18)
[2, 4, 6, 8]
>>>
```

In this case, the list comprehension defines the expression as `x*2`, which multiplies each data value in the original list by 2.

You can make the expression as complex as you like. Python just applies the expression—whatever it is—to the new data values in the new list. You can also use list comprehensions with string functions and values, as shown here:

```
>>> tuple19 = 'apples', 'bananas', 'oranges', 'pears'
>>> list19 = [fruit.upper() for fruit in tuple19]
>>> print(list19)
['APPLES', 'BANANAS', 'ORANGES', 'PEARS']
>>>
```

In this example, you apply the `upper()` function (see Hour 10) to the string values contained in the list.

# Working with Ranges

To close out this topic, there is one other Python data type you'll run into that can create multiple data elements. The `range` data type contains an immutable sequence of numbers that work a lot like a tuple but are a lot easier to create.

You create a new range value by using the `range()` method, which has the following format:

```
range(start, stop, step)
```

The `start` value determines the number where the range starts, and the `stop` value determines where the range stops (always one less than the `stop` value specified). The `step` value determines the increment value between values. The `stop` and `step` values are optional; if you leave them out, Python assumes a value of 0 for `start` and 1 for `step`.

The `range` data type is a bit odd to work with: You can't reference it directly, such as to print it out. You can only reference the individual data values contained in the range, like this:

```
>>> range1 = range(5)
>>> print(range1)
range(0, 5)
>>> print(range1[2])
2
>>> for x in range1:
    print(x)


0
1
2
3
4
>>>
```

When you try to print the `range1` variable, Python just returns the `range` object, showing the `start` and `stop` values. However, you can print the `range1[2]` value, which references the third data value in the range.

The `range` data type comes in most handy in the `for` statement, as shown in the preceding example. You can easily iterate through a range of values in the `for` loop by just specifying the range.

WATCH OUT!

## The `range()` Change

In Python v2, the `range()` function created a sequence of numbers as a standard list data type. Python v3 changed that to make the `range` data type separate from the `list` data type. Be careful if you run into any v2 code that assumes that `range` is `list`!

# Summary

In this hour, you took a look at the tuple and list data types in Python. Tuples allow you to reference multiple data values using a single variable. Tuple values are immutable, so once you create a tuple, you can't change it in your program code. Lists also contain multiple data values, but you can change, add, and delete values in lists. Python supports lots of functions to help you manipulate data using lists. They come in handy when you need to iterate through a data set of values in your scripts. List comprehensions allow you to create new lists based on values in another list, a tuple, or a range of values. You can define complex equations to manipulate the data as Python transfers it using a list comprehension, making it a very versatile tool in Python.

In the next hour, we'll turn our attention to yet another type of data storage in Python, using dictionaries and sets.

# Q&A

**Q. Can you use lists to perform matrix arithmetic?**

**A.** Not easily. Python doesn't have any built-in functions that can perform mathematical operations on list data values directly. You'd have to write your own code to iterate through the individual list values and perform the calculations.

Fortunately, there's the NumPy module (see Hour 5, "Using Arithmetic in Your Programs"), which provides a separate matrix object and functions to perform matrix math using those objects.

**Q. Most programming languages support associative arrays, matching a key to a value in an array. Do lists or tuples support this feature?**

**A.** No. Python uses a separate data type to support associative array features (see Hour 9, "Dictionaries and Sets"). Tuples and lists can only use numeric index values.

# Workshop

## Quiz

**1.** What does Python use to denote a list value?

    **a.** Parentheses

    **b.** Square brackets

    **c.** Braces

**2.** You can change a data value in a tuple, but not in a list. True or False.

**3.** What list comprehension statement should you use to quickly create a list of multiples of 3 up to 30?

## Answers

**1.** b. You'll need to get in the habit of remembering that Python uses parentheses for tuples, and square brackets for lists.

**2.** False. Python allows you to change the data values in a list, but tuple values remain constant, you can't change them!

**3.** `[x * 3 for x in range(11)]`. The comprehension uses the variable x to represent the numbers in the range. Each iteration multiplies the number by 3 before saving it in the range.

# Dictionaries and Sets

## What You'll Learn in This Hour:

- ▶ What a dictionary is
- ▶ How to populate a dictionary
- ▶ How to obtain information from a dictionary
- ▶ What a set is
- ▶ How to program with sets

In this hour, you will read about two additional Python collection types: dictionaries and sets. You will learn what they are, how to create them, how to fill them with data, how to manage them, and how to use them in Python scripts.

# Understanding Python Dictionary Terms

A *dictionary* is a simple structure, also called an *associative array*, that contains data. Each piece of data contained in a dictionary is called an *element*, an *entry*, or a *record*. Each element is broken up into two parts. One part is called the *value*. To locate a particular value in a dictionary, you use the other part of the dictionary element, the *key*. A key is immutable, and it is assigned only to one particular value in the dictionary. Thus, another name for a dictionary element is a *key/value pair*.

BY THE WAY

### When Is a Dictionary Not a Dictionary?

Don't let the term *dictionary* fool you. A dictionary in Python is not the same as a reference book containing word definitions found in the library. For one thing, a word in a dictionary reference may have multiple definitions. A key in a Python dictionary has only one value.

An example of a Python dictionary is a college, which has a list of student names and their associated student ID numbers. The college decides to build an array, where the student ID number would be the key and the student name would be the value. Each student ID would be assigned to only one student name. Thus, the key/value pair for this college dictionary would be the student ID number/student name.

# Exploring Dictionary Basics

Before you can start programming with dictionaries, you need to learn a few basics, such as how to access the data in a dictionary. Learning these basics will help when you read through the "Programming with Dictionaries" section of this hour.

## Creating a Dictionary

Creating and using dictionaries in Python is very simple. To create an empty dictionary, you just use the Python statement *dictionary_name = {}*.

In Listing 9.1, a dictionary called student is created. Then the type function is used on it. You can see that student is a dictionary (dict) type.

**LISTING 9.1**   Creating an Empty Dictionary

```
>>> student = {}
>>> type (student)
<class 'dict'>
>>>
```

## Populating a Dictionary

*Populating a dictionary* means putting keys and their associated values into the dictionary. To populate a dictionary, you use this syntax:

*dictionary_name = {key1:value1, key2:value2...}*

BY THE WAY

### No Need to Pre-Create

You don't have to create an empty dictionary before you start to populate it. You can create it and populate it all with one command. Just issue the command to populate the dictionary, and Python automatically creates the dictionary for you.

Listing 9.2 shows an example of populating a dictionary. Here the student dictionary is populated with three students. The key is the student ID number, such as 400A42, and the value is the student's name.

**LISTING 9.2** Populating a Dictionary

```
>>> student = {'400A42':'Paul Bohall','300A04':'Jason Jones','000B35':'Raz Pi'}
>>> student
{'300A04': 'Jason Jones', '000B35': 'Raz Pi', '400A42': 'Paul Bohall'}
>>>
```

Notice in Listing 9.2 that the three student key/value elements are between a pair of curly brackets. Each key/value element is set apart from the other elements by a comma. Also, both the key and the value are character strings and thus must have quotation marks around them.

BY THE WAY

### No Order in a Dictionary

The elements in a Python dictionary are not ordered. This is why you may put into a dictionary key/value pairs in a certain order, and then they end up being displayed in a different order! You can see this in Listing 9.2.

You can also add key/value pairs to a dictionary one at a time. The syntax for this method is *dictionary_name[key1] = value1*, as shown in Listing 9.3.

**LISTING 9.3** Populating a Dictionary One Pair at a Time

```
>>> student['000B35'] = 'Raz Pi'
>>> student
{'300A04': 'Jason Jones', '000B35': 'Raz Pi', '400A42': 'Paul Bohall'}
>>>
```

There are a couple important items to note about key/value pairs:

- ▶ The key cannot be a list.
- ▶ The key must be an immutable object.
- ▶ The key can be a string.
- ▶ The key can be a number (integer or floating point).
- ▶ The key can be a tuple.
- ▶ The key must belong to only one value (no duplicate keys allowed).

The rules concerning the value in a key/value pair are much more simple. Basically, a value can be anything.

# Obtaining Data from a Dictionary

Once a dictionary is populated, you can obtain and use the elements from it. To obtain a single dictionary value, you use this syntax:

```
dictionary_name[key]
```

Obviously, with this method, you need to know the *key* in order to obtain its associated *value*.

In Listing 9.4, a data value was obtained from the sample student dictionary. By using the associated value's key, the value was obtained.

**LISTING 9.4**   Obtaining a Dictionary Value via Its Key

```
>>> student['000B35']
'Raz Pi'
>>>
```

BY THE WAY

## Mappings

Key/value pairs are sometimes called *mappings*. This is because a single key maps directly to a particular value.

You need to know a few rules about looking up key/value pair elements:

▶ If you enter a dictionary lookup with a key that doesn't exist, you get a `KeyError` exception.

▶ When using a character string for a key, you must use the correct case.

▶ Only a key can be used to access the value. You cannot use a numeric index to access a key/value element because the elements in a dictionary are associative not positional.

To avoid receiving an error exception for nonexistent dictionary keys, you use the `get` operation. This is the basic syntax:

```
database_name.get(key, default)
```

When the `get` operation finds a key, it returns the associated value. When the `get` operation does not find a key, it returns the string listed in its optional default. Listing 9.5 shows an example of successfully locating a key in the dictionary and then an unsuccessful attempt.

**LISTING 9.5**   Using the Dictionary get Operation

```
>>> student.get('000B35','Not Found')
'Raz Pi'
>>> student.get('000B34','Not Found')
'Not Found'
>>>
```

Notice that when the unsuccessful attempt occurs, the second string (the *default*) is displayed. That is because the default allows you to create your own error message. If the default section is not included in the get operation and the key is not in the dictionary, you receive back the string "None".

You can also use a loop to obtain dictionary values. First, you get a list of the dictionary's keys by using the keys operation. The syntax is *dictionary_name*.keys (). Set the results of this operation as a value to a variable. You can then use a for loop to traverse the dictionary, as shown in Listing 9.6.

**LISTING 9.6**   Using a for Loop to Traverse a Dictionary

```
>>> key_list = student.keys()
>>> print (key_list)
dict_keys(['300A04', '000B35', '400A42'])
>>>
>>> for the_key in key_list:
...      print(the_key, end = ' ')
...      student[the_key]
...
300A04 'Jason Jones'
000B35 'Raz Pi'
400A42 'Paul Bohall'
>>>
```

Notice in Listing 9.6 that the students' ID numbers are not listed in order. Remember in a dictionary, the elements are unordered. You can solve this display problem by using the sorted function.

BY THE WAY

**Not a List**

For Listing 9.6, you might think that you could just enter the list operation key_list.sort (), but that will not work. The variable key_list is a dictionary key (dict_keys) object type and not a list. Thus, you cannot use a list operation on it!

In Listing 9.7, the `key_list` variable is sorted first. Then it is used as an iteration in the `for` loop, to traverse the dictionary.

**LISTING 9.7**    Using a Sorted Key List to Traverse a Directory

```
>>> key_list = student.keys()
>>> type (key_list)
<class 'dict_keys'>
>>>
>>> key_list = sorted(key_list)
>>> type (key_list)
<class 'list'>
>>>
>>> for the_key in key_list:
...     print(the_key, end = ' ')
...     student[the_key]
...
000B35 'Raz Pi'
300A04 'Jason Jones'
400A42 'Paul Bohall'
>>>
```

Notice in Listing 9.7 that the variable `key_list` changes object types after it is sorted! It starts out as `dict_keys` and becomes a `list` object type.

## Updating a Dictionary

Remember that keys are immutable, so they cannot be changed. However, you can update a key's associated value. The syntax for doing so is *database_name*[*key*]=*value*. In Listing 9.8, the student's name associated with the student ID number (key) `'000B35'` is changed.

**LISTING 9.8**    Updating a Dictionary Element

```
>>> student['000B35']   #Element shown before the change.
'Raz Pi'
>>> student['000B35'] = 'Raz Berry Pi'
>>>
>>> student['000B35']   #Element shown after the change.
'Raz Berry Pi'
>>>
```

In Listing 9.8, a particular key is updated to a new value. However, if the key does not already exist, a new key/value pair is created. Therefore, you can use this method not only to update a dictionary but add new elements as well.

When you need to delete a key/value pair from a dictionary, you use the following syntax:

```
del dictionary_name[key]
```

However, if the key/value pair does not exist in the dictionary, the `del` operation throws an error exception. Listing 9.9 uses an `if` statement to make sure the key does exist, before deleting the element.

**LISTING 9.9**    Deleting a Dictionary Element

```
>>> if '400A42' in student:
...     del student['400A42']
...
>>> student
{'300A04': 'Jason Jones', '000B35': 'Raz Berry Pi'}
>>>
```

BY THE WAY

### `has_key` **No Longer Available**

The dictionary operation `has_key` allowed you to determine whether a particular key existed in a dictionary. For Python v3, this dictionary operation is no longer available. Instead, you now use the `if` statement as shown in Listing 9.9.

# Managing a Dictionary

In addition to the ones you've already seen this hour, a few other dictionary operations may prove useful when you're using a Python dictionary. Table 9.1 lists them, as well as a few of the ones you've already seen this hour.

**TABLE 9.1**    Python Dictionary Management Operations

| Operation | Function |
| --- | --- |
| `len(dictionary_name)` | Returns the number of elements in a dictionary. |
| `dictionary_name.keys ()` | Returns the current keys in the dictionary (no values returned). |
| `dictionary_name.values ()` | Returns the current values in the dictionary (no keys shown). |
| `dictionary_name.items ()` | Returns a tuple that contains the dictionary's key/value pairs. |
| `dictionary_name.update (other_dictionary)` | Compares `dictionary` to `other_dictionary` and adds to the `dictionary` any key/value pairs that exist in `other_dictionary` but are missing from the `dictionary`. Finally, updates any key/value pairs in the `dictionary` to the matching key/value pairs in `other_dictionary`. |
| `dictionary_name.clear ()` | Removes all the dictionary's elements. |

Now that you have an idea of the various dictionary operations, you can learn about how to program using dictionaries.

# Programming with Dictionaries

Put on your weather researcher hat: You are going to do some weather data processing using a dictionary. In this part of the hour, you will be looking at three different Python scripts that use dictionaries to store and then analyze weather data.

The first script, script0901.py, populates a dictionary with daily record high temps (Fahrenheit) in Indianapolis, Indiana, for the month of May. In Listing 9.10, you can see the dictionary used to store the collected data. Each dictionary element's key is the day of the month in May. Each key's associated value is the logged record high temperature for that day.

**LISTING 9.10**  The script0901.py Script

```
1: pi@raspberrypi ~ $ cat py3prog/script0901.py
2: # script0901.py - Populate the Record High Temps Dictionary
3: # Author: Blum and Bresnahan
4: # Date:  May
5: ###########################################################
6: #
7: # Populate dictionary for Record High Temps (F) during May in Indianapolis
8: print ()
9: print ("Enter the record high temps (F) for May in Indianapolis...")
10: #
11: may_high_temp = {}              #Create empty dictionary
12: #
13: for may_date in range (1, 31 + 1):  #Loop to enter temps
14: #
15:     # Obtain record high temp for date
16:     prompt = "Record high for May " + str(may_date) + ": "
17:     record_high = int(input(prompt))
18:     # Put element in dictionary
19:     may_high_temp[may_date] = record_high
20: #
21: ###########################################################
22: # Display Record High Temps Dictionary
23: #
24: print ()
25: print ("Record High Temperatures (F) in Indianapolis during Race Month")
26: #
27: date_keys = may_high_temp.keys()    #Obtain list of element keys
28: #
29: for may_date in date_keys:          #Loop to display key/value pairs
30:     print ("May", may_date, end = ': ')
```

```
31:        print (may_high_temp[may_date])
32: #
33: #########################################################
34: pi@raspberrypi ~ $
```

On line 11, the empty dictionary `may_high_temp` is created. Then, using a `for` loop, the 31 days of high temperatures (Fahrenheit) are entered into the dictionary on lines 13–19. The elements in the dictionary are then pulled out and displayed one by one, using another `for` loop on lines 29–31. Notice that to make the values display in the right order, they are retrieved using their key value on line 27.

Listing 9.11 shows partial output from this Python script being run. The data being entered is record high daily temperatures (Fahrenheit) for May in Indianapolis. After this data is entered, it is then displayed back out to the screen.

**LISTING 9.11**   Output of `script0901.py`

```
pi@raspberrypi ~ $ python3 py3prog/script0901.py

Enter the record high temps (F) for May in Indianapolis...
Record high for May 1: 88
Record high for May 2: 85
Record high for May 3: 88
...
Record high for May 29: 90
Record high for May 30: 92
Record high for May 31: 90

Record High Temperatures (F) in Indianapolis during Race Month
May 1: 88
May 2: 85
May 3: 88
...
May 29: 90
May 30: 92
May 31: 90
pi@raspberrypi ~ $
```

Now your dictionary is loaded with record temperatures, in Fahrenheit. To convert the temperatures to Celsius, you can make a small change to the original script.

Listing 9.12 shows part of the new script. The script user has to enter the Fahrenheit temperature data only one time. Using a single loop on lines 11–17, the following happens to the temperature:

▶ It is stored in a new dictionary for Fahrenheit temperatures.

▶ It is converted to Celsius.

▶ It is stored in another new dictionary for Celsius temperatures.

**LISTING 9.12**    The `script0902.py` Script

```
1: pi@raspberrypi ~ $ cat py3prog/script0902.py
2:
...
3: # Create the Celsius version of the dictionary
4: #
5: may_high_temp_c = {}                 #Create empty dictionary
6: #
7: may_high_temp_c.update(may_high_temp)    #Create deep copy
8: #
9: date_keys = may_high_temp_c.keys()      #Obtain list of element keys
10: #
11: for may_date in date_keys:            #Loop to convert F to C
12: #
13:       high_temp_f = may_high_temp_c[may_date]  #Obtain Fahrenheit
14: #
15:       high_temp_c = (high_temp_f - 32) * 5 / 9 #Convert to Celsius
16: #
17:       may_high_temp_c[may_date] = high_temp_c  #Update dictionary
18: #
19: ##############################################################
20: # Display Record High Temps Dictionaries (Both F & C)
21: #
22: print ()
23: print ("Record High Temperatures in Indianapolis during Race Month")
24: #
25: date_keys = may_high_temp.keys()    #Obtain list of element keys
26: #
27: for may_date in date_keys:            #Loop to display key/value pairs
28:       print ("May", may_date, end = ': ')
29:       print (may_high_temp[may_date],"F", end = '\t')
30:       print ("{0:.1f}".format(may_high_temp_c[may_date]),"C")
...
```

In Listing 9.12, note that the `.update` dictionary operation is used on line 7. This operation performs a *deep copy* of the dictionary. A *deep copy* copies both the structure of an object and its elements. A *shallow copy*, on the other hand, copies only the structure of an object.

BY THE WAY

## The Inefficient Script

The `script0902.py` script is inefficient in how it has the user enter the data, then make a copy of the dictionary, and then reenter the Fahrenheit data into another dictionary. Keep in mind that these scripts are for learning purposes only. If they are to be used for non-educational purposes, they should be rewritten for efficiency's sake. In fact, rewriting them would be a good exercise for you to do as you learn Python programming!

In Listing 9.12, on line 15, the Fahrenheit temperature is converted to Celsius using a math equation. You learned about math in Python back in Hour 5, "Using Arithmetic in Your Programs." Once the conversion is made, the value in the `may_high_temp_c` dictionary is updated via an assignment option. Remember that when you use an existing key during a value assignment, the key simply has its associated value updated.

In Listing 9.13, you can see the output produced by the `script0902.py` script. Notice the Celsius temperature output in Listing 9.13 and then look back to line 30 in Listing 9.12. The `format` function, as you learned in Hour 5, enables the proper display of the calculated Celsius temperature.

## LISTING 9.13   Output of `script0902.py`

```
pi@raspberrypi ~ $ python3 py3prog/script0902.py

Enter the record high temps (F) for May in Indianapolis...
Record high for May 1: 88
...
Record High Temperatures in Indianapolis during Race Month
May 1: 88 F       31.1 C
May 2: 85 F       29.4 C
May 3: 88 F       31.1 C
...
pi@raspberrypi ~ $
```

Now you have a script that gets the temperature data into a dictionary and a script that calculates the Celsius values. Finally, you need a script that allows you to do some calculations on the temperatures for your pretend weather research.

Listing 9.14 shows part of `script0903.py`. This Python script takes the temperature data in a dictionary and calculates the maximum, the minimum, and the mode of the high temperatures.

**LISTING 9.14**    The `script0903.py` Script

```
1: pi@raspberrypi ~ $ cat py3prog/script09023.py
2:
...
3: # Determine Maximum, Minimum, and Mode Temps
4: #
5: temp_list = may_high_temp.values()
6: max_temp = max(temp_list)          #Determine maximum high temp
7: min_temp = min(temp_list)          #Determine minimum high temp
8: #
9: # Determine mode (most common) high temp ###
10: #
11: # Import Counter function
12: from collections import Counter
13: #
14: # Count temps and take the most frequent (mode) temperature
15: mode_list = Counter(temp_list).most_common(1)
16: #
17: # Extract mode high temp from 2-dimensional mode list
18: mode_temp = mode_list[0][0]
19: #
20: print ()
21: print ("Maximum high temp in May:\t", max_temp,"F")
22: print ("Minimum high temp in May:\t", min_temp,"F")
23: print ("Mode high temp in May:\t\t", mode_temp,"F")
...
```

Calculating the maximum and minimum temperatures is easy. You simply grab the values from the dictionary, on line 5, using the `.values` operation. Next you use the built-in `max` or `min` function on the values. Determining the mode of the high temperatures takes a little more work.

DID YOU KNOW

**What Is Mode?**

In a list of values, the mode is the value that occurs the most often. Thus, in the list 1, 2, 3, 3, 3, the number 3 is the list's mode.

In order to determine the most common temperature, you must import the non-built-in `Counter` function, as shown on line 12 in Listing 9.14. Using this function's `.most_common` operation on the list of temperatures values returns a two-dimensional sorted list object. (You learned about two-dimensional lists in Hour 8, "Using Lists and Tuples.") The temperature mode is the first item in the dimensional list, as shown being extracted on line 18. Listing 9.15 shows the output of this script.

**LISTING 9.15**   Output of `script0903.py`

```
pi@raspberrypi ~ $ python3 py3prog/script0903.py

Enter the record high temps (F) for May in Indianapolis...
Record high for May 1: 88
Record high for May 2: 85
Record high for May 3: 88
...
Maximum high temp in May:      92 F
Minimum high temp in May:      85 F
Mode high temp in May:         89 F
pi@raspberrypi ~ $
```

By using a dictionary in the Python scripts, you can temporarily store the temperature data so calculations can be performed upon them. Using a dictionary also allows for easy access to any of the stored temperatures via their key, which is the day of the month.

# Understanding Python Sets

A *set* is a collection of elements. Unlike the elements in a dictionary, a set's elements consist only of values. There is no key. There are two important things to know about Python sets:

- ▶ The elements in a set are unordered.

- ▶ Each element is unique.

Because a set's elements are unordered, you cannot access the set's data via an index, as you can in a list. However, similarly to a list, a set can contain different data types. (You learned about lists in Hour 8.) Using a set is much more efficient than using a list.

DID YOU KNOW

**That's Cold!**

A set's data elements are mutable and thus can be changed. There is another type of set, called a *frozenset*. A frozenset is immutable, and thus its elements cannot be changed. Its values are, in essence, "frozen" as they are.

# Exploring Set Basics

To create an empty set in Python, you use the built-in `set` function, which has the following syntax:

*set_name* = set()

In Listing 9.16, a set called `students_in_108` is created for the "108 Python Set Fundamentals" class. The `type` function is then used on it. You can see that the `students_in_108` is a set object type.

**LISTING 9.16**   Creating an Empty Set

```
>>> students_in_108 = set()
>>> type (students_in_108)
<class 'set'>
>>>
```

# Populating a Set

To add a single element to a set, you use the `.add` operation, which has the following syntax:

*set_name*.add *(element)*

To add the value elements to the `students_in_108` set, you enter them one at a time and press Enter after each, as shown in Listing 9.17.

**LISTING 9.17**   Populating a Set with the `.add` Operation

```
1: >>> students_in_108.add('Raz Pi')
2: >>> students_in_108.add('Jason Jones')
3: >>> students_in_108.add('Paul Bohall')
4: >>> students_in_108
5: {'Paul Bohall', 'Raz Pi', 'Jason Jones'}
6: >>>
```

To display the current elements in a set, you just type the set's name, as shown on line 4 in Listing 9.17. You can see that the elements are unordered, as you would expect in a set.

Using a less tedious method, you can create and populate a set all in one command. To do so, you use the following syntax:

*set_name*([*element1, element2, ... elementn*])

In Listing 9.18, a new set is created for the "133 Python Programming" class.

**LISTING 9.18**   Populating a Set with One Command

```
>>> students_in_133 = set(['Raz Pi', 'Benny Lora', 'Jody Sanchez'])
>>> students_in_133
{'Raz Pi', 'Benny Lora', 'Jody Sanchez'}
>>>
```

y must be between brackets. In this example, the ele-
lso be integers, floating-point numbers, lists, tuples,

## from a Set

cs calls *set theory*. You can easily determine what
s are different from one another, whether one ele-

gs to a particular set. As shown in Listing 9.19, you
element's membership.

### ent Membership

```
t Fundamentals' class.")

ass.")

ass.
```

i does have membership in the

A *set union* is where all the elements from two sets are combined to create a third set. You do not create a set union by using the + operand. Rather, you use the following syntax:

```
set_name#1.union(set_name#2)
```

Listing 9.20 is an example of combining sets into a union.

### LISTING 9.20    Performing a Set Union

```
>>> students_union = students_in_108.union(students_in_133)
>>> students_in_108
{'Paul Bohall', 'Raz Pi', 'Jason Jones'}
>>> students_in_133
```

```
{'Raz Pi', 'Benny Lora', 'Jody Sanchez'}
>>> students_union
{'Paul Bohall', 'Raz Pi', 'Jason Jones', 'Benny Lora', 'Jody Sanchez'}
>>>
```

The .union set operative adds the set members together. However, remember that every element in a set must be unique. So, even though the student Raz Pi was in both sets, he is listed only one time in the union set, students_union.

## Set Intersection

A *set intersection* contains set members that are also members in both a first and a second set. For example, in Listing 9.20, the student Raz Pi is in both sets students_in_108 and students_in_133. Therefore, an intersection of those two sets produces a set containing that student, Raz Pi (see Listing 9.21).

**LISTING 9.21** Performing a Set Intersection

```
>>> students_inter = students_in_108.intersection(students_in_133)
>>> students_in_108
{'Paul Bohall', 'Raz Pi', 'Jason Jones'}
>>> students_in_133
{'Raz Pi', 'Benny Lora', 'Jody Sanchez'}
>>> students_inter
{'Raz Pi'}
>>>
```

## Set Difference

A *set difference*, also called a *set complement*, is a created third set that contains items in the first set that are not in the second set. In essence, you subtract the second set from the first set, and the set difference is whatever is left.

Listing 9.22 shows an example of set difference. Using again the set of students in the 108 and 133 classes, the students_in_108 set has subtracted from it the students_in_133 set. This removes only the Raz Pi student. Thus, the resulting difference set contains Jason Jones and Paul Bohall.

**LISTING 9.22** Performing a Set Difference

```
>>> students_dif = students_in_108.difference(students_in_133)
>>> students_in_108
{'Paul Bohall', 'Raz Pi', 'Jason Jones'}
>>> students_in_133
```

```
{'Raz Pi', 'Benny Lora', 'Jody Sanchez'}
>>> students_dif
{'Paul Bohall', 'Jason Jones'}
>>>
```

Notice in Listing 9.22 that even though there are students in the `students_in_133` set that are not in the `students_in_108` set, subtracting them has no ill effects. Using the difference operator, you can subtract set elements that do not exist in the original set without throwing an error exception.

## Symmetric Set Difference

A *symmetric set difference* is a created third set that contains only elements that are solely in one set or the other. Thus, looking at the student example, a symmetric set difference would contain all the set elements from both `students_in_108` and `students_in_133`, except for Raz Pi. Since Raz Pi is in both sets, he would be excluded from the symmetric set difference, as shown in Listing 9.23.

**LISTING 9.23**    Performing a Symmetric Set Difference

```
>>> students_symdif = students_in_108.symmetric_difference(students_in_133)
>>> students_in_108
{'Paul Bohall', 'Raz Pi', 'Jason Jones'}
>>> students_in_133
{'Raz Pi', 'Benny Lora', 'Jody Sanchez'}
>>> students_symdif
{'Paul Bohall', 'Jason Jones', 'Benny Lora', 'Jody Sanchez'}
>>>
```

## Traversing a Set

Using a loop to obtain elements from a set is very easy because the set itself can be used for iteration. Listing 9.24 shows an example of this.

**LISTING 9.24**    Showing a Set Traversed

```
>>> for the_student in students_in_133:
...     print (the_student)
...
Raz Pi
Benny Lora
Jody Sanchez
>>>
```

Notice that the `for` loop has no problems traversing the set. However, due to the unordered nature of sets, you can potentially end up with an unordered display.

BY THE WAY

### A Sort Would Change the Type

You could use the `sorted` function to sort a set. However, be aware that `sorted` will convert the set to a list object type!

# Modifying a Set

A set is not immutable, and thus it can be changed. *Updating a set* does not mean changing individual elements within a set. For example, going back to the student set example, in `students_in_108`, the element `'Paul Bohall'` cannot be updated to be `'Sam Bohall'` because sets have no indexing capabilities. Updating a set actually means conducting a mass addition to the set.

To perform an update on a set, you use this syntax:

`set_name.update([element(s)_to_add])`

In Listing 9.25, a mass addition is made to the `students_in_108` set, using the `update` operation. As you can see, two additional elements are added to the set in a mass add.

**LISTING 9.25**   Updating a Set

```
>>> students_in_108
{'Paul Bohall', 'Raz Pi', 'Jason Jones'}
>>> students_in_108.update(['Alan Griffith','Otis McCallum'])
>>> students_in_108
{'Otis McCallum', 'Paul Bohall', 'Raz Pi', 'Jason Jones', 'Alan Griffith'}
>>>
```

You can also delete elements from a set. There are two set operations available for deleting elements. The first is the `remove` operation, which has this syntax:

`set_name.remove([element(s)_to_remove])`

The other is the `discard` operation, which has this syntax:

`set_name.discard([element(s)_to_discard])`

Listing 9.26 uses the `remove` operation to remove two students from the `students_in_108` set.

**LISTING 9.26**    Deleting Elements from a Set

```
>>> students_in_108
{'Otis McCallum', 'Paul Bohall', 'Raz Pi', 'Jason Jones', 'Alan Griffith'}
>>> students_in_108.remove('Alan Griffith')
>>> students_in_108.remove('Otis McCallum')
>>> students_in_108
{'Paul Bohall', 'Raz Pi', 'Jason Jones'}
>>>
```

The primary difference between doing a `remove` operation and a `discard` operation on a set has to do with missing elements. As shown in Listing 9.27, if an element does not exist in the set and you attempt remove it, an error exception is thrown. With a `discard` operation, no error exception is given.

**LISTING 9.27**    The Difference Between `remove` and `discard`

```
>>> students_in_108
{'Paul Bohall', 'Raz Pi', 'Jason Jones'}
>>> students_in_108.discard('Alan Griffith')
>>> students_in_108.remove('Otis McCallum')
Traceback (most recent call last):
  File "<stdin>", line 1, in <module>
KeyError: 'Otis McCallum'
>>>
```

BY THE WAY

**No En Masse**

You cannot remove or discard en masse, as you can with the `update` operation. You have to remove or discard set elements one at a time.

# Programming with Sets

In this section, you'll do some more weather temperature research—this time using sets. Using record high temperatures (Fahrenheit) in May in Indianapolis, you will build two sets and then do some analysis on them using a few set operations.

To build the first temperature set, highMayTemp2012, the set is initialized and then populated via a for loop. This is shown in Listing 9.28, which displays part of the Python script
script0904.py.

**LISTING 9.28**    Populating a Set with the script0904.py Script

```
1: pi@raspberrypi ~ $ cat py3prog/script09024.py
...
2: # Populate set with High Temps (F) during May 2012 in Indianapolis
3: print ()
4: print ("Enter the high temps (F) for May 2012 in Indianapolis...")
5: #
6: highMayTemp2012 = set()                #Create empty set
7: #
8: for may_date in range (1, 31 + 1):  #Loop to enter temps
9: #
10:       # Obtain high temp for date
11:       prompt = "High temperature (F) May " + str(may_date) + " 2012: "
12:       high_temp = int(input(prompt))
13:       # Put element in set
14:       highMayTemp2012.add(high_temp)
15: #
16: print ()
17: print ("The high temperatures (F) for May 2012 in a set are:")
18: print (highMayTemp2012)
19: #
...
```

DID YOU KNOW

**A Camel in the Case**

In Python script script0904.py, the set name highMayTemp2012 is using a variable style of
naming called *camel Case*. camel Case variable names, popular with Python script writers, start out
with lowercase, and then subsequent words in the name start with uppercase characters. This helps
to add clarity to a script. And, supposedly, the name looks like a camel with several humps.

Nothing is too exciting in script094.py so far. You have seen how to gather data into a set
before. In this script, a second set, not shown in Listing 9.28, is also built just like the first one,
except the set name is highMayTemp2011.

BY THE WAY

### Actual Temperature Data

The temperature data used in this hour is actual May temperatures for Indianapolis, Indiana. It was derived from www.almanac.com/weather/history/IN/Indianapolis.

Now that the necessary data is loaded into the scripts, you can do a little set mathematics to provide some analysis. First, you can compare the 2012's and 2011's May high temperatures by using a set intersection. The intersection should show the May high temperatures shared by both years. In Listing 9.29, the Python statement needed to accomplish this set intersection is shown on line 5.

**LISTING 9.29**  Setting an Intersection with the `script0904.py` Script

```
1: pi@raspberrypi ~ $ cat py3prog/script09024.py
...
2: # Determine Shared High Temps for May
3: #
4: # Find intersetion of high temp sets
5: shared_temps = highMayTemp2011.intersection(highMayTemp2012)
6: #
7: # Print out determined data
8: print ()
9: print ("High Temps (F) Shared by May 2012 & May 2011")
10: print (sorted(shared_temps))
...
```

Set mathematics performed on the temperature data also allows you to see which month was cooler (May 2011 or May 2012). To accomplish this, a set difference must be performed on the data. Listing 9.30 shows the code from `scrip0904.py`, which performs a set difference.

**LISTING 9.30**  Setting a Difference with the `script0904.py` Script

```
1: pi@raspberrypi ~ $ cat py3prog/script09024.py
...
2: # Determine Which Month was Cooler - May 2011 or May 2012
3: #
4: # Find difference of high temp sets
5: diff_temps2012 = highMayTemp2012.difference(highMayTemp2011)
6: diff_temps2011 = highMayTemp2011.difference(highMayTemp2012)
7: #
8: # Print out determined data
9: print ()
```

```
10: print ("Which month do you think was cooler?")
11: print ("May 2012:", sorted(diff_temps2012))
12: print ("            or")
13: print ("May 2011:", sorted(diff_temps2011))
...
```

Two difference sets are built, as shown on lines 5 and 6 in Listing 9.30. Both of these difference sets are then printed out, so the script user can determine which month was cooler.

Now that you have seen `scripts0904.py`'s construction, take a look at the output produced when the script is run. Listing 9.31 shows the final results for the intersection and difference calculations for the sets.

**LISTING 9.31    Output of `script0904.py`**

```
pi@raspberrypi ~ $ python3 py3prog/script0904.py

Enter the high temps (F) for May 2012 in Indianapolis...
High temperature (F) May 1 2012: 79
High temperature (F) May 2 2012: 84
High temperature (F) May 3 2012: 85
...
The high temperatures (F) for May 2012 in a set are:
{70, 72, 74, 75, 76, 78, 79, 80, 81, 83, 84, 85, 86, 87, 88, 90, 91, 92}
...
The high temperatures (F) for May 2011 in a set are:
{65, 68, 69, 70, 71, 72, 74, 77, 79, 80, 82, 84, 85, 87, 88, 57, 90, 59, 60}
...
High Temps (F) Shared by May 2012 & May 2011
[70, 72, 74, 79, 80, 84, 85, 87, 88, 90]

Which month do you think was cooler?
May 2012: [75, 76, 78, 81, 83, 86, 91, 92]
              or
May 2011: [57, 59, 60, 65, 68, 69, 71, 77, 82]
```

Notice that even though 31 days of data was entered, the sets are pretty small. Remember that each data element in a set must be unique, so any duplicate temperatures are eliminated.

You have seen that using set mathematics can help you draw conclusions about data. According to the Python script results, which month was cooler in Indianapolis: May 2012 or May 2011?

# Summary

In this hour, you got to try out two data storage object types: dictionaries and sets. You learned how to create both empty dictionaries and sets as well as how to populate them. Also, you were introduced to concepts such as obtaining data from, updating, and managing both dictionaries and sets. Finally, you saw some practical examples of using these data collection types in building a Python script. In Hour 10, you will learn about another object type: strings.

# Q&A

**Q. What is a dictionary pop operation?**

**A.** A dictionary pop operation is a method of removing a key/value pair. It is similar to the now retired `has_key` operation, in that you can specify what is to be returned, if the key is not found in the dictionary.

**Q. How can you determine whether one set is a subset of another set?**

**A.** You can use the `.issubset` operation to do this for Python sets. You can also determine whether a set is a superset of another by using the `.issuperset` operation.

**Q. Why are there no Try It Yourself sections in this hour?**

**A.** Unfortunately, there isn't enough room in this hour to include one. However, you can make your own practice exercise by going back to the "Programming with Dictionaries" and "Programming with Sets" sections and trying out the scripts shown there. Make it interesting by finding weather data pertaining to your part of the world and using it instead of the Indianapolis data.

# Workshop

## Quiz

1. A dictionary key may have one or more values associated with it at the same time. True or false?

2. What is the most common data item in a data list called?

3. Which set operation would you perform if you wanted a set that contained only members that are in two distinct sets?

   a. Union

   b. Difference

   c. Intersection

# Answers

1. False. A dictionary key can have only one value associated with it. Each key has one value and no more.

2. The mode is the most common data item in a data list.

3. Answer c is correct. Performing an intersection operation on two sets produces a third set that contains only members that are in both set 1 and set 2.

# HOUR 10
# Working with Strings

**What You'll Learn in This Hour:**

▶ How to create strings
▶ Working with string functions
▶ Formatting strings for output

One of the strong points of the Python programming language is its ability to work with text. Python makes manipulating, searching, and formatting text data almost painless. This hour explores how to create and work with text strings in your Python scripts.

# The Basics of Using Strings

Before we dive too deeply into the Python text world, let's take a look at the basics of working with text. For starters, Python handles text data as a string data type. The following sections outline how to use Python to create and work with string values and how to add text-handling features to your Python scripts.

## String Formats

Unfortunately, how Python handles string values has drastically changed in version 3. Previous versions of Python stored strings in ASCII format, which uses a single byte value for each character.

Python v3 changed that, and Python now uses Unicode format to store strings. The Unicode format uses 2 bytes to store each character, so it can accommodate a lot more text characters than the ASCII format does. This enables it to support many different languages, making it more popular in the world programming community.

### Using ASCII in Python v3

You can still work with ASCII characters and ASCII code values in Python v3. You can store ASCII string characters as binary data by storing the raw ASCII code value as a binary value, as in this example:

```
>>> binarystring = b'This is an ASCII string value'
>>> print(binarystring)
b'This is an ASCII string value'
>>> print(binarystring[1])
104
>>>
```

Because Python stores the string value as binary data, if you try to directly access an individual letter, you'll get the binary code for that letter. You can use the `chr()` function to convert the ASCII code into the corresponding string value, like this:

```
>>> print(chr(binarystring[1]))
h
>>>
```

If you're working with the English language in your scripts, the Python v3 change to Unicode format isn't readily apparent. You still store your text values the same way as in previous versions, and you retrieve them the same way, too. However, with Unicode you now have access to a wider variety of special characters that you can accommodate in your scripts!

# Creating Strings

Creating string values in Python is pretty straightforward. You just use a simple assignment statement to create a value and assign it to a variable. However, with string values, you must use quotes around the data to delineate the start and end of the string value, as in this example:

```
>>> string1 = 'This is a test string'
>>> print(string1)
This is a test string
>>>
```

You can use either single or double quotes to delineate a string value, but it's become somewhat standard in the Python community to use single quotes, unless there are quotes inside the text value itself.

If the text value includes single quotes, you can use double quotes to define the string beginning and end:

```
>>> string2 = "This'll work when defining a string"
>>> print(string2)
This'll work when defining a string
>>>
```

Or you can *escape* the quotes by placing a backslash in front of the quotes in the string value:

```
>>> string3 = 'This\'ll also work when defining a string'
>>> print(string3)
This'll also work when defining a string
>>>
```

The backslash isn't part of the string value; it just tells Python that the single quote in the data is part of the value. The same technique also works for embedding double quotes inside the string value.

You can break up long string values onto separate lines in your program or in the IDLE interface by adding a backslash at the end of the line and continuing the string on the next line. Python glues the two lines together to create a single string value, as shown here:

```
>>> string4 = 'This is a long string value \
that spans multiple lines.'
>>> print(string4)
This is a long string value that spans multiple lines.
>>>
```

There's another method for creating long string values, called *triple quotes*. With the triple quotes method, you place three single or double quotes in a row to define the start of the string, and then you place three single or double quotes in a row to define the end of the string, as shown in this example:

```
>>> string5 = '''This is another long string
value that will span multiple
lines in the output'''
>>> print(string5)
This is another long string
value that will span multiple
lines in the output
>>>
```

Notice though that with the triple-quotes method, the string value preserves any newlines that are added to the text. This can come in handy if you need to store text that has embedded newline characters that you want to display.

## Handling String Values

After you assign a string value to a variable, you can use the value as a whole, or you can work with parts of the string value.

As you've seen from the print() examples so far, to reference the whole string value, you just reference the string by specifying the variable name. You can also retrieve a subset of the string text stored in the variable by using a few different Python techniques.

Python treats string values somewhat like tuple values (see Hour 8, "Lists and Tuples"). You can reference an individual character in a string by using an index value, as shown here:

```
>>> string6 = 'This is a test string'
>>> print(string6[5])
i
>>>
```

However, as with tuples, Python won't let you change an individual character in the string by using the index. Here's an example:

```
>>> string6[5] = 'a'
Traceback (most recent call last):
  File "<pyshell#16>", line 1, in <module>
    string6[5] = 'a'
TypeError: 'str' object does not support item assignment
>>>
```

Also very much like with tuples, you can use slicing to retrieve a larger subset of characters from a string. Here's how:

```
>>> print(string6[5:7])
is
>>>
```

Slicing is a powerful tool to have when you're trying to extract specific data from string values, such as if you're trying to scrape data values from a webpage's content. Besides slicing, Python also supports lots of string functions to help you manipulate string values for just about every application. The next section takes a look at some of the most useful string functions that you'll run into as you code your Python scripts.

# Using Functions to Manipulate Strings

Python's popularity in working with strings is mostly due to the plethora of functions available for working with string data. The following sections walk through the most common string functions that you'll use as you work with strings in your Python scripts. Since there are so many string functions to choose from, the following sections split them up into categories to simplify a bit.

## Altering String Values

Python provides a handful of functions that manipulate either the text or the text format in string values. Table 10.1 shows the string functions that can come in handy when you need to manipulate string values.

**TABLE 10.1**   String-Manipulation Functions

| Function | Description |
|---|---|
| capitalize() | Makes the first letter in the string uppercase and the rest lowercase. |
| casefold() | Changes all characters to lowercase but also accommodates special characters found in some languages. |
| center(*width*[,*char*]) | Centers the string value within *width* spaces, using either spaces or *char* characters. |
| encode(*encoding*, *errors*) | Returns an alternate encoding of the string, using the encoding specified. |
| expandtabs([*tabsize*]) | Replaces each tab with the specified number of spaces. |
| ljust(*width*[,*char*]) | Left-justifies the string within *width* spaces or *char* characters. |
| lower() | Converts all characters to lowercase. |
| lstrip([*chars*]) | Removes leading whitespace characters or any characters specified in the *chars* string. |
| replace(*old*, *new*[,*count*]) | Replaces the substring *old* with the substring *new*. If *count* is specified, only the first *count* occurrences are replaced. |
| rjust(*width*[,*char*]) | Right-justifies the string within *width* spaces or *char* characters. |
| rstrip([*chars*]) | Removes trailing whitespace characters or any characters specified in the *chars* string. |
| strip([*chars*]) | Removes leading and trailing whitespace characters or any characters specified in the *chars* string. |
| swapcase() | Reverses the case of all characters in the string. |
| title() | Converts the first character of each word to uppercase and all other characters to lowercase. |
| translate(*map*) | Converts characters based on a predefined character map stored in a dictionary value. |
| upper() | Converts all characters to uppercase. |
| zfill(*width*) | Left-fills the string with zeros to create *width* characters. |

The string-manipulation functions don't change the value of the original string; they return a new string value. If you want to use the result in your script, you have to assign it to another variable, as in this example:

```
>>> string7 = 'Rich is working on the problem'
>>> string8 = string7.replace('Rich', 'Christine')
>>> print(string7)
Rich is working on the problem
>>> print(string8)
Christine is working on the problem
>>>
```

The `replace()` function changes the string text and returns the result to the `string8` variable. The original `string7` value remains the same.

## Splitting Strings

Another useful function in string manipulation is the ability to split strings into separate substrings. This comes in handy when you're trying to parse string values to look for words. Table 10.2 shows the Python string-splitting functions that are available.

**TABLE 10.2**   String-Splitting Functions

| Function | Description |
| --- | --- |
| `partition(char)` | Splits the string at the first occurrence of the character specified. |
| `rpetition(char)` | Splits the string at the last occurrence of the character specified. |
| `rsplit(char[, max])` | Returns a list of substrings, split at the specified character in the string. If a *max* value is specified, only the *max* rightmost substrings are split out. |
| `split(char[, max])` | Returns a list of substrings, split at the specified character in the string. If a *max* value is specified, only the *max* substrings are split out. |
| `splitlines([keepends])` | Splits string into a list of lines, split at line boundaries. If `keepends` is specified, the line breaks are included in the substrings. |

If you don't specify a split character, the split functions use any type of whitespace character as the split character. In the following example, the result is a list value, with each data value being a separate word in the original string:

```
>>> string9 = 'This is a test string used for splitting'
>>> list1 = string9.split()
>>> print(list1)
['This', 'is', 'a', 'test', 'string', 'used', 'for', 'splitting']
>>>
```

This is a great tool for breaking out individual words from a text string for manipulation.

Splitting strings can be somewhat of an art form, and sometimes it takes some experimenting to get it just right.

## Joining Strings

The opposite of splitting out string values into a list is joining them, which you do via the join() function. The join() function allows you to reassemble all the data values in a list back into a string value.

The join() function is a bit quirky, but it's extremely versatile, and will come in handy if you have to manipulate strings.

The join() function uses a single parameter, which is the list or tuple that you want to join into a string. However, that doesn't tell the join() function what character to use to separate the different list values. You need to define a string value that the join() method applies to. To see how this works, you can take a quick look at the join() function in action. Here are some additional actions taken on the list1 variable created in the previous example:

```
>>> list1[7] = 'joining'
>>> string10 = ' '.join(list1)
>>> print(string10)
This is a test string used for joining
>>>
```

This example shows a few different things about strings. First, it replaces the list1 data value at index 7 with a new word. Then it uses the join() function to reassemble the list back into a string value. The two single quotes surround a space character, so the join() function adds a space character between the data values in the list when it creates the string value. Printing the new string value shows that the list values were reassembled, including the updated value, using the space character. This is a tricky way to modify words within a text string.

## Testing Strings

A vital function in string manipulation is to have the ability to test string values for specific conditions. Python provides several string-testing functions that help out with that; Table 10.3 shows them.

**TABLE 10.3**   String-Testing Functions

| Function | Description |
|---|---|
| endswith(*chars*[,*start*[, *end*]]) | Returns True if the string ends with the specified characters. You can specify an optional starting index and ending index for a slice. |
| isalnum() | Returns True if the string contains only numbers and letters. |
| isalpha() | Returns True if the string contains only letters. |
| isdecimal() | Returns True if the string contains only decimal characters. |
| isdigit() | Returns True if the string contains only numbers. |
| isidentifier() | Returns True if the string is a valid Python identifier. |
| islower() | Returns True if the string contains only lowercase characters. |
| isnumeric() | Returns True if the string contains only numeric characters. |
| isprintable() | Returns True if the string contains only printable characters. |
| isspace() | Returns True if the string contains only whitespace characters. |
| istitle() | Returns True if the string is in title format. |
| isupper() | Returns True if the string contains only uppercase characters. |
| startswith(*chars*[,*start*[,*end*]]) | Returns True if the string starts with the specified characters. You can specify an optional starting index and ending index for a slice. |

The string-testing functions help when you need to validate input data that your scripts receive. If a script requests a numeric value from the user, it's a good idea to test what value the user enters before actually using it in your code! You can try this out by creating a test script.

**Test Strings**

Try adding a string-testing feature to a small script by following these steps:

1. Open your favorite text editor and add the following code:

```
#!/usr/bin/python3

choice = input('Please enter your age: ')
if (choice.isdigit()):
    print('Your age is ', choice)
else:
    print('Sorry, that is not a valid age')
```

2. Save the file as `script1001.py` in your Python code folder.

3. From the command prompt, run the program:

```
python3 sscript1001.py
```

The `script1001.py` script uses the `isdigit()` string function to test the string value that the `input()` function returns. If the string contains an invalid digit, the script produces a message telling the user about the error:

```
pi@raspberry script% python3 script1001.py
Please enter your age: 34
Your age is  34
pi@raspberry script% python3 script1001.py
Please enter your age: Rich
Sorry, that is not a valid age
pi@raspberry script% python3 script1001.py
Please enter your age: 12g5
Sorry, that is not a valid age
pi@raspberry script%
```

You can also try using the other string-testing functions in the same manner to see how they validate different types of text that you enter from the prompt.

# Searching Strings

Yet another common string function is searching for a specific value within a string. Python provides a couple functions to help out with this.

If you only need to know if a substring value is contained within a string value, you can use the in operator. The in operator returns a True value if the string contains the substring value, and it returns a False value if not. Here's an example:

```
>>> string12 = 'This is a test string to use for searching'
>>> 'test' in string12
True
>>> 'testing' in string12
False
>>>
```

If you need to know exactly where in a string the substring is found, you need to use either the find() or rfind() functions.

The find() function returns the index location for the start of the found substring, as shown here:

```
>>> string12.find('test')
10
>>>
```

The result from the find() function shows that the string 'test' starts at position 10 in the string value. (Strings start at index position 0.) If the substring value isn't in the string, the find() function returns a value of -1:

```
>>> string12.find('tester')
-1
>>>
```

The find() function searches the entire string unless you specify a start value and an end value to define a slice, as in this example:

```
>>> string12.find('test', 12, 20)
-1
>>>
```

It's also important to know that the find() function returns only the location of the first occurrence of the substring value, like this:

```
>>> string13 = 'This is a test of using a test string for searching'
>>> string13.find('test')
10
>>>
```

You can use the rfind() function to start the search from the right side of the string:

```
>>> string13.rfind('test')
26
>>>
```

Yet another searching function is the index() function. It performs the same function as find(), but instead of returning -1 if the substring isn't found, it returns a ValueError error, as shown here:

```
>>> string13.index('tester')
Traceback (most recent call last):
  File "<pyshell#9>", line 1, in <module>
    string13.index('tester')
ValueError: substring not found
>>>
```

The benefit of retuning an error instead of a value is that you can catch the error as a code exception (see Hour 17, "Exception Handling") and have your script act accordingly.

If you'd like to just count the number of occurrences of a substring value within a string, you use the count() function, as shown here:

```
>>> string13.count('test')
2
>>>
```

Between the find(), index(), and count() functions, you have a full arsenal of tools to help you search for data within your strings.

# Formatting Strings for Output

Python includes a powerful method of formatting the output that your script displays. The format() function allows you to declare exactly how you want your output to look. This section walks through how the format() function works and how you can use it to customize how your script output looks.

WATCH OUT!

### Python Change in String Formatting

The way Python defines formatting codes for the format() function drastically changed between version 2 and 3. Since this book focuses on Python v3, we only cover the v3 format() function formatting codes. If you need to use the Python v2 formatting method, refer to the Python documentation at www.python.org.

## The format() Function

The format() function is the most complicated of the built-in Python string functions. However, once you get the hang of it, you'll find yourself using it in lots of places in your scripts to help make your output more user friendly.

This is the syntax for the `format()` function:

`string.format(expression)`

There are two parts to using the `format()` function. The `string` component is the output string that you want to display, and the `expression` component defines what variables to embed in the output.

In the output string, you also need to embed placeholders within the string where you want the variable values from the expression to appear. There are two types of placeholders you can use:

▶ Positional placeholders

▶ Named placeholders

The next two sections discuss each of these types of placeholders.

## Positional Placeholders

Positional placeholders create spots in the output string to insert variable values by using numeric index values representing the order of the variables in the expression. To identify the placeholders, you put the index value within braces inside the string text. While this might sound confusing, it's actually pretty straightforward. Here's an example:

```
>>> test1 = 10
>>> test2 = 20
>>> result = test1 + test2
>>> print('The result of adding {0} and {1} is {2}'.format(test1, test2,
result))
The result of adding 10 and 20 is 30
>>>
```

Python inserts the value of each variable in the expression list in its associated positional placeholder. The `test1` variable value is placed in the `{0}` location, the `test2` variable value is placed in the `{1}` location, and the `result` variable value is placed in the `{2}` location.

## Named Placeholders

Instead of using index values, for the named placeholder method, you assign names to each variable value you want placed in the output string. You assign the names to each of the replacement values in the expression list and then use the names in the placeholders in the output string, as in this example:

```
>>> vegetable = 'carrots'
>>> print('My favorite vegetable is {veggie}'.format(veggie=vegetable))
My favorite vegetable is carrots
>>>
```

Python replaces the {veggie} named placeholder with the value assigned to the veggie name in the format() expression. If you have more than one named value in the expression, you just separate them using commas, as shown here:

```
>>> vegetable = 'carrots'
>>> fruit = 'bananas'
>>> print('Fruit: {fruit}, Veggie: {veggie}'.format(fruit=fruit,
 veggie=vegetable))
Fruit: bananas, Veggie: carrots
>>>
```

You can also assign string and numeric values directly to the named placeholder, as in this example:

```
>>> print('My favorite fruit is a {fruit}.'.format(fruit='banana'))
My favorite fruit is a banana.
>>>
```

You might be thinking that so far all this does is add an extra layer of complexity to displaying string values, with no additional purpose. However, the true power of the format() function comes in its formatting capabilities. The next section examines those capabilities.

## Formatting Numbers

The true power of the format() function comes into play when you need to display numeric values in your output. By default, Python treats numeric values as strings in the output generated by the print() function. That can lead to some pretty ugly printouts, as there's no control over things such as how many decimal places to display or whether to use scientific notation to display large values.

The format() function provides a wide array of formatting codes for you to specify exactly how Python displays the values. You just place the formatting codes within the placeholder braces in the string value, separated by a colon from the placeholder number or name.

You can use different formatting codes, based on the type of data you want to display. Here's a quick example to demonstrate:

```
>>> total = 3.4999999
>>> print('The total is {0:.2f}'.format(total))
The total is 3.50
>>>
```

This formatting code tells Python to round the floating-point value to two decimal places for you. Now that's handy! The following sections walk through the different codes you can use, based on the data type of the value you need to display.

## Integer Values

Displaying integer values doesn't usually involve too much formatting. By default, Python just displays integer values using the decimal format, which is usually just fine.

However, you can spice things up by specifying formatting codes to have Python convert the integer value to another base (such as octal or hexadecimal) for the display automatically. Table 10.4 lists the integer-formatting codes that are available.

**TABLE 10.4**  Integer-Formatting Codes

| Code | Description |
| --- | --- |
| b | Displays the number in binary format. |
| c | Converts the integer to a Unicode character before printing. |
| d | Displays the number in decimal format. |
| o | Displays the number in octal format. |
| x | Displays the number in hex format with lowercase letters. |
| X | Displays the number in hex format with uppercase letters. |
| n | Displays the number with numeric separators. |

There's nothing tricky about any of these codes. You just include them in the placeholder to output the integer value in that format, as shown here:

```
>>> test1 = 154
>>> print('Binary: {0:b}'.format(test1))
Binary: 10011010
>>> print('Octal: {0:o}'.format(test1))
Octal: 232
>>> print('Hex: {0:x}'.format(test1))
Hex: 9a
>>>
```

Now you're starting to see some of the built-in power of using the `format()` function for your output!

## Floating-Point Values

Displaying floating-point values can be somewhat of a pain. Not only do you have to worry about small values with several places past the decimal point, you may also have to worry about very large numbers. To get your floating-point values to display in a user-friendly manner, you can use the floating-point formatting codes for the `format()` function. Table 10.5 shows what's available for you to use.

**TABLE 10.5**   Floating-Point Formatting Codes

| Code | Description |
|------|-------------|
| e | Displays the value using scientific notation. |
| E | Displays the value using scientific notation with an uppercase *E*. |
| f | Displays the value as a fixed-point number. |
| F | Displays the value as a fixed-point number but uses uppercase for NAN and INF. |
| g | Uses no formatting. |
| G | Uses no formatting, unless the value gets too large, and then it uses scientific notation. |
| n | Uses no formatting but applies number separator characters. |
| % | Displays the value as a percentage, multiplying it by 100 and displaying it in fixed format. |

With floating-point values, besides the formatting code, you can also specify the number of decimal places Python should round the value to. Here's an example of that:

```
>>> test1 = 10
>>> test2 = 3
>>> result = test1 / test2
>>> print(result)
3.3333333333333335
>>> print('The result is {0:.2f}'.format(result))
The result is 3.33
>>>
```

Without the format() function, the print() function displays the result variable value with the repeating decimal places. The .2f format tells Python to round the value to two decimal places, using a fixed-point format.

## Sign Formatting

The format() function provides a way for you to define how Python handles the sign in a number.

The plus sign (+) tells Python that a sign should be used for both positive and negative numbers in the output. The negative sign (–) tells Python that a sign should be used only for negative numbers. The default is to use the sign only for negative numbers.

Here are a few examples of using the sign-formatting codes:

```
>>> test1 = 45
>>> print(test1)
45
>>> print('{0:+}'.format(test1))
+45
>>> test2 = -12.56
>>> print(test2)
-12.56
>>> print('{0:+.2f}'.format(test2))
-12.56
>>>
```

If you need your numeric columns to line up in the output, you can use a space for the sign formatting. The space indicates that a leading space should be used on positive numbers, and a minus sign should be used on negative numbers.

## Positional Formatting

If you have to work with lining up numbers in columns, there are a few other formatting codes you can use to help out. Table 10.6 describes the tools that are available to help align the numbers in your output.

**TABLE 10.6**  Positional-Formatting Codes

| Code | Description |
| --- | --- |
| < | Left-aligns the value (the default). |
| > | Right-aligns the value. |
| = | Places padding between the sign and the digits. |
| ^ | Centers the value. |

With the left-align, right-align, and center formats, you specify the number of spaces reserved for the number before the positional format code. Python then positions the number accordingly within that space area, as you can see here:

```
>>> print('The result is {0:>10d}'.format(test1))
The result is         45
>>>
```

Python reserves 10 spaces for the output of the numeric value and then right-aligns the value within that space area.

With all these formatting options, you should be able to create custom reports with numeric data in no time!

# Summary

This hour explores how Python handles text strings, and what functions you have available for working with them. You can use slicing to extract substrings out of a larger string at a specific location, or you can use the string-splitting functions to extract substrings based on a separation character. You can also use some search functions to search through a string to find a substring value. Finally, some handy string formatting functions help you format any output strings that your Python scripts produce.

In the next hour, we'll explore how to use files with your Python scripts. It's important to know how to store and retrieve data from your scripts, and using plain files is the easiest way to do that!

# Q&A

**Q. Does Python support searching for text in strings using regular expressions?**

**A.** Yes. Regular expressions are complicated enough to have their own hour (see Hour 16, "Regular Expressions").

**Q. Can you embed nonprintable and other characters in Python string values?**

**A.** Yes, you can use Unicode escape encoding to embed any Unicode character using its numeric code. Just precede the code with a \u. For example, the Unicode code for a space is 0020, so to embed it in a string you use, do this:

```
>>> print('This\u0020is\u0020a\u0020test')
This is a test
>>>
```

# Workshop

## Quiz

1. What Python string function should you use to exchange a word in a string with another word?

   **a.** swapcase()

   **b.** split()

   **c.** replace()

   **d.** find()

2. The `format()` function can display decimal values in hexadecimal or binary formats. True or false?

3. What `format()` function formatting code should you use to display a monetary value that requires two decimal places?

## Answers

1. c. The replace() function allows us to search for a specific string value and replace it with another string value within a larger string value.

2. True. You can specify whether to use decimal, hexadecimal, or binary formats when you display variable values using the format() function.

3. `.2f`. The "f" tells Python to display the value as a floating point number, and the ".2" tells it to display only two decimal places of the floating point value. This is exactly what you need for displaying monetary values!

# Using Files

## What You'll Learn in This Hour:

▶ File types that Python can handle

▶ How to open a file

▶ Reading a file's data

▶ Writing data to a file

Storing strings, lists, dictionaries, and so on in memory is fine for small Python scripts. However, when writing large scripts, you need to store the data in files. In this hour, you will explore how to use various files in your Python scripts.

# Understanding Linux File Structures

Python can deal with various operating systems' file structures. It can also handle the input and output for text files, binary files, compressed files, and so on. If you want a language with great file-handling capabilities that is also cross-platform, Python is your language.

Table 11.1 lists a few file types that Python can handle. Keep in mind that this is not a complete list!

**TABLE 11.1** A Few File Types Python Can Process

| Data | File Type | Description |
| --- | --- | --- |
| Binary digits | Binary | Binary data that is for program use and is not readable via a text editor. This data is often pickled in Python. |
| Compressed | zip, bzip2, gzip, tar | Data that has been compressed using a compression utility such as gzip. |
| Numeric | Text | Number data, which is stored as character strings. |

| Data | File Type | Description |
|------|-----------|-------------|
| String | Text | Character strings stored as either UTF-8 or ASCII. |
| XML | XML | Extensible Markup Language (XML) data. |
| Comma separated | Text | Comma-separated values (CSV) for use in other applications, such as a database. |

Notice in Table 11.1 that the file types can overlap. For example, a numeric text file can be compressed after it is created. The primary purpose of this table is to show that Python is extremely flexible in its ability to handle various file formats.

BY THE WAY

### Overwhelming?

Don't feel overwhelmed by the different file types in Table 11.1. The focus this hour is on handling text files. (You can breathe a sigh of relief now.)

The various file types that Python can handle "live" in various places in the Raspbian directory structure. Their type or purpose dictates their placement within the structure.

## Looking at Linux Directories

The Linux directory structure is called an *upside-down tree* because the top of the directory structure is called the *root*. Figure 11.1 shows the top root directory (/) with the subdirectories right beneath it.

**FIGURE 11.1**
The subdirectories in the top root directory (/).

Each subdirectory stores particular files according to their purpose. Directory names are written in two ways: as an absolute directory reference or as a relative directory reference.

An *absolute directory reference* always begins with the root directory. For example, when you log in to your Raspberry Pi using the `pi` account, you are in the directory `/home/pi`. This is an absolute directory reference because it starts with the root directory (`/`).

BY THE WAY

## Memory Trick

One way to remember that an absolute directory reference begins with the root directory (`/`) is a simple memory sentence, like this: "Absolute directories absolutely begin with the root directory."

A *relative directory reference* does not begin with the root directory (`/`). Instead, it denotes a directory relative to where your present working directory is now. Back in Hour 2, "Understanding the Raspbian Linux Distribution," you learned that a *present working directory* is where you are currently located in the directory structure. You can see the present working directory by using the `pwd` shell command. Listing 11.1 shows an example of using a relative directory reference.

**LISTING 11.1**    A Relative Directory Reference Example

```
pi@raspberrypi ~ $ pwd
/home/pi
pi@raspberrypi ~ $ ls py3prog
sample.py       script0402.py   script0702.py   script0901.py   script0903.py
script0401.py   script0701.py   script0703.py   script0902.py   script0904.py
pi@raspberrypi ~ $
```

In Listing 11.1, the `pwd` command is used to show the present working directory of the user `pi`. You can see that the present working directory is `/home/pi`, which is an absolute directory reference. Then the command `ls py3prog` is entered, in order to display the Python scripts currently located within the `py3prog` subdirectory. The `ls` command uses a relative directory reference. To use an absolute directory reference, the command would be `ls /home/pi/py3prog`.

To help you learn about using files in Python, the rest of this book uses the directories shown in Table 11.2.

**TABLE 11.2**    Python Directories for This Book

| Directory | What Is Stored There |
|---|---|
| `/home/pi/py3prog` | Python scripts |
| `/home/pi/temp` | Temporary data files |
| `/home/pi/data` | Permanent data files |

Test yourself here. In Table 11.2, which type of directory reference is used: absolute or relative?

# Managing Files and Directories via Python

You learned how to create directories in Hour 2. You were the one who created the /home/pi/ py3prog directory by using the mkdir shell command. Now you will learn how to manage files and make directories by using a Python program!

Python comes with a multiplatform function called os. The os function allows you to conduct various operating system functions, such as creating directories. Table 11.3 lists some of the os methods you can use to manage files and directories in Python.

**TABLE 11.3**   A Few os Function Methods

| Method | Description |
|---|---|
| os.chdir('*directory_name*') | Changes your present working directory to *directory_name*. |
| os.getcwd() | Provides the present working directory's absolute directory reference. |
| os.listdir('*directory_name*') | Provides the files and subdirectories located in *directory_name*. If no *directory_name* is provided, it returns the files and subdirectories located in the present working directory. |
| os.mkdir('*directory_name*') | Creates a new directory. |
| os.remove('*file_name*') | Deletes *file_name* from your present working directory. It will not remove directories or subdirectories. There are no "Are you sure?" questions provided. |
| os.rename('*from_file*','*to_file*') | Renames a file from the name *from_file* to the name *to_file* in your present working directory. |
| os.rmdir('*directory_name*') | Deletes the directory *directory_name*. It will not delete the directory if it contains any files. |

The os function is not a built-in Python function. Therefore, you need to issue the Python statement import os before you can use the methods listed in Table 11.3.

DID YOU KNOW

**More os, Please**

The os function has a great deal more methods than shown here. To learn about these methods, go to docs.python.org/3/library/os.html.

Listing 11.2 shows a few of the os function's methods being used. You can see on line 8 that the os function is imported, and a new subdirectory is created on line 11. The present working directory is then changed to the newly created subdirectory on line 12.

**LISTING 11.2** Using the os Function

```
1: pi@raspberrypi ~ $ pwd
2: /home/pi
3: pi@raspberrypi ~ $ python3
4: Python 3.2.3 (default, Jan 28 2013, 11:47:15)
5: [GCC 4.6.3] on linux2
6: Type "help", "copyright", "credits" or "license" for more information.
7: >>>
8: >>> import os
9: >>> os.getcwd()
10: '/home/pi'
11: >>> os.mkdir('MyNewDir')
12: >>> os.chdir('MyNewDir')
13: >>> os.getcwd()
14: '/home/pi/MyNewDir'
15: >>>
```

Handling these methods within Python allows you to manage directories from within your scripts. You can create and use files within these directories.

# Opening a File

To access a file in a Python script, you use the built-in open function. This is the basic syntax for using this function:

*filename_variable* = open (*filename, options*)

Several options can be used in the open function, as shown in Table 11.4.

**TABLE 11.4** open Function Options

| Option | Description |
| --- | --- |
| mode | Designates the mode, such as read. |
| buffering | Designates the buffering policy. |
| encoding | Specifies the process to be used for putting characters into a specific format. |
| errors | Specifies how encoding/decoding errors are handled. |
| newline | Designates how to handle newlines in files. |
| closefd | Specifies when to close a file descriptor when a file is closed. |
| opener | Designates a created function to be called. |

Which options are used typically depends on which file type (see Table 11.1) you are opening. For learning purposes here, the focus is on the mode option of the open function.

# Designating the Open Mode

For the mode option in the open function, several modes can be designated, as shown in Table 11.5.

**TABLE 11.5**   The open Function mode Designations

| Mode | Description |
| --- | --- |
| a | Opens a file for writing to the end of the file, if the file already exists. If the file does not exist, it is created. |
| r | Opens a file for reading. |
| w | Opens a file for writing. It erases the file's current contents, if it already exists. If the file does not exist, it is created. |

For all the options a b and/or a + can be tacked onto the end. For example, the r option can be r+, rb, or rb+. A b tacked onto a mode indicates that the file is binary. Thus, the mode wb indicates that a binary file is being open to be written to. The + tacked onto a mode indicates two things. One is that the file pointer will be at the beginning of the file. The other is that the file is open for both reading and writing/appending.

DID YOU KNOW

**What Is a File Pointer?**

Think of a file pointer as a place keeper. It keeps your current place in the file for you as your script reads (or writes) data from the file.

In Listing 11.3, a file called May2012TempF.txt is opened. Before it is opened, a few of the os functions are used to navigate to the file's location.

**LISTING 11.3**   Opening the Temperature File

```
>>> import os
>>> os.chdir('/home/pi/data')
>>> os.getcwd()
'/home/pi/data'
>>>
>>> temp_data_file = open('May2012TempF.txt','r')
>>>
```

Once the file is opened, methods on the file can be done using the variable `temp_data_file`. Notice in Listing 11.3 that when the `open` function was used, both the file's name and the mode arguments were passed as strings. You can also use variables as arguments, if desired.

# Using File Object Methods

You can act on an opened file by using its variable name. The file's variable name in Listing 11.3 is `temp_data_file`. This variable name, called a *file object*, has methods associated with it. For example, once a file is open, you can check various file attributes. Table 11.6 shows a few of file object methods for checking file attributes.

**TABLE 11.6**   File Object Methods for File Attributes

| Method | Description |
| --- | --- |
| *filename_variable*.closed | Returns `True` if the file is closed. Returns `False` if the file is not closed. |
| *filename_variable*.mode | Returns the mode (character string) with which the file is/was opened. |
| *filename_variable*.name | Returns the file name with which the file is/was opened. |

Going back to the example used in Listing 11.3, you can see in Listing 11.4 these file object methods being used to determine current file attributes.

**LISTING 11.4**   Determining File Attributes

```
...
>>> os.getcwd()
'/home/pi/data'
>>> temp_data_file = open('May2012TempF.txt', 'r')
>>>
>>> temp_data_file.closed
False
>>> temp_data_file.mode
'r'
>>> temp_data_file.name
'May2012TempF.txt'
>>>
```

Notice that the result of the `.name` method returns the name used in the `open` function. In this case, the file name used does not need to include an absolute directory reference. This is because the file is located in the present working directory. In Listing 11.5, the file to be opened is not in the present working directory, so a slight change has to be made in the `open` argument.

**LISTING 11.5**   Opening a File Using Absolute Directory Reference

```
...
>>> os.getcwd()
'/home/pi'
>>> temp_data_file = open('/home/pi/data/May2012TempF.txt', 'r')
>>> temp_data_file.name
'/home/pi/data/May2012TempF.txt'
>>>
```

Notice in Listing 11.5, that the absolute directory reference is used for the file name in the open function. When this is done, the file object method .name returns the entire file's name and its directory location. Therefore, you can see that these file object methods are based on the attributes used in the open function.

Now that you know how to open a file, you should learn how to read one. Reading files is the next item on this hour's agenda.

# Reading a File

To read a file, of course, you must first use the open function to open the file. The mode chosen must allow the file to be read, which the r mode does. After the file is opened, you can read an entire file into your Python script in one statement, or you can read the file line by line.

## Reading an Entire File

You can read a text file's entire contents into a variable by using the file object method .read, as shown in Listing 11.6.

**LISTING 11.6**   Reading an Entire File into a Variable

```
...
1: >>> temp_data_file = open('/home/pi/data/May2012TempF.txt', 'r')
2: >>>
3: >>> temp_data = temp_data_file.read()
4: >>> type (temp_data)
5: <class 'str'>
6: >>> print (temp_data)
7: 1 79
8: 2 84
9: 3 85
10: ...
11: 29 92
12: 30 81
13: 31 76
```

```
14: >>> print(temp_data[0])
15: 1
16: >>> print(temp_data[0:4])
17: 1 79
18: >>>
```

In Listing 11.6, you can see that the variable `temp_data` is used to receive the entire file contents from the `.read` method on line 3. The data comes in as a string into the `temp_data` variable, as shown on lines 4 and 5, using the `type` function. You can then access the file data now stored in the variable by using string slicing. This is shown on lines 14 through 17. (You learned about string slicing in Hour 10, "Working with Strings.")

# Reading a File Line by Line

With Python, you can have a file read line by line. To read a file line by line into a Python script, you use the `.readline` method.

BY THE WAY

## What Is Considered a Line?

From Python's point of view, a text file *line* is a string of characters of any length that is terminated by the newline escape sequence, `\n`.

In Listing 11.7, the `.readline` method is used on the temperature file, `/home/pi/data/May2012TempF.txt`. A `for` loop is used to iterate through the file, line by line, printing out each read file line.

**LISTING 11.7**   Reading a File Line by Line

```
>>> temp_data_file = open('/home/pi/data/May2012TempF.txt', 'r')
>>>
>>> for the_date in range (1, 31 + 1):
...     temp_data = temp_data_file.readline()
...     print (temp_data,end = '')
...
1 79
2 84
3 85
...
29 92
30 81
31 76
>>>
```

Notice in Listing 11.7 that when printing out the temperature data, the print functions newline (\n) has to be suppressed by using end=''. This is needed because the data already has a \n character on the end of each line. If you do not suppress the print function's newline (\n), your output will be double spaced.

DID YOU KNOW

### Stripping Off the Newline!

There may be times when you need to remove the newline characters that are read into your Python script by the .readline method. You can achieve this by using the strip method. Since the newline character is on the right side of the line string, you more specifically use the .rstrip method. If this method were used in Listing 11.7, it would look like this:
temp_data = temp_data.rstrip('\n').

Reading a file line by line, as you would expect, reads the file in sequential order. Python maintains a file pointer that keeps track of its current position in a file. You can see this file pointer getting updated after each line read by using the .tell method. In Listing 11.8, the .tell method is performed right after the file is opened on line 2. You can see that the file pointer is set to 0, or the beginning of the file.

**LISTING 11.8**    Following the File Pointer Using .tell

```
 1: >>> temp_data_file = open('/home/pi/data/May2012TempF.txt', 'r')
 2: >>> temp_data_file.tell()
 3: 0
 4: >>> for the_date in range (1, 31 + 1):
 5: ...        temp_data = temp_data_file.readline()
 6: ...        print (temp_data,end = '')
 7: ...        temp_data_file.tell()
 8: ...
 9: 1 79
10: 5
11: 2 84
12: 10
13: 3 85
14: 15
15: ...
16: 29 92
17: 165
18: 30 81
19: 171
20: 31 76
21: 177
22: >>>
```

So that you can see the process of the file pointer, another .tell method is embedded in Listing 11.8, inside the for loop on line 7. After each line is read and printed out, the file pointer is updated to the next character to read.

# Reading a File Nonsequentially

You can actually use the file pointer from Listing 11.8 to directly access data within the file. This requires you to know where the data is located, as with string slicing. To do this, you need to use the .seek and .read methods.

For a text file, the .read method has the following basic syntax:

```
filename_variable.read(number_of_characters)
```

In Listing 11.9, you can see in line 1 that the .read method is used on temp_data_file to read in the first four characters of the file. Using the .tell method on line 4, the file pointer is now pointing at character number 4. The next .read, on line 6, grabs only one character, which is the newline (\n) escape sequence. This is why the subsequent print command on line 7 prints out two blank lines (lines 8 and 9).

**LISTING 11.9**   Reading File Data Using .read

```
1: >>> temp_data = temp_data_file.read(4)
2: >>> print (temp_data)
3: 1 79
4: >>> temp_data_file.tell()
5: 4
6: >>> temp_data = temp_data_file.read(1)
7: >>> print (temp_data)
8:
9:
10: >>> temp_data_file.tell()
11: 5
12: >>> temp_data = temp_data_file.read(4)
13: >>> print (temp_data)
14: 2 84
15: >>> temp_data_file.tell()
16: 9
17: >>>
```

To reposition the file pointer back to the beginning of the file, you need to use the .seek method. The .seek method has the following basic syntax:

```
filename_variable.seek(position number)
```

The *position number* for the start of the file is 0. Listing 11.10 shows an example of using .seek and .read to read a file nonsequentially.

**LISTING 11.10**   Repositioning the File Pointer Using .seek

```
1: >>> temp_data_file.seek(0)
2: 0
3: >>> temp_data = temp_data_file.read(4)
4: >>> print (temp_data)
5: 1 79
6: >>> temp_data_file.seek(25)
7: 25
8: >>> temp_data = temp_data_file.read(4)
9: >>> print (temp_data)
10: 6 84
11: >>>
```

Notice in Listing 11.10 that after the file pointer is position at 0 on line 1, the .read method is set to read the next four characters in the file. This is very similar to how the .readline method works in Listing 11.7. However, because only four characters are read in this example, the new-line (\n) does not need to be suppressed in the print function on line 4.

You need to notice one more item to note in Listing 11.10 before you can move on to the Try It Yourself section. On line 6, the file pointer is positioned at character 25. This allows the next .read method to read the four characters starting at that position on line 9. Thus, using the .seek and .read methods allows you to read a file nonsequentially.

▼ TRY IT YOURSELF

**Open a File and Read It Line by Line**

In the following steps, you will create a file outside Python. After it is created, you will enter the Python interactive shell environment, open the created file, and read it into Python line by line. In these steps, you will try out another more elegant way to read through a file. And, hopefully, have a little fun along the way. Here's what you do:

1. If you have not already done so, power up your Raspberry Pi and log in to the system.

2. If you do not have the LXDE GUI started automatically at boot, start it now by typing startx and pressing Enter.

3. Open the LXTerminal by double-clicking the LXTerminal icon.

4. At the shell prompt, type mkdir /home/pi/data and press the Enter key. The shell command creates the data subdirectory needed for storing your permanent Python data files.

5. You need to create a new blank file in your new directory. Do this by typing `touch /home/pi/data/friends.txt` at the shell prompt and pressing Enter.

6. Double-check that the file is there by typing `ls data` and pressing Enter. You should see the file `friends.txt` listed. Notice that when you enter the `ls` command, you use a relative directory reference of `data` instead of an absolute directory reference of `/home/pi/data`.

7. Using a shell command in order to write records to the `friends.txt` file, type `echo "Chris" > /home/pi/data/friends.txt` and press Enter.

8. Another shell command will be used to append another friend to the bottom of the file, friends.txt. Type `echo "Zach" >> /home/pi/data/friends.txt` and press Enter. (Note that two greater-than signs (`>>`) are used this time.)

9. Type `echo "Karl" >> /home/pi/data/friends.txt` and press Enter.

10. Type `echo "Zoe" >> /home/pi/data/friends.txt` and press Enter.

11. Type `echo "Simon" >> /home/pi/data/friends.txt` and press Enter. (Yes, this is tedious. But it will help you appreciate writing to a file later this hour all the more.)

12. Type `echo "John" >> /home/pi/data/friends.txt` and press Enter.

13. Type `echo "Anton" >> /home/pi/data/friends.txt` and press Enter. (Do you recognize these names yet?)

14. Finally, you are all done creating your friends.txt file! Take a look at its contents by typing `cat data/friends.txt` and pressing Enter. Don't worry if there are typos in your file. Just note them, so they won't cause you confusion later on in this section.

15. Type `pwd` and press Enter. Take note of your present working directory.

16. Open the Python interactive shell by typing `python3` at the shell prompt and pressing Enter.

17. At the Python interactive shell prompt, `>>>`, type `import os` and press Enter to import the `os` function into your Python shell.

18. Type `os.getcwd()` and press Enter. You should see the same present working directory displayed that you saw in step 15.

19. To move down into the data subdirectory, type `os.chdir('data')` and press Enter.

20. Type `os.listdir()` and press Enter. Do you see the file `friends.txt` listed in the output of this command? You should see it!

21. You will create a file object and open your `friends.txt` file. Type `my_friends_file = open ('friends.txt','r')` at the prompt and press Enter.

22. Create a `for` loop to read the friends file you just opened, line by line, by typing in `for my_friend in my_friends_file:` and pressing Enter. Wait a minute! Where is the `range` statement? Don't worry. You can elegantly loop through the `friends.txt` file by using this `for` loop structure. Just wait and see.

23. Press the Tab key one time and then type `print (my_friend, end=' ')`. That's right! There is no readline method included in this loop.

24. To see this nice little loop read through the `friends.txt` file, press the Enter key two times. You should see results similar to the output in Listing 11.11.

**LISTING 11.11**   Reading `friends.txt` Line by Line

```
Chris
Zach
Karl
Zoe
Simon
John
Anton
>>>
```

25. Press Ctrl+D to exit the Python interactive shell.

26. If you want to power down your Raspberry Pi now, type `sudo poweroff` and press the Enter key.

The new style of `for` loop you used in these steps does not require any read methods. This is because you can use a file object variable as a method for iterating through the file.

One item that has been very sloppily handled so far in this book is closing files. Notice in step 25 that you just quit out of the Python interactive shell, without doing any proper file closure. In the next section, you will learn how to properly close a file.

# Closing a File

When a file is opened in almost any program, it is considered good form to close it before you exit that program. This is true in Python scripting as well. The general syntax for closing an opened file is:

```
filename_variable.close ()
```

In Listing 11.12, the temperature file is opened and then closed. Whether or not the file is opened is tested two times by the `.closed` method.

**LISTING 11.12    Closing the Temperature File**

```
>>> temp_data_file = open('/home/pi/data/May2012TempF.txt','r')
>>> temp_data_file.closed
False
>>> temp_data_file.close()  # Close the temperature data file.
>>> temp_data_file.closed
True
>>>
```

Python automatically closes a file if its file name variable is reassigned to another file. However, when you're writing to a file, closing a file can be critical. This is due to the fact that the operating system buffers the write methods in memory. The data is written to the file only when the buffer reaches a certain level. When a file is closed in Python, the buffer in memory is automatically written to the file, whether it is full or not. Not properly closing a file could leave your file's data in a very interesting state!

You can see why it is considered good form to properly close a file, especially if it is being written to. This leads to the next topic of this hour: writing to a file.

# Writing to a File

A file can be opened for writing only, or it can be opened to be read and written. The open mode for writing a text file is either w or a, depending on whether you want to create a new file or append to an old one. Adding a + at the end of the w or a open mode allows you to both read and write to the file.

## Creating and Writing to a New File

Listing 11.13 shows a new text file opened using the open function. In this case, a new file is needed to hold new temperature data.

**LISTING 11.13    Opening a File for Creation and Writing**

```
pi@raspberrypi ~ $ ls /home/pi/data
friends.txt  May2012TempF.txt
pi@raspberrypi ~ $
pi@raspberrypi ~ $ python3
Python 3.2.3 (default, Jan 28 2013, 11:47:15)
[GCC 4.6.3] on linux2
Type "help", "copyright", "credits" or "license" for more information.
>>>
>>> import os
>>> os.listdir('/home/pi/data')
['friends.txt', 'May2012TempF.txt']
```

```
>>>
>>> ctemp_data_file = open('/home/pi/data/May2012TempC.txt', 'w')
>>>
>>> os.listdir('/home/pi/data')
['friends.txt', 'May2012TempC.txt', 'May2012TempF.txt']
>>>
```

You can see in Listing 11.13 that the file `May2012TempC.txt` was nonexistent before the `open` function was used. Once the file was opened, using the `w` mode, the text file was created.

WATCH OUT!

### Write Mode Removes

Keep in mind that if a file is opened in write mode, `w`, using the `open` function, and it already exists, the entire file's contents will be erased! To preserve a preexisting file, use the append, `a`, mode to open a file. See the next section in this hour, "Writing to a Preexisting File," for how to accomplish appending to a file.

Once the file is properly opened in the correct mode, you can begin to write data to it by using the `.write` method. Again using the example from Hour 9, "Dictionaries and Sets," now you can read a file containing temperatures in Fahrenheit, convert the temperatures to Celsius, and then write the data to a new file.

In Listing 11.14, the Fahrenheit file `May2012TempF.txt` is open for reading, so the temperatures in it can be converted to Celsius. The Celsius file `May2012TempC.txt` is opened for writing. Included on line 5 is a `for` loop for reading the Fahrenheit temperatures from the Fahrenheit file. Notice that the `for` loop is the more elegant version of a file reading loop that you used in the Try It Yourself section of this hour.

**LISTING 11.14**   Writing to the Celsius Temperature File

```
 1: >>> ftemp_data_file = open ('/home/pi/data/May2012TempF.txt', 'r')
 2: >>> ctemp_data_file = open ('/home/pi/data/May2012TempC.txt', 'w')
 3: >>>
 4: >>> date_count = 0
 5: >>> for ftemp_data in ftemp_data_file:
 6: ...      ftemp_data = ftemp_data.rstrip('\n')
 7: ...      ftemp_data = ftemp_data[2:len(ftemp_data)]
 8: ...      ftemp_data = ftemp_data.lstrip(' ')
 9: ...      ftemp_data = int(ftemp_data)
10: ...       ctemp_data = round((ftemp_data - 32) * 5/9, 2)
11: ...       date_count += 1
12: ...       ctemp_data = str(date_count) + ' ' + str(ctemp_data) + '\n'
13: ...       ctemp_data_file.write(ctemp_data)
```

```
14: ...
15: 8
16: 8
17: 8
18: ...
19: 9
20: 9
21: 9
22: >>> ftemp_data_file.close()
23: >>> ctemp_data_file.close()
```

After the Fahrenheit temperature is read into the variable ftemp_data, some processing is needed to pull out the temperature from the read-in data. First, on line 6, the newline escape sequence is stripped off. On line 7, the temperature is obtained using string slicing. (You learned about string slicing in Hour 10.) Before calculations start, any preceding blank spaces are stripped off in line 8, and the temperature data character string is turned into an integer on line 9.

BY THE WAY

## A Number Is a String

When reading in data from a text file in Python, numbers are not typed as numeric, such as integer or floating point. Instead, Python assigns them the character string type (str).

Also, when writing a number to a text file, you need to perform a conversion. A number must be converted from its numeric type to a character string before it is written to a text file.

The temperature is converted from Fahrenheit to Celsius on line 10 of Listing 11.14. Before the data can be written to the new Celsius text file, it must be converted from floating point to character string, which happens on line 12. A character string with all the needed data is created on line 12 because the .write method can accept only one argument. Notice also on line 12 that a newline escape sequence is added at the end of the string to act as a data separator. Now the data can be written to the new file on line 13 in Listing 11.14.

BY THE WAY

## What Are Those Numbers?

In Listing 11.14, you can see a partial string of 8s and 9s after the .write method in the code. This is because the .write method displays to output how many characters it wrote to a text file or how many bytes it wrote to a binary file.

After all the data has been read from the Fahrenheit file, processed, and written out to the Celsius file, the files should be closed. Remember that closing files is important when writing to a file, and it's considered good form. The files are closed on lines 22 and 23 in Listing 11.14.

So does everything work in this example? The Fahrenheit temperature data is properly read in, converted, and written out to the Celsius file, as shown in Listing 11.15.

**LISTING 11.15**   The Celsius Temperature File

```
pi@raspberrypi ~ $ cat /home/pi/data/May2012TempC.txt
1 26.11
2 28.89
3 29.44
4 29.44
...
28 33.33
29 33.33
30 27.22
31 24.44
pi@raspberrypi ~ $
```

Note that each Celsius temperature has the day of the month it was recorded and is displayed in floating-point format. Also notice that each temperature is on its own line in the text file. This is because the newline \n escape sequence was tacked onto the end of each Celsius temperature data string in Listing 11.14, line 12.

# Writing to a Preexisting File

You tell Python that written data will be appended to a preexisting file via the open function. After that, writing data to a preexisting file using the .write method is no different than using the .write method for writing data to a new file.

In Listing 11.16, the Celsius temperature file is opened. This file was just filled with data in the last section of this hour.

**LISTING 11.16**   Opening a File to Append to It

```
>>> ctemp_data_file = open ('/home/pi/data/May2012TempC.txt', 'a')
>>>
```

The open statement's mode setting keeps the file's data from being overwritten. The mode is set to a, which does two things. First, it preserves the current data. Second, the file pointer is set to point at the end of the file. Thus, any .write methods that occur, start writing at the file's bottom, and no data is lost.

## Open a File and Write to It

In the last Try it Yourself, you created a file outside Python. In the following steps, you get to create a file inside Python! Before you can complete these steps, you need to have completed the last Try It Yourself in this hour. If you haven't, go back and do it now! When you're ready, follow these steps:

1. If you have not already done so, power up your Raspberry Pi and log in to the system.

2. If you do not have the LXDE GUI started automatically at boot, start it now by typing `startx` and pressing Enter.

3. Open the LXTerminal by double-clicking the LXTerminal icon.

4. Open the Python interactive shell by typing `python3` at the shell prompt and pressing Enter.

5. To open the file needed, at the Python interactive shell prompt, `>>>`, type `my_friends_file = open ('/home/pi/data/friends.txt','w+')` and press Enter. Wait! Won't this delete the contents of the `friends.txt` file you created in the last Try It Yourself section? Yes, you are correct. Opening the file using the `w` mode will indeed clear out the `friends.txt` file. But, don't worry. You will be rebuilding it. This is why you use the `+`. After you fill the file with data, you will be reading it.

6. Create a `for` loop to create the friends file, using keyboard input by typing `for friend_count in range(1, 7+1):` and pressing Enter.

7. Press the Tab key one time to properly indent under the `for` loop. Now that you are properly indented, type `my_friend = input("Friend's name: ")` and press Enter. (You learned about getting keyboard input into a Python script in Hour 4, "Understanding Python Basics." Go back there to refresh your memory if needed.)

8. Press the Tab key one time to properly indent under the `for` loop. Now that you are properly indented, type `my_friend = my_friend + '\n'` and press Enter. Python will now add the needed newline escape sequence to the end of each friend's name.

9. Press the Tab key one time to maintain the proper indentation under the `for` loop. Type `my_friends_file.write(my_friend)` and press Enter. This Python statement causes the name to be written out to the `friends.txt` file.

10. Press the Enter key two times to kick off your loop for filling the `friends.txt` file with data.

**11.** Each time the loop asks you for a friend's name, enter a name from this list and then press Enter:

Chris
Zack
Karl
Zoe
Simon
John
Anton

Don't let it throw you that a number displays after each name you enter. Remember that the `.write` method displays the number of characters it wrote to a file. (Have you figured out who these people are yet?)

**12.** So that you can now read the file from the beginning, reset the file pointer to the start of the file by typing `my_friends_file.seek(0)` and pressing Enter.

**13.** Create a `for` loop to read the newly populated friends file by typing `for my_friend in my_friends_file:` and pressing Enter. Notice that there is no need to close and reopen the file. This is because in step 5, the file was opened with the mode `w+`, which allows you to write to the file and read it.

**14.** Press the Tab key one time and then type `print (my_friend, end=' ')`. This causes the data read into the Python script from the file to be displayed to your screen.

**15.** Press the Enter key two times. You should see the friends' names from the `friends.txt` file displayed to the screen.

**16.** Close the `friends.txt` file by typing `my_friends_file.close()` and pressing the Enter key. Now the file is properly closed.

**17.** Just to double-check, type `my_friends_file.closed` and press the Enter key. If the file is truly closed, you should receive back the word `True`. If you get `False`, repeat step 16.

**18.** Press Ctrl+D to exit the Python interactive shell.

**19.** If you want to power down your Raspberry Pi now, type `sudo poweroff` and press the Enter key.

Creating the /home/pi/data/friends.txt file this time was much easier than it was in the last Try It Yourself section! By now you should have a good handle on how to open, close, read, and write data files.

# Summary

In this hour, you read about using files in Python. You saw how to open a file and how to close a file. Also, you were introduced to Python statements and structures that allow you to read from a file and write to a file. You got to try out both writing to a file in the dash shell and reading it, and then writing to a file within Python and reading it. In Hour 12, "Creating Functions," you will investigate how to create your own Python functions, which means you are starting to move into more advanced Python concepts.

# Q&A

**Q.** Which directory reference is used in Table 11.2?

**A.** The directories shown in Table 11.2 use an absolute directory reference. Remember that an absolute directory reference absolutely starts with the root directory. Therefore, /home/pi/ py3prog, /home/pi/temp, and /home/pi/data are all absolute directory references.

**Q.** What is pickling?

**A.** Pickling is a method of preserving vegetables, such as cucumbers, in a salty vinegar solution. But you are probably asking about pickling data, mentioned in Table 11.1, and not vegetables.

Pickling is a method of transforming a Python object, such as a dictionary, into a series of bytes for storage into a file. Turning an object into a series of bytes is called *serializing* an object. You need to import the pickle function to perform pickling of objects.

The advantage of pickling is that it allows you to quickly and easily handle objects. A pickled object can be read from a file into a single variable. The disadvantage is that pickling has no security measures built around it. Therefore, you could cause a system to be compromised by using it.

**Q.** How can I keep the .write method from displaying the number of characters it has written during the running of a Python script?

**A.** You can use a "cheat" method. Import the non-built-in function sys. Then before using your .write method, redirect the terminal output by typing in the following statement: sys.stdout = open ('/dev/null', 'w'). This sends the characters written as output to "nowhere." However, be very careful using a feature like this! You will also send any error messages to "nowhere" as well!

**Q.** I can't figure it out. Who are those people in the friends.txt file?

**A.** Here's a hint: They are Sci-Fi related.

# Workshop

## Quiz

1. The `os` function is a built-in function. True or false?

2. Which file object methods can be used to read a text file in a random-access manner?

3. To open a file to be both read and written, which mode should be used in the `open` function statement?

    a. `w+`

    b. `r&w`

    c. `r+`

## Answers

1. False. The `os` function is not a built-in function. You must import it to use its various methods. To import the `os` function, you use the Python statement `import os`.

2. The `.read` and `.seek` file object methods can be used to read a text file in a random-access (nonsequential) manner. You use the `.seek` method to put the file pointer in the correct position in the file. You use the `.read` method to read a particular number of characters from the file.

3. This is a trick question! Both answers a and c are correct. Remember that tacking on a + to either `r` or `w` will allow the file to have the other option done to it as well. Therefore, `w+` lets you read and write to a file, and `r+` also lets you read and write to a file.

# Creating Functions

---

**What You'll Learn in This Hour:**

- ▶ How to create your own functions
- ▶ Retrieving data from functions
- ▶ Passing data to functions
- ▶ Using lists with functions
- ▶ Using functions in your Python scripts

Often while writing Python scripts, you'll find yourself using the same code in multiple locations. With just a small code snippet, that's usually not a big deal. However, rewriting large chunks of code multiple times in your Python scripts can get tiring. Python helps you out by supporting user-defined functions. You can encapsulate your Python code into a function and then use it as many times as you want, anywhere in your script. This hour walks you through the process of creating your own Python functions and demonstrates how to use them in other Python script applications.

## Utilizing Python Functions in Your Programs

As you start writing more complex Python scripts, you'll find yourself reusing parts of code that perform specific tasks. Sometimes it's something simple, such as displaying a text message and retrieving an answer from the script users. Other times it's a complicated calculation that's used multiple times in a script as part of a larger process.

In each of these situations, writing the same blocks of code over and over again in your script can get tiresome. It would be nice to just write the block of code once and then be able to refer to that block of code other places in your script, without having to rewrite it.

Python provides a feature that allows you to do just that. *Functions* are blocks of script code that you assign names to, and then you can reuse them anywhere in your code. Any time you need to use that block of code in your script, you simply use the name you assigned to the function; this is referred to as *calling the function*. The following sections describe how to create and use functions in your Python scripts.

## Creating a Function

To create a function in Python, you use the `def` keyword followed by the name of the function, with parentheses, as shown here:

```
def name():
```

Note the colon at the end of the statement. By now you should recognize that this means there's more code associated with the statement. You just place any code that you want in your function under the function statement, indented, like this:

```
def myfunction():
    statement1
    statement2
    statement3
    statement4
```

With Python, there's no "end of function" type of delimiter statement. When you're done with the statements contained within the function, you just place the next code statement back on the left margin.

## Using Functions

To use a function in a Python script, you specify the function name on a line, just as you would any other Python statement. Listing 12.1 shows the script1201.py program, which demonstrates how to define and use a function in a sample Python script.

**LISTING 12.1**    Defining and Using Functions in a Script

```
#!/usr/bin/python3

def func1():
    print('This is an example of a function')

count = 1
while(count <= 5):
    func1()
    count = count + 1

print('This is the end of the loop')
func1()
print('Now this is the end of the script')
```

The code in the `script1201.py` script defines a function called `func1()`, which prints out a line to let you know it ran. The script then calls the `func1()` function from inside a `while` loop, so the function runs five times. When the loop finishes, the code prints out a line, calls the function one more time, and then prints out another line to indicate the end of the script.

When you run the script, you should see this output:

```
pi@raspberrypi ~ $ python3 script1201.py
This is an example of a function
This is an example of a function
This is an example of a function
This is an example of a function
This is an example of a function
This is the end of the loop
This is an example of a function
Now this is the end of the script
pi@raspberrypi ~ $
```

The function definition doesn't have to be the first thing in your Python script, but be careful. If you attempt to use a function before it's defined, you get an error message. Listing 12.2 shows an example of this with the script1202.py program.

**LISTING 12.2**    Trying to Use a Function Before It's Defined

```python
#!/usr/bin/python3

count = 1
print('This line comes before the function definition')

def func1():
    print('This is an example of a function')

while(count <= 5):
    func1()
    count = count + 1

print('This is the end of the loop')
func2()
print('Now this is the end of the script')

def func2():
    print('This is an example of a misplaced function')
```

When you run the `script1202.py` function, you should get an error message:

```
pi@raspberrypi ~ $ python3 script1202.py
This line comes before the function definition
This is an example of a function
```

```
This is an example of a function
This is an example of a function
This is an example of a function
This is an example of a function
This is the end of the loop
Traceback (most recent call last):
  File "script1202.py", line 14, in <module>
    func2()
NameError: name 'func2' is not defined
pi@raspberrypi ~ $
```

The first function, func1(), is defined after a couple of statements in the script, which is perfectly fine. When the func1() function is used in the script, Python knows where to find it.

However, the script attempts to use the func2() function before it is defined. Because the func2() function isn't defined yet when the script reaches the place where you use it, you get an error message.

You also need to be careful about function names. Each function name must be unique, or you have problems. If you redefine a function, the new definition overrides the original function definition, without producing any error messages. Take a look at the script1203.py script in Listing 12.3 for an example of this.

**LISTING 12.3**   Trying to Redefine a Function

```
#!/usr/bin/python3

def func1():
   print('This is the first definition of the function name')

func1()

def func1():
   print('This is a repeat of the same function name')
func1()
print('This is the end of the script')
```

When you run the script1203.py script, you should get this output:

```
pi@raspberrypi ~ $ python3 script1203.py
This is the first definition of the function name
This is a repeat of the same function name
This is the end of the script
pi@raspberrypi ~ $
```

The original definition of the `func1()` function works fine, but after the second definition of the `func1()` function, any subsequent uses of the function use the second definition instead of the first one.

# Returning a Value

So far the functions that you've used have just output a string and ended. Python uses the `return` statement to exit a function with a specific value. With the `return` statement, you can specify a value that the function returns back to the main program after it finishes, and then it uses that value back in the main program.

The `return` statement must be the last statement in the function definition, as shown here:

```
def func2():
    statement1
    statement2
    return value
```

In the main program, you can assign the value returned by the function to a variable and then use it in your code. Listing 12.4 shows the `script1204.py` script, which demonstrates how to do this.

**LISTING 12.4    Returning a Value from a Function**

```
#!/usr/bin/python3

def dbl():
  value = int(input('Enter a value: '))
  print('doubling the value')
  result = value * 2
  return result

x = dbl()
print('The new value is ', x)
```

In the `script1204.py` code, you define a function called `dbl()` that prompts for a number, converts the answer to an integer, and then multiples it by 2 and returns it.

Then in the application part of the code, you call the `dbl()` function and assign its output to the variable x. If you run the `script1204.py` script and enter a value at the prompt, you should see these results:

```
pi@raspberrypi ~ $ python3 script1204.py
Enter a value: 10
doubling the value
The new value is  20
pi@raspberrypi ~ $
```

---

BY THE WAY

**Returning Values**

In this example, the function returns an integer value, but you can also return strings, floating-point values, and even other Python objects!

---

# Passing Values to Functions

You might have noticed in the functions defined so far this hour that they have all created their own data. However, most functions don't operate in a vacuum but require information from the main program to process. The following sections discuss how you can ensure that information gets to the Python functions you create.

## Passing Arguments

You pass values into a function from your main program by using arguments. Arguments are values enclosed within the function parentheses, like this:

```
result = funct3(10, 50)
```

To retrieve the argument values in your Python functions, you define parameters in the function definition. Parameters are variables you place in the function definition to receive the argument values when the main program calls the function.

Here's an example of defining a function that uses parameters:

```
>>> def addem(a, b):
    result = a + b
    return result

>>>
```

The addem() function defines two parameters. The a variable receives the first argument value, and the b variable receives the second argument value. You can then use the a and b variables anywhere within the function code.

Now when you call the addem() function from your Python code, you must pass two argument values:

```
>>> total = addem(10, 50)
>>> print(total)
60
>>>
```

If you don't provide any arguments, or if you provide the incorrect number of arguments, you get an error message from Python, like this:

```
>>> total = addem()
Traceback (most recent call last):
  File "<pyshell#7>", line 1, in <module>
    total = addem()
TypeError: addem() missing 2 required positional arguments: 'a' and 'b'
>>> total = addem(10, 20, 30)
Traceback (most recent call last):
  File "<pyshell#8>", line 1, in <module>
    total = addem(10, 20, 30)
TypeError: addem() takes 2 positional arguments but 3 were given
>>>
```

If you pass string values as arguments, it's important to remember to use quotes around the values, as shown here:

```
>>> def greeting(name):
        print('Welcome', name)


>>> greeting('Rich')
Welcome Rich
>>>
```

If you don't place the quotes around the string value, Python thinks you're trying to pass a variable, as shown in this example:

```
>>> greeting(Barbara)
Traceback (most recent call last):
  File "<pyshell#13>", line 1, in <module>
    greeting(Barbara)
NameError: name 'Barbara' is not defined
>>>
```

This brings up a good point: You can use variables as arguments when calling a function, as shown here:

```
>>> x = 100
>>> y = 200
>>> total = addem(x, y)
>>> print(total)
300
>>>
```

Inside the `addem()` function, Python retrieves the value stored in the x variable in the main program and stores it in the a variable inside the function. Likewise, Python retrieves the value stored in the y variable in the main program and stores it in the b variable inside the function.

WATCH OUT!

**Positional Parameters**

Be careful when passing arguments to your Python functions. Python matches the argument values in the same order that you define them in the function parameters. These are called *positional parameters*.

## Setting Default Parameter Values

Python allows you to set default values assigned to parameters if no arguments are provided when the main program calls the function. You just set the default values inside the function definition, like this:

```
>>> def area(width = 10, height = 20):
        area = width * height
        return area

>>>
```

The `area()` function definition defines default values for the two parameters. If you call the `area()` function with arguments, Python uses those arguments in the function and overrides the default values, as shown here:

```
>>> total = area(15, 30)
>>> print(total)
450
>>>
```

However, if you call the function without any arguments, instead of giving you an error message, Python uses the default values assigned to the parameters, like this:

```
>>> total2 = area()
>>> print(total2)
200
>>>
```

If you specify just one argument, Python uses it for the first parameter and takes the default for the second parameter, as in this example:

```
>>> area(15)
300
>>>
```

If you want to define a value for the second parameter but not the first, you have to define the argument value by name, like this:

```
>>> area(height=15)
150
>>>
```

You don't have to declare default values for all the parameters. You can mix and match which ones have default values, as shown here:

```
>>> def area2(width, height = 20):
        area2 = width * height
        print('The width is:', width)
        print('The height is:', height)
        print('The area is:', area2)


>>>
```

With this definition, the width parameter is required, but the height parameter is optional. If you call the area2() function with just one argument, Python assigns it to the width variable, as shown in this example:

```
>>> area2(10)
The width is: 10
The height is: 20
The area is: 200
>>>
```

If you call the area2() function with no parameters, you get an error message for the missing required parameter, as shown here:

```
>>> area2()
Traceback (most recent call last):
  File "<pyshell#28>", line 1, in <module>
    area2()
TypeError: area2() missing 1 required positional argument: 'width'
>>>
```

Now you have more control over how to use functions in other Python scripts.

## WATCH OUT!

### Ordering of Parameters

When you list the parameters used in your functions, make sure you place any required parameters first, before parameters that have default values. Otherwise, Python gets confused and does not know which arguments match with which parameters!

# Dealing with a Variable Number of Arguments

In some situations, you might not have a set number of parameters for a function. Instead, a function may require a variable number of parameters. You can accommodate this by using the following special format to define the parameters:

```
def func3(*args):
```

When you place an asterisk in front of the variable name, the variable becomes a tuple value, containing all the values passed as arguments when the function is called.

You can retrieve the individual parameter values by using indexes of the variable. In this example, Python assigns the first parameter to the args[0] variable, the second parameter to the args[1] variable, and so on for all the arguments passed to the function.

Listing 12.5 shows the script1205.py script, which demonstrates using this method to retrieve multiple parameter values.

**LISTING 12.5**   **Retrieving Multiple Parameters**

```python
#!/usr/bin/python3

def perimeter(*args):
    sides = len(args)
    print('There are', sides, 'sides to the object')
    total = 0
    for i in range(0, sides):
        total = total + args[i]
    return total

object1 = perimeter(2, 3, 4)
print('The perimeter of object1 is:', object1)
object2 = perimeter(10, 20, 10, 20)
print('The perimeter of object2 is:', object2)
object3 = perimeter(10, 10, 10, 10, 10, 10, 10, 10)
print('The perimeter of object3 is:', object3)
```

In the script1205.py code, the perimeter() function uses the *args parameter variable to define the parameters for the function. Because you don't know how many arguments are used when the function is called, the code uses the len() function to find out how many values are in the args tuple.

While you could just use the for() loop to directly iterate through the args tuple, this example demonstrates retrieving each value individually. The script1205.py code uses a for() loop to iterate through a range from 0 to the number of values that the tuple contains and then uses the arg[i] variable to reference each value directly. When the for loop is complete, the perimeter() function returns the final value.

This example shows three different examples of using the perimeter() function, each with a different number of arguments. In each case, the perimeter() function totals the argument values and returns the result. When you run the script1205.py script, you should see the following output:

```
pi@raspberrypi ~$ python3 script1205.py
There are 3 sides to the object
The perimeter of object1 is: 9
There are 4 sides to the object
The perimeter of object2 is: 60
There are 8 sides to the object
The perimeter of object3 is: 80
pi@raspberrypi ~$
```

BY THE WAY

### The args Variable

The examples in this section use the variable args to represent the tuple of the argument values. This is not a requirement; you can use any variable name you choose. However, it's become somewhat of a de facto standard in Python to use the args variable name in this situation.

## Retrieving Values Using Dictionaries

You can use a dictionary variable to retrieve the argument values passed to a function. To do this, you place two asterisks (**) before the dictionary variable name in the function definition parameter:

```
def func5(**kwargs):
```

When you place the two asterisks in front of the kwargs variable, it becomes a dictionary variable. When you call the func5() function, you must specify a keyword and value pair for each argument:

```
func5(one = 1, two = 2, three = 3)
```

To retrieve the values, you use the kwargs['one'], kwargs['two'], and kwargs['three'] variables.

Listing 12.6 shows an example of using this method in the script1206.py Python script.

**LISTING 12.6**   Using Dictionaries in Functions

```
#!/usr/bin/python3

def volume(**kwargs):
    radius = kwargs['radius']
    height = kwargs['height']
    print('The radius is:', radius)
    print('The height is:', height)
    total = 3.14159 * radius * radius * height
    return total

object1 = volume(radius = 5, height = 30)
print('The volume of object1 is:', object1)
```

The `script1206.py` code demonstrates how the `kwargs` variable becomes a dictionary variable, using the keywords you specify when you call the function. The `kwargs['radius']` variable contains the value set to the radius value in the function call, and the `kwargs['height']` variable contains the value set to the height value in the function call.

BY THE WAY

**The `kwargs` Variable**

You can use any variable name for the dictionary variable, but the `kwargs` variable name has become a de facto standard for defining dictionary parameters in Python coding.

# Handling Variables in a Function

As you can probably tell by now, handling variables in Python functions can be rather complex. To make things even more complicated, there are two different types of variables that you can use inside Python functions:

▶ Local variables

▶ Global variables

These two types of variables behave somewhat differently in your program code, so it's important to know just how they work. The following sections break down the differences between using local and global variables in your Python scripts.

# Local Variables

*Local variables* are variables that you create inside a function. Because you create the variables inside the function, you can only access them inside the function. Outside the function, the rest of the script code doesn't recognize them. Listing 12.7 shows the script1207.py program, which demonstrates this principle.

**LISTING 12.7    Working with Local Variables in a Function**

```
#!/usr/bin/python3

def area3(width, height):
    total = width * height
    print('Inside the area3() function, the value of width is:',width)
    print('Inside the area3() function, the value of height is:',height)
    return total

object1 = area3(10, 40)
print('Outside the function, the value of width is:', width)
print('Outside the function, the value of height is:', height)
print('The area is:', object1)
```

The `script1207.py` code defines the `area3()` function with two parameters and then uses those parameters inside the function to calculate the area. However, if you try to access those variables outside the function, you get an error message, like this:

```
pi@raspberrypi ~% python3 script1207.py
Inside the area3() function, the value of width is: 10
Inside the area3() function, the value of height is: 40
Traceback (most recent call last):
  File "C:/Python33/script1207.py", line 11, in <module>
    print('Outside the function, the value of width is:', width)
NameError: name 'width' is not defined
pi@raspberrypi ~%
```

The code starts out just fine, passing two arguments to the `area3()` function, which completes without a problem. However, when the code tries to access the `width` variable outside the `area3()` function, Python produces an error message, indicating that the `width` variable is not defined.

# Global Variables

*Global variables* are variables you can use anywhere in your program code, including inside functions. Values assigned to a global variable in the main program are accessible in the function code, but there's a catch: While the function can read the global variables, by default it can't change them. Listing 12.8 shows the script1208.py program, which is an example of how this can go wrong in your Python scripts.

**LISTING 12.8**   Global Variables Causing Problems

```
#!/usr/bin/python3

width = 10
height = 60
total = 0

def area4():
    total = width * height
    print('Inside the function the total is:', total)

area4()
print('Outside the function the total is:', total)
```

The `script1208.py` code shown in Listing 12.8 defines three global variables: `width`, `height`, and `total`. You can read them in the `area4()` function just fine, as shown by the output of the `print()` statement. However, if you expect the total variable to still be set when the code exits the `area4()` function, you have a problem, as shown here:

```
pi@raspberrypi ~$ python3 script1208.py
Inside the function the total is: 600
Outside the function the total is: 0
pi@raspberrypi ~$
```

When you try to read the `total` variable in the main program, the value is set back to the global value assignment, not the value that was changed inside the `area4()` function!

There is a solution to this problem. To tell Python that the function is trying to access a global variable, you need to add the `global` keyword to define the variable:

```
global total
```

This equates the variable named `total` inside the function to the variable named `total` defined in the main program. Listing 12.9 shows a corrected example of using this principle with the script1209.py program.

**LISTING 12.9**   Properly Using Global Variables

```
#!/usr/bin/python3

width = 10
height = 60
total = 0
```

```
def area5():
    global total
    total = width * height
    print('Inside the function the total is:', total)

area5()
print('Outside the function the total is:', total)
```

Adding the one `global` statement causes the code to run correctly now, as shown here:

```
pi@raspberrypi ~$ python3 script1209.py
Inside the function the total is: 600
Outside the function the total is: 600
pi@raspberrypi ~$
```

WATCH OUT!

### Using Global Variables

You may be tempted to use global variables to pass values to a function and retrieve values from a function. While that certainly works, it's somewhat frowned upon in Python programming circles. The idea is to make a function as self-contained as possible, so you can use it in other programs without lots of extraneous coding. Requiring global variables for the function to work complicates reusing the function in other programs. If you stick with parameters and return values, your functions can easily be reused in any program where you need them!

# Using Lists with Functions

When you pass values as arguments to functions, Python passes the actual value, and not the variable location in memory; this is called *passing by reference*. However, there's an exception to this.

If you pass a mutable object (such as a list or dictionary variable), the function can make changes to the object itself. That may seem a bit odd, but it can come in handy.

Listing 12.10 shows the `script1210.py` script, which demonstrates passing a list to a function that modifies the list.

### LISTING 12.10   Passing a List Value to a Function

```
#!/usr/bin/python3

def modlist(x):
    x.append('Jason')
```

```
mylist = ['Rich', 'Christine']
print('The list before the function call:', mylist)
modlist(mylist)
print('The list after the function call:', mylist)
```

The `script1210.py` code creates a function named `modlist()`, which appends a value to the list passed as the function parameter.

The code then tests the `modlist()` function by creating a list called `mylist`, calling the `modlist()` function, and displaying the value of the `mylist` list variable, as shown here:

```
pi@raspberrypi ~$ python3 script1210.py
The list before the function call: ['Rich', 'Christine']
The list after the function call: ['Rich', 'Christine', 'Jason']
pi@raspberrypi ~$
```

As you can see from the output, the `modlist()` function modifies the `mylist` list variable, which maintains its value in the main program code.

# Using Recursion with Functions

A popular use of functions is in a process called *recursion*. In recursion, you solve an algorithm by repeatedly breaking the algorithm into subsets until you reach a core definition value.

The factorial algorithm is a classic example of recursion. The *factorial* of a number is defined as the result of multiplying all the numbers up to and including that number. So, for example, the factorial of 5 is 120, as shown here:

```
5! = 1 * 2 * 3 * 4 * 5 = 120
```

By definition, the factorial of 0 is equal to 1. Notice that to find the factorial of 5, you just multiply 5 by the factorial of 4, and to find the factorial of 4, you multiply 4 by the factorial of 3. You continue on until you get to the factorial of 0, which, by definition, is 1. This is a perfect example of using recursion in your functions.

▼ TRY IT YOURSELF

### Creating a Factorial Function Using Recursion

To use recursion, you need to define an endpoint in the function so that it doesn't get stuck in a loop. For the factorial function, the endpoint is the factorial of 0:

```
if (num == 0):
    return 1
```

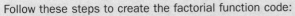

Follow these steps to create the factorial function code:

1.  Create the file script1211.py, and open it in your editor program. Here's the code to use for the file:

```
#!/usr/bin/python3

def factorial(num):
    if (num == 0):
        return 1
    else:
        return num * factorial(num - 1)

result = factorial(5)
print('The factorial of 5 is', result)
```

2.  Save the file, then run the program.

    When you run the `script1211.py` program, you get this output:

```
pi@raspberrypi ~$ python3 script1211.py
The factorial of 5 is 120
pi@raspberypi ~$
```

The `factorial()` function first checks whether the parameter value is 0. If it is, it returns the default definition value of 1. If the parameter value isn't 0, it runs a new calculation, returning the number multiplied by the factorial of one less than the number. So the `factorial()` function calls itself, each time with a lower number, until it gets to the 0 value.

# Summary

In this hour, you learned how to create and use your own functions in Python. You use the `def` keyword to define your function code, and then you can just reference your function anywhere in your script code. You can return a value from the function back to the main program that called it, and you can pass values from the main program into the function. You also learned how to work with variables in functions. Any variable that you define inside a function can be used only inside the function, while variables you define outside the function can be used inside the function.

In the next hour, we'll turn our attention to modules. Python lets us use modules to package our functions, as well as use functions from others!

# Q&A

**Q.** **Can I group all my function definitions together into a single file and then just reference that file in my Python scripts?**

**A.** Yes, that technique is called using a module, and it's covered in Hour 13, "Working with Modules"!

**Q.** **What if you write a function that uses recursion that doesn't have an endpoint, and your program gets stuck in an infinite loop?**

**A.** Python will continue to iterate through the functions until you manually stop the program by sending a SIGINT signal to the program (using the Ctrl-C key combination).

**Q.** **If both the function and the main program can read and process global variables, why do I need to pass parameters to a function? Can't I just use global variables?**

**A.** The idea of the function is that it should be as self-contained as possible. That way you can easily copy functions between programs. When the function uses global variables, that means the other programs would also need to define the global variables. With parameters, all of the data required for the function is self-contained in the function call!

# Workshop

## Quiz

**1.** What do you call the variables that are defined to receive values passed to a function?

    **a.** Arguments

    **b.** Parameters

    **c.** Global variables

    **d.** Recursion

**2.** A function can never reference itself. If it did, you'd get an endless loop. True or false?

**3.** You must define a function before you can use it in your Python script. True or false?

## Answers

**1.** b. Parameters. The parameters help keep the function self-contained; all of the data required for the function are passed as parameters from the main program.

**2.** False. You can use recursion to reference a function inside itself. However, there must be a predefined endpoint in the function; otherwise, it will get stuck in an infinite loop.

**3.** True. If the Python interpreter sees a function in use in your code, it must have the definition in memory to know how to process it. The Python interpreter can't read ahead in the code to find the function definition.

# Working with Modules

## What You'll Learn in This Hour:

▶ What a module is

▶ Standard Python modules

▶ What a module contains

▶ How to create a custom module

Keeping your Python scripts a reasonable size helps you use and manage them. Modules can help you keep a Python script to a reasonable size. In this hour, you will learn about modules: how to create them and how to use them in your Python scripts.

## Introducing Module Concepts

A *module* is a collection of functions. You learned about writing your own functions in Hour 12, "Creating Functions," and started using functions, such as `print`, in Hour 4, "Understanding Python Basics." A module is external to a Python script and has to be imported using the `import` statement. Once it is imported, the function or functions the module contains are available for you to use within your script. The module you import may be a standard module, such as the `os` module. A module may also be a user-created module, which is a module that contains functions a user (such as you) wrote.

BY THE WAY

### Term Confusion

Because a module contains functions, you often see the terms *function* and *module* used interchangeably in Python books and documentation. Also, a function within a module is sometimes called a *method* or an *operation*.

A module must be imported before a function in its collection can be used. However, there are three flavors of Python modules:

- ▶ **Python functions stored in a `.py` file**—These modules can be either locally available (via the Python standard library) or downloaded from somewhere else.

- ▶ **C programs that are dynamically loaded into the Python interpreter**—These modules are built-in.

- ▶ **C programs that are linked with the Python interpreter**—These modules are built-in.

To determine a particular module's flavor, you start by viewing the modules housed in files that end with `.py`. At the shell prompt, issue the command `ls /usr/lib/python*/*.py` on your Raspbian system. You then see the various versions of Python installed, along with each version's modules stored in files.

Viewing linked modules is a little trickier. You need to import the `sys` module to display a list of the built-in modules, as shown in Listing 13.1. Notice that the `math` module is listed here.

**LISTING 13.1** List of Built-in Python Modules

```
>>> import sys
>>> sys.builtin_module_names
('__main__', '_ast', '_bisect', '_codecs', '_collections', '_datetime',
'_elementtree', '_functools', '_heapq', '_io', '_locale', '_pickle',
'_posixsubprocess', '_random', '_socket', '_sre', '_string', '_struct',
'_symtable', '_thread', '_warnings', '_weakref', 'array', 'atexit',
'binascii', 'builtins', 'errno', 'fcntl', 'gc', 'grp', 'imp', 'itertools',
'marshal', 'math', 'operator', 'posix', 'pwd', 'pyexpat', 'select', 'signal',
'spwd', 'sys', 'syslog', 'time', 'unicodedata', 'xxsubtype', 'zipimport',
'zlib')
>>>
```

Compare the `os` module and the `math` module, which have been used in previous hours. The `os` module's functions are Python statements stored in a `.py` file. The `math` module is written in the C programming language and is linked with the Python interpreter. Even though they are different flavors of modules, as shown in Listing 13.2, both must be imported before their functions can be used.

**LISTING 13.2** Importing Different-Flavored Modules

```
1: >>> math.factorial(5)
2: Traceback (most recent call last):
3:   File "<stdin>", line 1, in <module>
4: NameError: name 'math' is not defined
5: >>> import math
```

```
6: >>> math.factorial(5)
7: 120
8: >>> os.getcwd()
9: Traceback (most recent call last):
10:    File "<stdin>", line 1, in <module>
11: NameError: name 'os' is not defined
12: >>> import os
13: >>> os.getcwd()
14: '/home/pi'
15: >>>
```

This is the standard syntax for using functions within modules:

*module.function*

To understand this syntax, look at line 6 in Listing 13.2. You can see that the module is `math`, and the function (sometimes called a *method* or an *operation*) is `factorial`.

---

DID YOU KNOW

### A Group of Modules

You can also gather a collection of modules in Python. This is called a *package*.

---

Using and creating Python modules have some strong benefits. They include manageability and reusability. The term *module* is derived from the theory of "modular programming." A script broken up into small code chunks is easy to track and manage. Large scripts can be unwieldy and difficult to debug. Small scripts that import in reusable code "chunks" (that is, modules) are much easier to handle. Also, when you use modules, you don't have to reinvent functions for each script.

# Exploring Standard Modules

The Python standard library, which contains hundreds of modules, is included when Python is installed on your system. One of the Python catch phrases is "Python comes with batteries included." That catch-phrase applies to the Python standard library, with all its prewritten functions housed in modules.

Most of the standard library modules can be loosely fit into general categories. (Many of the modules could fit into multiple categories.) Those categories include the following:

- ▶ Character strings processing
- ▶ Data compression and backup
- ▶ Database management

- ▸ Date and time tools

- ▸ File I/O and format processing

- ▸ Game development tools

- ▸ Graphics utilities

- ▸ Internationalization utilities

- ▸ Internet I/O and format processing

- ▸ Interprocess communication

- ▸ Multimedia tools

- ▸ Network management

- ▸ Platform-specific commands

- ▸ Python script development tools

- ▸ Scientific (including math) utilities

- ▸ Security management

- ▸ Web development tools

With so many modules full of functions, it makes sense to search through the standard library for what you need before you decide to write your own. To determine what modules are loaded in your Python standard library, use the `help` function, as shown in Listing 13.3. The listing displays only a small portion of the entire module list because it is just too long to show the entire list in this book!

**LISTING 13.3**    Using `help` to Find Modules

```
>>> help('modules')

Please wait a moment while I gather a list of all available modules...

CDROM               audioop           imp              shlex
DLFCN               base64            importlib        shutil
IN                  bdb               inspect          signal
RPi                 binascii          io               site
TYPES               binhex            itertools        sitecustomize
...
Enter any module name to get more help.  Or, type "modules spam" to search
for modules whose descriptions contain the word "spam".

>>>
```

Notice the standard library module, RPi, in Listing 13.3. This is a Raspberry Pi–specific Python module, which contains functions to control the General Purpose Input/Output (GPIO) on the Pi. You will have to wait until Hour 24, "Working with Advanced Pi/Python Projects," to learn about that module!

**More Modules, Please**

One of the great things about the Python community is its willingness to share. You can find all kinds of user-created Python modules on the Internet to supplement the standard library. Besides just using your favorite web search engine, take a look at the Python Package Index (PyPi), at pypi.python.org/pypi.

Reading a module's name may give you a clue about what type of functions it contains. However, the best way to find out what is inside takes a little more work, as you will see in the next section.

# Learning About Python Modules

To learn about a module and see a description of its various functions, you use the `help` function. For example, suppose you want help on the `calendar` module. In Listing 13.4, the `calendar` module is imported, and the `help` function is used on it.

**LISTING 13.4**   Using `help` to See Module Descriptions

```
>>> import calendar
>>> help(calendar)

Help on module calendar:

NAME
    calendar - Calendar printing functions

MODULE REFERENCE
    http://docs.python.org/3.2/library/calendar

    The following documentation is automatically generated from the Python
    source files.  It may be incomplete, incorrect or include features that
...
CLASSES
    builtins.ValueError(builtins.Exception)
        IllegalMonthError

...
```

You can see that lots of information is provided by the help function, although it is a little cryptic for those new to Python. A nice site to visit when you need a little more help on modules than you get from help is docs.python.org/3/py-modindex.html.

BY THE WAY

**Different help Endings**

Sometimes using help in the Python interactive shell just puts you back at the shell prompt, as when you use the command help('modules'). Other times, such as when issuing the command help(calendar), you need to use special keys to navigate and quit the help function. This is covered in Hour 3, "Setting Up a Programming Environment." Review the section "Learning About the Python Interactive Shell" in that hour if you need a refresher.

For a quick list of functions, contained within a module, you can use the dir function. For example, Listing 13.5 shows a list of the functions the calendar module provides.

**LISTING 13.5**   Using dir to List Functions Within a Module

```
>>> import calendar
>>> dir(calendar)
['Calendar', 'EPOCH', 'FRIDAY', 'February', 'HTMLCalendar',
 'IllegalMonthError', 'IllegalWeekdayError', 'January', 'LocaleHTMLCalendar',
 'LocaleTextCalendar', 'MONDAY', 'SATURDAY', 'SUNDAY', 'THURSDAY', 'TUESDAY',
 'TextCalendar', 'WEDNESDAY', '_EPOCH_ORD', '__all__', '__builtins__',
 '__cached__', '__doc__', '__file__', '__name__', '__package__', '_colwidth',
 '_locale', '_localized_day', '_localized_month', '_spacing', 'c', 'calendar',
 'datetime', 'day_abbr', 'day_name', 'different_locale', 'error',
 'firstweekday', 'format', 'formatstring', 'isleap', 'leapdays', 'main',
 'mdays', 'month', 'month_abbr', 'month_name', 'monthcalendar', 'monthrange',
 'prcal', 'prmonth', 'prweek', 'setfirstweekday', 'sys', 'timegm', 'week',
 'weekday', 'weekheader']
>>>
```

Notice that one of the functions that calendar provides is prcal. You can get help on a particular function, such as prcal, without digging through all the other module help information. You use the syntax help(*module.function*), as shown in Listing 13.6.

**LISTING 13.6**   Using help to See How to Use a Function

```
>>> import calendar
>>> help(calendar.prcal)

Help on method pryear in module calendar:
```

```
pryear(self, theyear, w=0, l=0, c=6, m=3) method of calendar.TextCalendar instance
    Print a year's calendar.
(END)
>>>
```

Listing 13.6 shows a quick description of the `prcal` function and the arguments it accepts. Notice also that `help` calls `prcal` a "method" of the `calendar` module. Remember that *method* is another term for a function stored within a module.

## WATCH OUT!

### Do I Need to Import a Module to Use `help` on It?

You may have stumbled onto the fact that you do not have to import a module in order to get help on it or use the `dir` function on it. For example, without importing the `calendar` function, you can issue the command `help('calendar')`. However, the information you get this way may be lacking. It is always a best practice to import a module before performing any function on it, including `help` or `dir`.

Using the information provided by `help`, you can try out the `calendar` module's `prcal` method, as shown in Listing 13.7.

## LISTING 13.7  Using the `prcal` Function of `calendar`

```
>>> import calendar
>>> calendar.prcal(2014)
                        2014

        January                  February                   March
Mo Tu We Th Fr Sa Su     Mo Tu We Th Fr Sa Su      Mo Tu We Th Fr Sa Su
       1  2  3  4  5                     1  2                      1  2
 6  7  8  9 10 11 12      3  4  5  6  7  8  9       3  4  5  6  7  8  9
13 14 15 16 17 18 19     10 11 12 13 14 15 16      10 11 12 13 14 15 16
20 21 22 23 24 25 26     17 18 19 20 21 22 23      17 18 19 20 21 22 23
27 28 29 30 31           24 25 26 27 28            24 25 26 27 28 29 30
                                                   31


         April                     May                      June
Mo Tu We Th Fr Sa Su     Mo Tu We Th Fr Sa Su      Mo Tu We Th Fr Sa Su
    1  2  3  4  5  6          1  2  3  4                            1
...
```

Look back to Listing 13.6 and compare the `help` syntax description to the actual use of the syntax in Listing 13.7. You can see that `help` just gives you a push in the right direction. Trying it out for yourself will help you clearly understand how to use a module's function.

## ▼ TRY IT YOURSELF

### Explore the Modules on Your Raspberry Pi

In the following steps, you are going to explore the Python modules currently available on your Raspberry Pi. You will get to try a new method of looking for the `.py` module files, and you are going to get to do a little Easter egg hunting. Follow these steps:

1. If you have not already done so, power up your Raspberry Pi and log in to the system.

2. If you do not have the LXDE graphical interface started automatically at boot, start it now by typing `startx` and pressing Enter.

3. Open the LXTerminal by double-clicking the LXTerminal icon.

4. Open the Python interactive shell by typing `python3` at the shell prompt and pressing Enter.

5. At the Python interactive shell prompt, `>>>`, type `import os` and press Enter. Python imports the `os` module into your Python shell.

6. Type `import time` and press Enter to import the `time` module and all its functions into your Python shell.

7. Look to see if the `os` module has a `.py` file by typing the command `os.__file__`. Note that before and after the file in the command are two underscores (_). Press Enter to see if an absolute directory reference and a file ending in `.py` are shown. You should see that this module does have a `.py` file.

8. Look to see if the `time` module has a `.py` file by typing the command `time.__file__` and press Enter. You should receive an error because the `time` module does not have a `.py` file. (It's okay that this step generates an error!) Since the `time` module does not have a `.py` file, it may be built-in.

9. To determine whether the `time` module is built-in, import the `sys` module by typing `import sys` and pressing Enter.

10. Produce a list of the built-in modules by typing `sys.builtin_module_names` and pressing Enter. Do you see the `time` module listed in the output of the Python statement? You should!

11. Get a little help on the `time` module by typing `help(time)` and pressing Enter. In the `help` documentation, what are the two representations of time? Can you find this information? One is called the epoch, and the other is a tuple.

12. Still looking at the `time` module's `help` information, locate the `time()` function description. Which time representation does it use: epoch or the tuple?

13. Press Q to exit the `time` module's `help` information.

**14.** To import the `antigravity` module, type `import antigravity` and press Enter. You might need to wait a few minutes, until you see an Easter egg! You should see the Midori web browser open up, with a little surprise on the inside. Read the webpage that appears and then hover your mouse over it to see the secret message. Funny!

BY THE WAY

**What Is an Easter Egg?**

An *Easter egg* is a secret message or hidden surprise. You have to know exactly the right commands or keystrokes to find it.

**15.** Close the Midori web browser by clicking the X on the upper-right side of the window.

**16.** Press Ctrl+D to exit the Python interactive shell.

**17.** If you want to power down your Raspberry Pi now, type `sudo poweroff` and press the Enter key.

Understanding how to look at the various modules and their functions will be useful for you in the years to come. This will be especially true as new modules become part of the standard library. And didn't you have fun seeing an Easter egg?

# Creating Custom Modules

When you have created several functions, you might want to reuse them in your Python scripts. It can be helpful to create a custom module to house your functions. Generally speaking, it takes about seven steps to create a custom module:

**1.** Create or gather functions that it makes sense to put together.

**2.** Determine a module name.

**3.** Create the custom module in a test directory.

**4.** Test the custom module.

**5.** Move the module to a production directory.

**6.** Check the path and modify it if needed.

**7.** Test the production custom module.

A few of these steps require more work than they might seem like they'd need. The following sections examine all the steps and give you directions on successfully accomplishing them.

## Creating or Gathering Functions That Go Together

What functions to gather into a module is relative. It comes down to what makes sense and will be the most productive for you. However, keep in mind that you might want to distribute your functions to others, so take some time here. A logical gathering of functions can serve the greater community.

As an example in the following sections, you will gather two of the functions created in Hour 12 to create a module. They are dbl() and addem().

## Determining a Module Name

Naming a module is not a trivial step. Python has a few rules for naming modules, and if you do not follow them, your module won't work. For example, a module name cannot be a Python keyword. In Hour 4, the "Python Keywords" section explores this topic concerning variables. The same rules apply to modules.

It is also a standard practice to name custom modules using very short names, with all the characters in lowercase. Additional rules focus on the file name that holds the module's functions:

▶ Custom module file names must end in .py. If they don't, Python cannot import the modules.

▶ A file name cannot contain a dot (.) except right before the file name extension. For example, my.module.py is not a legal module file name. However, mymodule.py is a legal module file name because you need the last dot (.) to denote the file's extension.

▶ The file name must match the module name. Therefore, if the module's name is ni, the file name must be ni.py.

You could put the two functions, dbl() and addem() into a module called arith, which is short for arithmetic. This module name follows all the module name rules.

## Creating the Custom Module in a Test Directory

The file name for your new module should be arith.py in order to properly follow naming conventions. In Listing 13.8, you can see that the module resides in /home/pi/py3prog as its test directory.

**LISTING 13.8**  The arith.py Module File

```
1: pi@raspberrypi ~ $ cat /home/pi/py3prog/arith.py
2: def dbl(value):
3:     result=value * 2
4:     return result
5:
6: def addem(a, b):
```

```
7:      result=a + b
8:      return result
9: pi@raspberrypi ~ $
```

The `arith` module is a very simple one. It contains only the two functions from Hour 12. In Listing 13.8, you can see that the `dbl()` function starts on line 2 and ends on line 4. The `addem()` function starts on line 6 and ends on line 8.

BY THE WAY

### Simple or Complex Modules

A module can be as simple or as complex as you choose. The custom modules shown here are very simple. If desired, you can add help utilities to your modules, include Python version directives, import other modules, and so on.

To create a custom module, you can use your favorite text editor or the Python IDLE editor. Be sure to save the module to a location you have designated as a *test directory*. Often, this is simply a subdirectory of your home login directory, as shown in Listing 13.8.

## Testing the Custom Module

To test your module, you need to ensure that your present working directory is in the same location where the module file resides. In Listing 13.9, the `os` module is used to show the present working directory, /home/pi, on line 2. This is not the test directory which was designated earlier, and thus the `import` of the `arith` module on line 4 does not work.

**LISTING 13.9** A Test of the `arith` Custom Module

```
1: >>> import os
2: >>> os.getcwd()
3: '/home/pi'
4: >>> import arith
5: Traceback (most recent call last):
6:   File "<stdin>", line 1, in <module>
7: ImportError: No module named arith
8: >>>
9: >>> os.chdir('/home/pi/py3prog')
10: >>> import arith
11: >>>
12: >>> arith.dbl(10)
13: 20
14: >>> arith.addem(5,37)
15: 42
16: >>>
```

After the test directory is reached on line 9, using the os.chdir statement, the import arith on line 10 works fine. Both functions in the arith module are tested, and no problems result.

BY THE WAY

**Multiple Imports?**

It doesn't hurt anything if you accidently import a module a second (or third) time. When the import command is issued, Python first checks whether the module is already imported. It the module is already imported, then Python does nothing (and doesn't complain, either!).

# Moving a Module to a Production Directory

When you have any problems resolved in your module, you should move it to a production directory location. This is an optional step. If you are the only user of the Raspberry Pi and really don't care where you keep your Python code, you can skip this step. However, good form dictates that a module should be in a standard production directory.

A *production directory* is a directory where modules can be accessed by all the Python script users. You can see the current production directories by using the sys module, as shown in Listing 13.10.

**LISTING 13.10**    Production Directories Used by Python

```
>>> import sys
>>> sys.path
['', '/usr/lib/python3.2', '/usr/lib/python3.2/plat-linux2',
 '/usr/lib/python3.2/lib-dynload', '/usr/local/lib/python3.2/dist-packages',
 '/usr/lib/python3/dist-packages']
>>>
```

Notice in Listing 13.10 that the first directory shown is just two single quotes with no directory name in between the quotes. When Python sees this, it searches your present working directory for the module. This is why Listing 13.9 is able to use the os.chdir function to switch to /home/pi/py3prog and test your module.

Also, notice in Listing 13.10 that /usr/lib/python3.2 is listed as a location in the path. This is where the standard Python library modules' .py files, are located for this Python version. You looked at this directory back in the "Introducing Module Concepts" section of this hour.

### The Path

Many programming languages and operating systems use the term *path*. A path is a list of directories a program searches for other programs, libraries, components, and so on. For modules, Python searches the path directories for modules (actually, it searches for their `.py` files) that have been requested to be imported.

Following standards is always good form. Also, it protects any modules you create from being unintentionally removed. For example, if you put your custom modules with the other standard Python modules in /usr/lib/python*version*, when the Python software is upgraded, your modules could be deleted.

Table 13.1 shows the standard Python module directories for a Debian-based operating system. Remember from Hour 2 that Raspbian is a Debian-based Linux distribution.

**TABLE 13.1** Standard Python Module Directories

| Directory | Description |
|---|---|
| /usr/lib/python*version* | Contains Python modules that come standard with Python. |
| /usr/local/lib/python*version*/dist-packages | Contains public third-party Python modules that have been downloaded and installed with a package installer. |
| /usr/local/lib/python*version*/site-packages | Your local custom production-quality modules. |

According to Table 13.1, when you are ready to move your module into a production directory, it needs to go into the /usr/local/lib/python*version*/site-packages directory. The steps needed to make this move have to be performed at Raspbian's dash shell. Line 1 of Listing 13.11 checks the directory /usr/local/lib/python3.2 to see if the site-packages subdirectory exists. In this case, it does not. Only the dist-packages subdirectory exists, as you see on line 2. Therefore, the site-packages subdirectory is created on line 3, using the sudo and mkdir commands you learned about in Hour 2. If your system already has the site-packages subdirectory, you do not need to use those commands.

**LISTING 13.11** Copying `arith.py` to the Production Directory

```
1: $ ls /usr/local/lib/python3.2/
2: dist-packages
3: $ sudo mkdir /usr/local/lib/python3.2/site-packages
4: $
```

```
5: $ cd /home/pi/py3prog
6: $
7: $ sudo cp arith.py /usr/local/lib/python3.2/site-packages/
8: $ cd
9: $
```

In Listing 13.11, once the directory is created, `cd` is used to change the present working directory to the location of the tested module on line 5. The `arith.py` module file is then copied to the production directory on line 7, using the `sudo` and `cp` commands.

BY THE WAY

### Don't Want a Copy?

If you do not want or need to keep a copy of your newly created module in your test directory, you can use the `mv` (move) command. Simply replace `cp` with the `mv` command, and the module is moved to the production directory, with no copy of it left in the test directory.

Now that the module has been copied to the production directory, you should be able to import the `arith` module, right? Well, look at Listing 13.12.

### LISTING 13.12    Module arith Not Found

```
$ python3
...
>>> import arith
Traceback (most recent call last):
  File "<stdin>", line 1, in <module>
ImportError: No module named arith
>>>
```

Obviously, there is a problem with Python finding the module. You can resolve this problem in the next step.

# Checking the Path and Modifying It, if Needed

It is a good idea to check your current Python path directories. As shown in Listing 13.10, the two commands to use are `import sys` and `sys.path`. Notice in Listing 13.13 that the production directory `/usr/local/lib/python3.2/site-packages` is not listed. This is why Python could not find the `arith` module file.

**LISTING 13.13    Checking the Python Path Directories**

```
>>> import sys
>>> sys.path
['', '/usr/lib/python3.2', '/usr/lib/python3.2/plat-linux2',
 '/usr/lib/python3.2/lib-dynload',
 '/usr/local/lib/python3.2/dist-packages',
 '/usr/lib/python3/dist-packages']
>>>
```

To modify the path, if needed, you use a function from the sys module, as shown in Listing 13.14. After the path is added, you can import the needed module in the added production directory.

**LISTING 13.14    Adding a New Directory to the Path**

```
>>> import sys
>>> sys.path.append('/usr/local/lib/python3.2/site-packages')
>>>
>>> sys.path
['', '/usr/lib/python3.2', '/usr/lib/python3.2/plat-linux2',
 '/usr/lib/python3.2/lib-dynload',
 '/usr/local/lib/python3.2/dist-packages',
 '/usr/lib/python3/dist-packages',
 '/usr/local/lib/python3.2/site-packages']
>>>
>>> import arith
>>>
```

As you can see in Listing 13.14, with the site-packages directory now added to the path, Python can find the arith module with no problems.

Keep in mind that the path resets back to its default every time your script finishes or you exit the Python interactive shell. This means you need to add the new path every time you want to import the new module.

DID YOU KNOW

**The Python Path**

You Linux gurus should know that there is an environment variable called PYTHONPATH that you can modify to include your new custom modules directory. By changing this environment variable and adding it to an environment file, you make the path permanently include the custom modules directory. There is then no need to use the sys module to change your path every time.

# Testing the Production Custom Module

Testing a production custom module is optional. However, it is always a good idea to test your module's functions after you have moved it to the production directory.

**LISTING 13.15**    Testing the Production `arith` Module

```
>>> import arith
>>> arith.dbl(20)
40
>>> arith.addem(7,35)
42
>>>
>>>
```

As you can see in Listing 13.15, the `arith` module is imported with no problem. Both the `.dbl` method and the `.addem` method run flawlessly.

BY THE WAY

## Importing Near the Beginning

A good rule of thumb for importing modules—custom or standard—into your Python script is to do it at the top of the script. Therefore, in Python scripts, put in all your `import` statements after the leading documentation lines.

▼ TRY IT YOURSELF

## Create and Use a Custom Module in Python

In the following steps, you are going to create a custom module. In essence, you will be following the seven steps just described to create it, test it, and then move it into production.

Pretend that you have two great functions. One, called `discountp`, that shows the actual price of an item if you give it the original price and the listed discount. The other function, `priceper`, shows the price per item when the prices at a store are shown for a group of items (for example, 3 for $11).

You decide to put these two useful functions into their own module so that you can use them in several scripts. Next, you determine that `shopper` would be a good module name. It is a nice, short name, and it is not a Python keyword. Next, you need to put the two functions into a file called `shopper.py`. This is where you start in the following steps:

1. If you have not already done so, power up your Raspberry Pi and log in to the system.

2. If you do not have the LXDE GUI started automatically at boot, start it now by typing `startx` and pressing Enter.

3. Open the LXTerminal by double-clicking the LXTerminal icon.

4. At the command-line prompt, type `nano py3prog/shopper.py` and press Enter. The command puts you into the nano text editor and creates the custom module `shopper.py` in the `/home/pi/py3prog` directory.

5. Type the following code into the nano editor window, pressing Enter at the end of each line:

BY THE WAY

## Be Careful!

Be sure to take your time here and avoid making typographical errors. You can make corrections by using the Delete key and the up- and down-arrow keys.

```
def discountp(percentage, price):
    return round(price -((percentage /100) * price),2)

def priceper(no_items, price):
    return round(price / no_items,2)
```

6. Double-check to make sure you have entered the code into the nano text editor window as shown above. Make any corrections needed.

7. Write out the information from the text editor to the module by pressing Ctrl+O. The module file name should appear along with the prompt `File name to write`. Press Enter to write out the contents to the `shopper.py` module.

8. Exit the nano text editor by pressing Ctrl+X.

9. Now that the `shopper.py` custom module is created, test it by typing `python3` and pressing Enter to enter the Python interactive shell.

10. Now import the `os` module by typing `import os` and pressing Enter. You need this module to change the present working directory within the Python interactive shell.

11. Type `os.chdir('/home/pi/py3prog')` and press Enter. Python changes your present working directory to `/home/pi/py3prog`, where the `shopper.py` module file is located. (You can double-check your present working directory by typing `os.cwd()` and pressing Enter.)

12. Now see if you can import your new custom module by typing `import shopper` and pressing Enter. Does Python give you an error? If so, you need to go back to step 4 and try again to create the file `shopper.py`. If you get no errors, you can continue on to the next test.

**13.** Test the function `priceper`, which is stored in the `shopper` module, by typing `shopper.priceper(3,12)` and pressing Enter. You should see the result 4. Test the rest of the functions in the module.

**14.** Press Ctrl+D to exit the Python interactive shell and begin the process of moving the `shopper` module to a production directory.

**15.** At the Raspbian shell prompt, type `python3 -V` and press Enter. This allows you to determine the current version of Python you have on Raspbian. (You learned about this is Hour 3.) You only need to know the first two numbers of the version number displayed (for example, 3.2) for the next step

**16.** You must check and see if the production directory currently exists on your Raspberry Pi by typing `ls /usr/local/lib/python`*`version`*`/site-packages` (where *version* is the version number from step 15, such as /usr/local/lib/python3.2/site-packages) and press Enter. If you do not get an error message after pressing Enter, you can skip step 17. If you get an error message, that is okay; simply fix the problem in step 17.

**17.** Type `sudo mkdir /usr/local/lib/python`*`version`*`/site-packages` and press Enter. Remember to type the version number you found in step 15 rather than *version*.

**18.** Move your custom module, `shopper`, to the `site-packages` directory by typing `sudo cp /home/pi/py3prog/shopper.py /usr/local/lib/python`*`version`*`/site-packages` and pressing Enter. Be careful here! The command may wrap on your screen (and that's okay). Again, remember to type the version number you found in step 15 rather than *version*.

**19.** Now you have finished moving the module to a production directory, type `python3` and press Enter to reenter the Python interactive shell.

**20.** At the interactive shell prompt, type `import sys` and press Enter to import the `sys` module.

**21.** Type `sys.path` and press Enter to display the directories currently in the Python path. Do you see the `/usr/local/lib/python`*`version`*`/site-packages` directory? If yes, you can skip step 22. If you don't see this directory, move on to step 22.

**22.** Type `sys.path.append('/usr/local/lib/python`*`version`*`/site-packages')` and press Enter to add the production directory to the Python path. Remember to type the version number you found in step 15 rather than *version*. With the production directory containing the `shopper.py` module file added to Python's path, you are ready to test the production custom module.

**23.** Import the custom shopper module by typing `import shopper` and pressing Enter.

**24.** Test the function `discount` by typing `shopper.discountp(100,10)` and pressing Enter. You should see the response 90. Congratulations! Your custom module has been tested and confirmed!

▼

**25.** If you want to power down your Raspberry Pi now, type `sudo poweroff` and press the Enter key.

BY THE WAY

### Proud of a Module You've Created?

You can be a part of the Python community by sharing modules you have created. Start the sharing process by reading about Distutils at docs.python.org/3/distutils/index.html. Also, take a look at github, github.com, where you can share your work.

Understanding how to create a custom module is very useful in your Python learning adventure. Just remember to follow the seven steps described in this hour, and you will be able to create lots of custom modules containing your homemade functions.

# Summary

In this hour, you read about finding and using modules in Python. You also read how to create your own custom modules. The various Python module flavors were covered, along with where to find modules stored in `.py` files on your Raspbian system. Also, you have learned where custom Python modules can be stored. You got to try out creating your own custom module and moving it to a production location. In Hour 14, "Exploring the World of Object-Oriented Programming," you will investigate object-oriented programing, including a concept called *classes*, which can also be stored in a Python module.

# Q&A

**Q.** I heard a Python expert talking about the "cheese shop." What is that?

**A.** The "cheese shop" is the Python Package Index (PyPi), where you can find lots of Python modules to download and install on your system. PyPi used to be called the "cheese shop," after a *Monty Python's Flying Circus* skit concerning a cheese shop that had no cheese. PyPi is full of cheese...err...modules and is located at pypi.python.org/pypi.

**Q.** I'm not a Linux guru, but I want to change the PYTHONPATH variable. How do I do this?

**A.** First, you must make a change to a file in the home directory—that is, the present working directory when you log into the Raspberry Pi. Edit the file `.profile` by typing `nano .profile` and pressing Enter. Navigate to the very bottom of the file and then add the following two lines:

```
PYTHONPATH=/usr/local/lib/pythonversion/site-packages
export PYTHONPATH
```

In place of *version*, type the version of Python on your Raspbian system, such as `/usr/local/lib/python3.2/site-packages`.

When you have these two lines, save the file by pressing Ctrl+O and then Enter. Then exit the editor by pressing Ctrl+X.

Test the change by logging out (typing `exit` and pressing Enter) logging back in, and jumping into the Python interactive shell (by typing `python3` and pressing Enter). Check the path variable by importing the `sys` module (by typing `import sys` and pressing Enter). Now see if the `/usr/local/lib/pythonversion/site-packages` directory is in the path by typing `sys.path` and pressing Enter.

**Q. How can I add some sort of help to my custom module?**

**A.** You can add help to your module by putting a triple-quoted string at the top of your custom module file. Between the quotes add a sentence or two about the purpose of the module and give a few examples of using the module's functions.

# Workshop

## Quiz

1. A Python script and a module are in essence the same thing because their file names both end in `.py`. True or false?

2. What must you do to a module if it is linked with the Python interpreter and you need to use it in a Python script?

3. In which directory should you store third-party public modules that have been downloaded and installed?

    **a.** `/usr/lib/pythonversion`

    **b.** `/usr/local/lib/pythonversion/dist-packages`

    **c.** `/usr/local/lib/pythonversion/site-packages`

## Answers

1. False. A Python script is designed to be run on its own. A Python module must be imported into a script or an interactive session before it can be used.

2. It must be imported before it can be used. It does not matter that the module is linked with the Python interpreter. All modules, no matter where they "live," must be imported before you can use the functions they contain.

3. This is a bit of a trick question. The correct answer is b, `/usr/local/lib/python version/dist-packages`. However, you don't move the modules there yourself. The package installer does it for you.

# Exploring the World of Object-Oriented Programming

---

**What You'll Learn in This Hour:**

► How to create object classes

► Defining attributes and methods in classes

► How to use classes in your Python scripts

► How to use class modules in your Python scripts

So far, all the Python scripts presented in this book have followed the procedural style of programming. With *procedural programming*, you create variables and functions within your code to perform certain procedures, such as storing values in variables and then checking them with structured statements. The data that you use and the functions you create are completely separate entities, with no specific relationship to one another. With *object-oriented programming*, on the other hand, variables and functions are grouped into common objects that you can use in any program. In this hour, you'll see just what object-oriented programming is and how to use it in your Python scripts.

# Understanding the Basics of Object-Oriented Programming

Before you can start working on object-oriented programming (commonly called OOP), you need to know how it works. OOP uses a completely different paradigm than the coding you've been doing so far in this book. OOP requires that you think differently about how your programs work and how you code them.

## What Is OOP?

With OOP, everything is related to objects. (I guess that's why they call it object-oriented!) Objects are the data you use in your applications, grouped together into a single entity.

For example, if you're writing a program that uses cars, you can create a `car` object that contains information on the car's weight, size, engine, and number of doors. If you're writing a program that tracks people, you might create a `person` object that contains information on each person's name, height, age, weight, and gender.

OOP uses classes to define objects. A *class* is the written definition in the program code that contains all the characteristics of the object, using variables and functions. The benefit of OOP is that once you create a class for an object, you can use that same class any time in any other application. Just plug in the class definition code and put it to use.

An OOP class has members, and there are two types of members:

▶ **Attributes**—Class attributes denote features of an object (such as the weight, engine, and number of doors of a car). A class can contain many attributes, with each attribute describing a different feature of the object.

▶ **Methods**—Methods are similar to the standard Python functions that you've been using. A method performs an operation, using the attributes in a class. For instance, you could create class methods to retrieve a specific person from a database, or change the address attribute for an existing person. Each method should be contained within a class and perform operations only in that class. The methods for one class shouldn't deal with attributes in other classes.

## Defining a Class

Defining a class in Python isn't too much different from defining a function. To define a new class, you use the `class` keyword, along with the class name, and a colon, followed by any statements contained in the class. Here's an example of a simple class definition:

```
>>> class Product:
    pass

>>>
```

The class name you choose must be unique within your program. And while it's not required, it's somewhat of a de facto standard in Python to start a class name with an uppercase letter.

The statements section for the class defines any attributes and methods that the class contains. Just as with functions, you must indent the class statements in the code. When you're done defining the class attributes and methods, you just place the next code statement on the left margin.

The `pass` statement shown in this example is special in Python. It's a statement that does nothing! You normally use the `pass` statement as a placeholder for code that you'll add in the future.

In this example, the `pass` statement creates an empty class definition. You can use the class in your Python code, but it won't do anything!

## Creating an Instance

The class definition defines the class, but it doesn't put the class to use. To use a class, you have to instantiate it. When you *instantiate* a class, you create what's called an *instance* of the class in your program. Each instance represents one occurrence of the object. To instantiate an object, you just call it by name, like this:

```
>>> prod1= Product()
>>>
```

The `prod1` variable is now an instance of the `Product` class. After you create an instance of a class, you can define attributes "on the fly," as shown here:

```
>>> prod1.description = 'carrot'
>>> print(prod1.description)
carrot
>>>
```

Each instance is a separate object in Python. If you create a second instance of the `Product` class, the attributes you define in that instance are separate from the attributes you define in the first instance, as shown in this example:

```
>>> prod2 = Product()
>>> prod2.description = "eggplant"
>>> print(prod2.description)
eggplant
>>> print(prod1.description)
carrot
>>>
```

Now you have two separate instances of the `Product` class: `prod1` and `prod2`. Each instance has its own attribute values.

## Default Attribute Values

Defining attributes on the fly as you use a class instance isn't good programming practice. It's better to define the class attributes inside the class definition code so that they're documented as part of the class. You can do that and set default values for the attributes all at the same time, like this:

```
>>> class Product:
    description = "new product"
    price = 1.00
    inventory = 10
>>>
```

Now when you instantiate a new instance of the `Product` class, the instance has values set for the description, price, and inventory attributes, as shown here:

```
>>> prod1 = Product()
>>> print('{0} - price: ${1:.2f}, inventory: {2:d}'.format(prod1.description,
prod1.price, prod1.inventory))
new product - price: $1.00, inventory: 10
>>>
```

After you create the new class instance, you can replace the existing values at any time, as in the following example:

```
>>> prod1.description = 'tomato'
>>> print('{0} - price: ${1:.2f}, inventory: {2:d}'.format(prod1.description,
prod1.price, prod1.inventory))
tomato - price: $1.00, inventory: 10
>>>
```

The description attribute of the `prod1` instance now has the new value that you assigned.

# Defining Class Methods

Classes consist of more than just a bunch of attributes. You also need to have methods for handling the attributes. There are three different types of class methods available:

▶ Mutator methods

▶ Accessor methods

▶ Helper methods

The following sections discuss the difference between mutator and accessor methods, and then you'll learn more about the helper methods.

## Mutator Methods

Mutator class methods are functions that change the value of an attribute. The most common type of mutator method is called a *setter*.

You use a setter method to set the value of an attribute in the class. While not required, it's somewhat of a standard convention in Python coding to name setter mutator methods starting with `set_`, as in this example:

```
def set_description(self, desc):
    self.__description = desc
```

The set_ methods use two parameters. The first parameter is a special value called self. The self parameter points the class to the current instance of the object. This parameter is required in all class methods as the first parameter.

The second parameter defines a value to set as the attribute value for the instance. Notice the assignment statement used in the method statement:

```
self.__description = desc
```

The attribute name is __description. Yet another de facto Python standard in OOP is to use two underscores at the start of the attribute name to indicate that it shouldn't be used by programs outside the class definition.

<hr>

BY THE WAY

**Private Attributes**

Some object-oriented programming languages provide a feature called *private attributes*. You can use private attributes inside the class definition but not outside the class in the program code. Python doesn't provide for private attributes; any attribute you define can be accessed from any program. The two-underscore naming convention just makes it obvious which attributes you prefer not to use outside the class definition.

<hr>

The self keyword used in the attribute assignment is used to reference the attribute to the current class instance.

You can also include other mutator methods that perform some type of calculations with the attributes. For example, you could create a buy_Product() method for the Product class that changes the inventory value of a product when a customer purchases it. That code would look something like this:

```
def buy_Product(self, amount):
    self.__inventory = self.__inventory - amount
```

The mutator method still requires the self keyword as the first parameter, and it uses a second parameter to provide data for the method. The assignment statement changes the __inventory attribute value for the instance by subtracting the value sent as the second parameter.

# Accessor Methods

Accessor methods are functions you use to access the attributes you define in a class. Creating special methods to retrieve the current attribute values helps create a standard for how other programs use your class object. These methods are often called *getters* because they retrieve the value of the attribute.

As with setters, it's somewhat of a standard convention to name getter accessor methods starting with get_, followed by the attribute name, as shown here:

```
def get_description(self):
    return self.__description
```

This is all there is to it! Accessor methods aren't overly complicated; they just return the current value of the attribute. Notice that even though no data is passed to the accessor method, you still need to include the self keyword as a parameter. Python uses this keyword internally to reference the class instance.

When you create your classes, you need to create one setter and one getter for each attribute you use in your class. Listing 14.1 shows the script1401.py program, which demonstrates creating setter and getter methods for the Product class attributes and then using them in a program.

**LISTING 14.1**   Using Setter and Getter Methods

```
 1: #!/usr/bin/python3
 2:
 3: class Product:
 4:     def set_description(self, desc):
 5:         self.__description = desc
 6:
 7:     def get_description(self):
 8:         return self.__description
 9:     def set_price(self, price):
10:         self.__price = price
11:
12:     def get_price(self):
13:         return self.__price
14:
15:     def set_inventory(self, inventory):
16:         self.__inventory = inventory
17:
18:     def get_inventory(self):
19:         return self.__inventory
20:
21: prod1 = Product()
22: prod1.set_description('carrot')
23: prod1.set_price(1.00)
24: prod1.set_inventory(10)
25:  print('{0} - price: ${1:.2f}, inventory: {2:d}'.format(
 prod1.get_description(),prod1.get_price(), prod1.get_inventory()))
```

After you instantiate an instance of the Product class (line 21), you need to use the setter methods to set the initial values (lines 22 through 24). To retrieve the attribute values (as in the print() statement in line 25), you just use the get_ methods for each attribute.

When you run the `script1401.py` program, you should see this output from the instance values:

```
pi@raspberrypi ~$ python3 script1401.py
carrot - price: $1.00, inventory: 10
pi@raspberrypi ~$
```

The program code contained in the `script1401.py` file creates an instance of the `Product` class; sets the `__description`, `__price`, and `__inventory` attribute values (using the appropriate setter methods); and then retrieves the attribute values using the getter methods.

So far, so good. But things are a bit cumbersome in terms of using the class methods. You had to go through a lot of work to create the class, set the initial values for the attributes, and then retrieve the attribute values using the getters. Fortunately, Python has some helper methods that help make your life easier!

# Adding Helper Methods to Your Code

Besides the accessor and mutator methods, there are a few other methods you can create for your classes that help make using classes much easier. The following sections go through some of the most common helper class methods that you'll want to use when working with classes in your Python programs.

## Constructors

It can get somewhat old trying to set attribute values using the `set_` mutator methods, especially if you have lots of attributes in a class. Using class constructor methods is a popular simple way to instantiate a new instance of a class with default values.

Python provides a special method called `__init__()` that it calls when you instantiate a new class instance. You can define the `__init__()` method in your class code to do any type of work when the class instance is created, including assign default values to attributes. The `__init__()` method requires parameters for each attribute that you want to define a value for, as shown here:

```
def __init__(self, description, price, inventory):
    self.__description = description
    self.__price = price
    self.__inventory = inventory
```

Now when you create the instance of the `Product` class, you must define the initial values directly in the class constructor:

```
>>> prod3 = Product('tomato', 2.00, 20)
>>> print('{0} - price: ${1:.2f}, inventory:
 {2:d}'.format(prod1.get_description(), prod1.get_price(),
```

```
 prod1.get_inventory()))
tomato – price: $2.00, inventory: 20
>>>
```

This makes creating a new instance of a class a lot easier! The only downside is that you must specify the default values, and if you don't, you get an error message, like this:

```
>>> prod1 = Product()
Traceback (most recent call last):
  File "<pyshell#31>", line 1, in <module>
    prod1 = Product()
TypeError: __init__() missing 3 required positional arguments: 'desc', 'price', and
'inventory'
>>>
```

To solve this problem, just as with Python functions, the constructor method lets you define default values in the parameters (see Hour 12, "Creating Functions"), as in this example:

```
def __init__(self, description = 'new product', price = 0, inventory = 0):
    self.__description = description
    self.__price = price
    self.__inventory = inventory
```

Now if you create a new instance of the Product class without specifying any default values, Python uses the default values you defined in the class constructor, as shown here:

```
>>> prod5 = Product()
>>> prod5.get_description()
'new product'
>>> prod5.get_price()
0.0
>>> prod5.get_inventory()
0
>>>
```

Now your class constructor is even more versatile!

## Customizing Output

The next thing to tackle is displaying the class instance. So far you've had to use the print() statement to display the individual attributes from the class instance, using the get_ methods. However, if you have to display your class data multiple times in your program code, that could get old. Python provides an easier way to do it.

You just need to define the __str__() helper method for your class to tell Python how to display the class object as a string value. Any time your program references the class instance as a string (such as when you use it in a print() statement), Python calls the __str__() method from the class definition. All you need to do is return a string value from the __str__() method that formats the class object attributes as strings. Here's what the __str__() method could look like for the Product class:

```
def __str__(self):
   return '{0} - price: ${1:.2f}, inventory:
 {2:d}'.format(self.__description, self.__price, self.__inventory)
```

Now, to display the class instance attribute values, you can just reference the instance variable in the print() statement, as shown here:

```
>>> prod6 = Product('banana', 1.50, 30)
>>> print(prod6)
banana - price: $1.50, inventory: 30
>>>
```

This is yet another method to make your life a lot easier!

## Deleting Class Instances

Handling memory management in Python programs is normally a lot easier than in other programming languages. By default, Python recognizes when a class instance is no longer in use and removes it from memory. However, there may be times when a program needs to do some type of "cleanup" work for the class before Python removes it from memory.

You can specify a helper method that Python automatically attempts to run just before it removes the instance from memory. Such methods are called *destructors*.

Destructors come in handy with a class that works with files to ensure that the files are properly closed before the class instance is removed.

You use the __del__() helper method to define any final statements to process before Python removes the class instance from memory:

```
def __del__(self):
   statements
```

The __del__() method doesn't allow you to pass any parameters into the method. All the statements that you specify in the method need to be self-contained and must not rely on any data from the main program.

WATCH OUT!

## Running Destructors

Python processes a class destructor any time it automatically removes a class instance from memory or when you use the del statement on the class instance. However, when the Python interpreter shuts down, there are no guarantees that Python will be able to run the descriptor class for any active class instances.

## Documenting the Class

Object-oriented classes are meant to be shared. Therefore, it's important that you document your Python classes so that anyone else who needs to use them knows what they do (and that may even include yourself, if you pick up some of your own Python code years later!).

While it's not exactly a method, Python provides the document string (called docstring) feature, which allows you to embed strings inside classes, functions, and methods to help document the code. You enclose the docstring in triple quotes to identify it in the class, function, or method definition. Also, the docstring must be the first item in the definition.

Here's an example of documenting the `Product` class:

```
class Product:
    """The Product class creates an instance of a product with three
 attributes - the product description, price, and inventory"""
```

To see the docstring for a class, just reference the special __doc__ attribute, as shown here:

```
>>> prod7 = Product()
>>> prod7.__doc__
The Product class creates an instance of a product with three attributes
- the product description, price, and inventory
>>>
```

You can also create a docstring value for each individual method inside the class, as shown here:

```
def get_description(self):
    """The description contains the product type"""
    return self.__description
```

To view a method's docstring, you just add the __doc__ attribute to the method in the instance, like this:

```
>>> prod8 = Product()
>>> prod8.get_description.__doc__
The description contains the product type
>>>
```

Now you have a way to share your comments on the class with others who may use your code in their own projects.

## The `property()` Helper Method

So far you have setter and getter methods defined to interface with the attributes you define for a class. However, it can get somewhat cumbersome trying to use the `set_` and `get_` methods all the time for each method. To solve this problem, Python provides the `property()` method.

The `property()` method creates a method that combines the setter and getter methods, along with the destructor and a docstring for an attribute, into a single method. Python calls the appropriate method, based on how you use the `property()` method in your code.

This is the syntax for defining the `property()` method in a class:

```
method = property(setter, getter, destructor, docstring)
```

You don't have to define all four parameters in the `property()` method. You can define a single parameter to create only a setter method; two methods to create only setter and getter methods, three methods for the setter, getter, and destructor; or all four parameters to create all four methods.

Here's an example of what you can add to the end of a `Product` class to create the `property()` methods for each attribute:

```
description = property(get_description, set_description)
price = property(get_price, set_price)
inventory = property(get_inventory, set_inventory)
```

Once you define the `property()` methods for the attributes, you can set or get the individual attributes by referencing their `property()` methods, as shown here:

```
>>> prod1 = Product('carrot', 1.00, 10)
>>> print(prod1)
carrot - price: $1.00, inventory: 10
>>> prod1.price = 1.50
>>> print('The new price is', prod1.price)
The new price is 1.50
>>>
```

The `prod1.price` property allows you to both set and retrieve the __price attribute value in your program code.

# Sharing Your Code with Class Modules

The whole point of creating Python object-oriented classes is that you can reuse the same code in any program that uses that object. If you combine the class definition code with your program code, sharing your class objects is more difficult.

The key to OOP in Python is to create separate modules for each object class. That way, you can just import the object class file that you need for any program that uses that type of object.

It's somewhat common practice to name the object class module the same name as the class. Doing so makes it easier to identify and import your classes. The following Try It Yourself walks through creating a module file for the `Product` class and then using it in a separate application script.

▼ TRY IT YOURSELF

## Create a Class Module

In the following steps, you'll create two Python script files. One file will contain the code that defines the `Product` class, and the other file will contain the script you'll run to use the class. Here's what you do:

**1.** Open a text editor and create the file `product.py`. Here's the code that you need to enter into the file:

```
#!/usr/bin/python3

class Product:

    def __init__(self, description, price, quantity):
        self.__description = description
        self.__price = price
        self.__inventory = quantity

    def set_description(self, description):
        self.__description = description

    def get_description(self):
        return self.__description

    description = property(get_description, set_description)

    def set_price(self, price):
        self.__price = price

    def get_price(self):
        return self.__price

    price = property(get_price, set_price)

    def set_inventory(self, inventory):
        self.__inventory = inventory

    def get_inventory(self):
        return self.__inventory

    inventory = property(get_inventory, set_inventory)

    def buy_Product(self, amount):
        self.__inventory = self.__inventory - amount
```

```
    def __str__(self):
        return '{0} - price: ${1:.2f}, inventory:
{2:d}'.format(self.__description, self.__price,
self.__inventory)
```

The `product.py` file incorporates all the attributes and methods for the `Product` class into a single module. You can now use your `Product` class in any of your Python scripts by simply importing the class from the `product.py` file.

**2.** Save the `product.py` file.

**3.** Open the text editor again and create the `script1403.py` file. Here's the code to enter into the file:

```
#!/usr/bin/python3
from product import Product

prod1 = Product('carrot', 1.25, 10)
print(prod1)

print('Buying 4 carrots...')
prod1.buy_Product(4)
print(prod1)
print('Changing the price to $1.50...')
prod1.price = 1.50
print(prod1)
```

The code first uses the `from` statement to reference the `Product` class from the `product.py` module file. (Make sure you have the `product.py` file in the same folder as the `script1403.py` file.)

**4.** Save the `script1403.py` file in the same folder where you saved the `product.py` file.

**5.** Run the `script1403.py` file from the command prompt.

Here's what you should see when you run the `script1403.py` file:

```
pi@raspberrypi ~$ python3 script1403.py
carrot - price: $1.25, inventory: 10
Buying 4 carrots...
carrot - price: $1.25, inventory: 6
Changing the price to $1.50...
carrot - price: $1.50, inventory: 6
pi@raspberrypi ~$
```

The script uses the `Product` class that you defined in the `product.py` file to create an instance of the `Product` class, uses the `buy_Product()` method to decrease the inventory value, and then uses the `set_price` accessor method to change the price of the product. This is starting to look like a real program!

# Summary

In this hour, you learned how to create and use object-oriented programming in Python. You can create object classes by using the `class` keyword, and then you define attributes and methods for the class. You also learned how to create access and mutator methods for your classes, as well as use many of the common helper methods to help make your coding job easier. Finally, you learned how to save a class definition in a separate code file and then import that class as a module in other Python scripts to use the class object.

In the next hour, we'll dig a little deeper into the object-oriented world and look at the topic of inheritance. That allows you to build new classes from existing classes!

# Q&A

**Q. Does Python support protected methods?**

**A.** No, Python doesn't support protected methods. You can, however, use the same idea as with private attributes and name your method starting with two underscores. The method is still publicly accessible, just not using the normal name.

**Q. Does Python support class inheritance?**

**A.** Yes, you can allow a class to inherit attributes and methods from another class. That's covered in Hour 15, "Employing Inheritance."

# Workshop

## Quiz

1. Which method should you define to create default values in a class constructor?

    **a.** `__del__()`

    **b.** `__init__()`

    **c.** `init()`

    **d.** `set_init()`

2. When you instantiate two instances of a class, you can share attribute values between the two instances. True or false?

3. How would you write an accessor method to set the value of a last name attribute?

# Answers

1. b. The __init__() special method allows you to pass parameters to the class constructor that you can use to define the default values for properties.

2. False. Separate instances of the same class are considered two separate objects, you can't share the same property values between them.

3. You can create a method called set_lastname() that accepts the name value as a single parameter, then assign that value to the self.__lastname property, like this:

```
def set_lastname(self, name):
        self.__lastname = name
```

# Employing Inheritance

**What You'll Learn in This Hour:**

▶ What subclasses are

▶ What inheritance is

▶ How to use inheritance in Python

▶ Inheritance in scripts

In this hour, you will learn about subclasses and inheritance, including how to create subclasses and how to use inheritance in scripts. Inheritance is the next step in understanding object-oriented programming in Python.

## Learning About the Class Problem

In Hour 14, "Exploring the World of Object-Oriented Programming," you read about object-oriented programming, classes, and class module files. Even with object-oriented programming, problems related to duplication of object data attributes and methods still exist. This is called the "class problem." This hour looks at the biological classification of animals and plants to help clarify the nature of the class problem.

Suppose you are creating a Python script for an insect scientist (entomologist). What makes an insect an insect? A very basic classification for an insect is that an animal must have the following in order to be considered an insect:

▶ No vertebral column (backbone)

▶ A chitinous exoskeleton (outside shell)

▶ A three-part body (head, thorax, and abdomen)

▶ Three pairs of jointed legs

▶ Compound eyes

▶ One pair of antennae

Using this information, you could create an insect object definition. It would include these characteristics as part of the object module.

But think about the ant. An ant is classified as part of the insect class because it has all the characteristics just listed. However, an ant also has these unique characteristics that not all other insects share:

▶ A narrow abdomen where it joins the thorax (looks like a tiny waist)

▶ At the narrow abdomen where it joins the thorax, a hump on top that is clearly separate from the rest of the abdomen

▶ Elbowed antennae, with a long first segment

Thus, to create an object definition for an ant, you would need to duplicate all the insect characteristics that are put into the insect object definition. In addition, you must add the characteristics that are specific to ants.

However, there are more insects than just ants. For example, the honey bee is also an insect. It shares the first characteristic specific to an ant (narrow abdomen). Therefore, to create a honey bee object definition, you would need to duplicate the characteristics from the insect object definition, duplicate the first ant characteristic from the ant object definition file, and then add the honey bee's unique characteristics. That is a lot of duplication!

The class problem is strongly demonstrated in the three object definitions we've just look at (insects, ants, and honey bees). And there are more insects besides honey bees and ants! To fix the inefficient duplication class problem, Python uses subclasses and inheritance.

# Understanding Subclasses and Inheritance

A *subclass* is an object definition. It has all the data attributes and methods of another class but includes additional attributes and methods specific to itself. These additional data attributes and methods make a subclass a specific version of a class. For example, an ant is a specific version of an insect.

A class whose data attributes and methods are used by a subclass is called a *superclass*. Using the insect example, the superclass would be `Insect`, and the subclass would be `Ant`. An ant has all the characteristics of an insect, as well as a few of its own, which are specific to ants.

BY THE WAY

### Object Class Terms

A *superclass*, also called a *base class*, is a class used in an object definition of a subclass. A *subclass*, which has all the data attributes and methods of a base class, as well as a few of its own, is also called a *derived class*.

Subclasses have what is called an "is a" relationship to their base class. For example, an ant (subclass) is an insect (base class). A honey bee (subclass) is an insect (base class). These are some other examples of "is a" relationships:

▶ A duck is a bird.

▶ Python is a programming language.

▶ A Raspberry Pi is a computer.

In order for a subclass object to gain the data attributes and methods of its base class, Python uses a process called *inheritance*. In Python, inheritance is more similar to inheriting genes from your biological parents than receiving a monetary inheritance.

*Inheritance* is the process by which a subclass may obtain a copy of the base class's data attributes and methods to include in its object class definition. The subclass object then creates its own data attributes and methods in its object class definition, to make itself a specialized version of the base class.

The example of ants and insects can be used to demonstrate inheritance. To keep it simple, only characteristics (data attributes) are used. But you could use behavior (methods) here, too! An insect base class object definition would contain the following data attributes:

▶ `backbone='none'`

▶ `exoskeleton='chitinous'`

▶ `body='three-part'`

▶ `jointed_leg_pairs=3`

▶ `eyes='compound'`

▶ `antennae_pair=1`

The ant object class definition would inherit all six of these insect data attributes. The following three data attributes would be added to make the subclass (ant object) a specialized version of the base class (insect object):

▶ `abdomen_thorax_width='narrow'`

▶ `abdomen_thorax_shape='humped'`

▶ `antennae='elbowed'`

The ant "is a" insect relationship would be maintained. Basically, the ant "inherits" the insect's object definition. There is no need to create duplicate data attributes and methods in the ant's object definition. Thus, the class problem is solved.

# Using Inheritance in Python

So what does inheritance look like in Python? This is the basic syntax for inheritance in a class object definition:

```
class classname:
      base class data attributes
      base class mutator methods
      base class accessor methods
      class classname (base class name):
            subclass data attributes
            subclass mutator methods
            subclass accessor methods
```

Listing 15.1 shows a bird class object definition stored in the object module /home/pi/py3prog/ birds.py. The Bird class is an overly simplified object definition for a bird. Notice that there are three immutable data attributes: feathers (line 7), bones (line 8), and eggs (line 9). The only mutable data attribute is sex (line 10) because a bird can be male, female, or unknown.

**LISTING 15.1**   Bird Object Definition File

```
 1: pi@raspberrypi ~ $ cat /home/pi/py3prog/birds.py
 2: # Bird base class
 3: #
 4: class Bird:
 5:     #Initialize Bird class data attributes
 6:     def __init__(self, sex):
 7:         self.__feathers = 'yes'       #Birds have feathers
 8:         self.__bones = 'hollow'       #Bird bones are hollow
 9:         self.__eggs = 'hard-shell'    #Bird eggs are hard-shell.
10:          self.__sex = sex                  #Male, female, or unknown.
11:
12:     #Mutator methods for Bird data attributes
13:     def set_sex(self, sex):      #Male, female, or unknown.
14:          self.__sex = sex
15:
16:     #Accessor methods for Bird data attributes
17:     def get_feathers(self):
18:         return self.__feathers
19:
20:     def get_bones(self):
21:         return self.__bones
22:
23:     def get_eggs(self):
24:          return self.__eggs
```

```
25:
26:    def get_sex(self):
27:        return self.__sex
28: pi@raspberrypi ~ $
```

Also notice in Listing 15.1 that there is one mutator method (lines 12 through 14) for the `Bird` class, and there are four accessor methods (lines 16 through 27). There's nothing too special here. Most of these items should look similar to the class definitions in Hour 14.

## Creating a Subclass

To add a subclass to the `Bird` base class, a barn swallow (also known as a European swallow) was chosen. For simplicity's sake, the `BarnSwallow` subclass is also overly simplified. Any ornithologist will recognize that there is much more to a barn swallow than is listed here!

To add the `BarnSwallow` subclass, the subclass must be declared using the class declaration, as shown here:

```
class BarnSwallow(Bird):
```

This class declaration allows you to define a subclass of `BarnSwallow` that inherits from its base class (`Bird`). Thus, the `BarnSwallow` subclass object definition inherits all the data attributes and methods from the `Bird` base class.

As with initializing a class, all the data attributes to be used in the `BarnSwallow` subclass are initialized. This includes both base class and subclass data items, as shown here:

```
def __init__(self, feathers, bones, eggs, sex,
             migratory, flock_size):
```

Within the initialization block, the `__init__` method of the `Bird` base class is used to initialize the inherited data attributes `feather`, `bones`, `eggs`, and `sex`. This needs to be done for inheritance purposes. You initialize the data attributes like this:

```
Bird.__init__(self, feathers, bones, eggs, sex)
```

Specialization of the `BarnSwallow` subclass can now begin. A barn swallow has the following specialized data attributes. One data attribute is immutable (`migratory`) and one is mutable (`flock_size`):

▶ **migratory**—Set to `yes` because a barn swallow is known for its large migratory range.

▶ **flock_size**—Indicates the number of birds seen in one sighting.

(Remember that this is an overly simplified example. A real barn swallow would have many more data attributes.)

These specialized data attributes are set using the following Python statements.

```
self.__migratory = 'yes'
self.__flock_size = flock_size
```

Since the first data attribute for the BarnSwallow subclass is immutable, the only mutator method needed is for flock_size. This is set as follows:

```
def set_flock_size(self,flock_size):
    self.__flock_size = flock_size
```

Finally, in the BarnSwallow subclass object definition, the accessor methods must be declared for the subclass data attributes. They are shown here:

```
def get_migratory(self):
    return self.__migratory
def get_flock_size(self, flock_size):
    return self.__flock_size
```

Once all the parts of the object definition have been determined, you can add the subclass to an object module file.

## Adding a Subclass to an Object Module File

The BarnSwallow subclass object definition can be stored in the same module file as the Bird base class (see Listing 15.2).

**LISTING 15.2**  The BarnSwallow Subclass in the Bird Object File

```
1: pi@raspberrypi ~ $ cat /home/pi/py3prog/birds.py
2: # Bird base class
3: #
4: class Bird:
5:     #Initialize Bird class data attributes
6:     def __init__(self, sex):
7: ...
8: #
9: # Barn Swallow subclass (base class: Bird)
10: #
11: class BarnSwallow(Bird):
12:
13:     #Initialize Barn Swallow data attributes & obtain Bird inheritance.
14:     def __init__(self, sex, flock_size):
15:
16:         #Obtain base class data attributes & methods (inheritance)
17:         Bird.__init__(self, sex)
```

```
18:
19:          #Initialize subclass data attributes
20:          self.__migratory = 'yes'        #Migratory bird.
21:          self.__flock_size = flock_size  #How many in flock.
22:
23:
24:      #Mutator methods for Barn Swallow data attributes
25:      def set_flock_size(self,flock_size):  #No. of birds in sighting
26:          self.__flock_size = flock_size
27:
28:      #Accessor methods for Barn Swallow data attributes
29:      def get_migratory(self):
30:          return self.__migratory
31:      def get_flock_size(self):
32:          return self.__flock_size
33: pi@raspberrypi ~ $
```

You can see a partial listing of the `Bird` base class object definition file on lines 2 through 7 of Listing 15.2. The `BarnSwallow` subclass object definition is on lines 8 through 32 of the `birds.py` object module file.

## WATCH OUT!

### Proper Indentation

Remember that you need to make sure you do the indentation properly for an object module file! If you do not indent object module blocks properly, Python gives you an error message, indicating that it cannot find a method or data attribute.

Inheritance allows you to use a subclass along with its base class in a module file. However, you are not limited to just one subclass in an object module file.

# Adding Additional Subclasses

You can add additional subclass object definitions to an object module file. For example, the South African cliff swallow is very similar to a barn swallow, but it is non-migratory.

Listing 15.3 adds the `SouthAfricanCliffSwallow` subclass. Again, it is an oversimplified version of a bird. However, you can see that the subclass object definition has its own place within the object module file. You could list every subclass of bird that exists in the file `birds.py`, if you wanted to.

**LISTING 15.3**   The `CliffSwallow` Subclass in the `Bird` Object File

```
 1: pi@raspberrypi ~ $ cat /home/pi/py3prog/birds.py
 2: # Bird base class
 3: #
 4: class Bird:
 5:     #Initialize Bird class data attributes
 6:     def __init__(self, sex):
 7: ...
 8:
 9: #
10: # Barn Swallow subclass (base class: Bird)
11: #
12: class BarnSwallow(Bird):
13:
14:     #Initialize Barn Swallow data attributes & obtain Bird inheritance.
15:     def __init__(self, sex, flock_size):
16: ...
17: #
18: # South Africa Cliff Swallow subclass (base class: Bird)
19: #
20: class SouthAfricaCliffSwallow(Bird):
21:
22:     #Initialize Cliff Swallow data attributes & obtain Bird inheritance.
23:     def __init__(self, sex, flock_size):
24:
25:         #Obtain base class data attributes & methods (inheritance)
26:         Bird.__init__(self, sex)
27:
28:         #Initialize subclass data attributes
29:         self.__migratory = 'no'           #Non-migratory bird.
30:         self.__flock_size = flock_size   #How many in flock.
31:
32:
33:     #Mutator methods for Cliff Swallow data attributes
34:     def set_flock_size(self,flock_size):  #No. of birds in sighting
35:         self.__flock_size = flock_size
36:
37:     #Accessor methods for Cliff Swallow data attributes
38:     def get_migratory(self):
39:         return self.__migratory
40:     def get_flock_size(self):
41:         return self.__flock_size
42: pi@raspberrypi ~ $
```

Remember that modularity is important when you're creating any program, including Python scripts. Thus, keeping all the bird subclasses in the same file as the `Bird` base class is not a good idea.

# Putting a Subclass in Its Own Object Module File

For better modularity, you can store a base class in one object module file and store each subclass in its own module file. In Listing 15.4, you can see the modified /home/pi/py3prog/birds.py object module file. It does not include the BarnSwallow or SouthAfricanCliffSwallow subclass.

**LISTING 15.4**  A Bird Base Class Object File

```
pi@raspberrypi ~ $ cat /home/pi/py3prog/birds.py
# Bird base class
#
class Bird:
    #Initialize Bird class data attributes
    def __init__(self, sex):
...
    def get_sex(self):
        return self.__sex
pi@raspberrypi ~ $
```

To put a subclass in its own object module file, you need to add an import Python statement to the file, as shown in Listing 15.5. Here the Bird base class is imported before the BarnSwallow subclass is defined.

**LISTING 15.5**  The BarnSwallow Subclass Object File

```
 1: pi@raspberrypi ~ $ cat /home/pi/py3prog/barnswallow.py
 2: #
 3: # BarnSwallow subclass (base class: Bird)
 4: #
 5: from birds import Bird    #import Bird base class
 6:
 7: class BarnSwallow(Bird):
 8:
 9:     #Initialize Barn Swallow data attributes & obtain Bird inheritance.
10:     def __init__(self, sex, flock_size):
11:
12:         #Obtain base class data attributes & methods (inheritance)
13:         Bird.__init__(self, sex)
14:
15:         #Initialize subclass data attributes
16:         self.__migratory = 'yes'        #Migratory bird.
17:         self.__flock_size = flock_size  #How many in flock.
18:
19:
20:     #Mutator methods for Barn Swallow data attributes
21:     def set_flock_size(self, flock_size):  #No. of birds in sighting
```

```
22:                 self.__flock_size = flock_size
23:
24:         #Accessor methods for Barn Swallow data attributes
25:         def get_migratory(self):
26:             return self.__migratory
27:         def get_flock_size(self):
28:             return self.__flock_size
29: pi@raspberrypi ~ $
```

You can see in Listing 15.5 that the `Bird` base class is imported on line 5. Notice that the `import` statement uses the `from` module_file_name `import` object_def format. It does so because the module file name is `bird.py` and the object definition is called `Bird`. After it is imported, the `BarnSwallow` subclass is defined on lines 7 through 28.

Once you have your object module files created—one containing the base class and others containing all the necessary subclasses—the next step is to use these files in Python scripts.

# Using Inheritance in Python Scripts

Using inheritance in a Python script is really not much different from using regular base class objects in a script. Both the `BarnSwallow` subclass and the `SouthAfricanCliffSwallow` subclass are used in `script1501.py`, along with their `Bird` base class. The script, as shown in Listing 15.6, simply goes through the objects and displays the immutable settings of each.

**LISTING 15.6**   Python Statements in `script1501.py`

```
1: pi@raspberrypi ~ $ cat /home/pi/py3prog/script1501.py
2: # script1501.py - Display Bird immutable data via Accessors
3: # Written by Blum and Bresnahan
4: #
5: ############# Import Modules #################
6: #
7: # Birds object file
8: from birds import Bird
9: #
10: # Barn Swallow object file
11: from barnswallow import BarnSwallow
12: #
13: # South Africa Cliff Swallow object file
14: from sacliffswallow import SouthAfricaCliffSwallow
15: #
16: def main ():
17:       ###### Create Variables & Object Instances ###
18:       #
19:       sex='unknown'  #Male, female, or unknown
```

```
20:        flock_size='0'
21:        #
22:        bird=Bird(sex)
23:        barn_swallow=BarnSwallow(sex,flock_size)
24:        sa_cliff_swallow=SouthAfricaCliffSwallow(sex,flock_size)
25:        #
26:        ########## Show Bird Characteristics ########
27:        #
28:        print("A bird has",end=' ')
29:        if bird.get_feathers() == 'yes':
30:            print("feathers,", end=' ')
31:        print("bones that are", bird.get_bones(), end=' ')
32:        print("and", bird.get_eggs(), "eggs.")
33:        #
34:        ###### Show Barn Swallow Characteristics #####
35:        #
36:        print()
37:        print("A barn swallow is a bird that", end=' ')
38:        if barn_swallow.get_migratory() == 'yes':
39:            print("is migratory.")
40:        else:
41:            print("is not migratory.")
42:        #
43:        ######## Show Cliff Swallow Characteristics ######
44:        #
45:        print()
46:        print("A cliff swallow is a bird that", end=' ')
47:        if sa_cliff_swallow.get_migratory() == 'yes':
48:            print("is migratory.")
49:        else:
50:            print("is not migratory.")
51: ########################################################
52: #
53: ########## Call the main function ######################
54: main()
```

In the script, the object module files are imported before the start of the main function declaration on lines 7 through 14. The variables sex and flock_size are to be used as arguments and thus are set to 'unknown' and 0, respectively, on lines 19 and 20.

In Listing 15.6, the object instances themselves are declared on lines 22 through 24. Finally, the accessors for each object are used to obtain the immutable values of each object class. They are printed to the screen on line 28 through line 50.

Listing 15.7 shows script1501.py in action. Both the base class and each subclass's immutable values are displayed.

**LISTING 15.7** Output of `script1501.py`

```
pi@raspberrypi ~ $ python3 /home/pi/py3prog/script1501.py
A bird has feathers, bones that are hollow and hard-shell eggs.

A barn swallow is a bird that is migratory.

A cliff swallow is a bird that is not migratory.
pi@raspberrypi ~ $
```

As you can see, the script runs fine. Both the `BarnSwallow` object and the `SouthAfricaCliffSwallow` object are able to inherit data attributes and methods within the script from the `Bird` object with no problems.

Many ornithology organizations around the world—such as Cornel's Great Backyard Bird Count, at www.birdsource.org/gbbc/—seek bird-sighting information. The `script1501.py` script was modified to include sighting information and renamed `script1502.py`.

Listing 15.8 shows the `script1502.py` script. It now includes methods to obtain flock size information.

**LISTING 15.8** Python Statements in `script1501.py`

```
1: pi@raspberrypi ~ $ cat /home/pi/py3prog/script1502.py
2: # script1502.py - Record a Swallow Sighting
3: # Written by Blum and Bresnahan
4: #
5: ########### Import Modules  #################
6: #
7: # Birds object file
8: from birds import Bird
9: #
10: # Barn Swallow object file
11: from barnswallow import BarnSwallow
12: #
13: # South Africa Cliff Swallow object file
14: from sacliffswallow import SouthAfricaCliffSwallow
15: #
16: # Import Date Time function
17: import datetime
18: #
19: ###############################################
20: def main ():      #Mainline
21:      ###### Create Variables & Object Instances ###
22:      #
23:      flock_size='0'       #Number of birds sighted
24:      sex='unknown'        #Male, female, or unknown
```

```
25:        species=''             #Barn or Cliff Swallow Object
26:        #
27:        barn_swallow=BarnSwallow(sex,flock_size)
28:        sa_cliff_swallow=SouthAfricaCliffSwallow(sex,flock_size)
29:        #
30:        ###### Instructions for Script User #########
31:        print()
32:        print("The following characteristics are listed")
33:        print("in order to help you determine what swallow")
34:        print("you have sighted.")
35:        print()
36:        #
37:        ###### Show Barn Swallow Characteristics #####
38:        #
39:        print("A barn swallow is a bird that", end=' ')
40:        if barn_swallow.get_migratory() == 'yes':
41:                print("is migratory.")
42:        else:
43:                print("is not migratory.")
44:        #
45:        ######## Show Cliff Swallow Characteristics ######
46:        #
47:        print("A cliff swallow is a bird that", end=' ')
48:        if sa_cliff_swallow.get_migratory() == 'yes':
49:                print("is migratory.")
50:        else:
51:                print("is not migratory.")
52:        #
53:        ######## Obtain Swallow Sighted ################
54:        print()
55:        print("Which did you see?")
56:        print("European/Barn Swallow    - 1")
57:        print("African Cliff Swallow    - 2")
58:        species = input("Type number & press Enter: ")
59:        print()
60:        #
61:        ######## Obtain Flock Size ################
62:        #
63:        flock_size=int(input("Approximately, how many did you see? "))
64:        #
65:        ###### Mutate Sighted Birds' Flock Size ####
66:        #
67:        if species == '1':
68:                barn_swallow.set_flock_size(flock_size)
69:        else:
70:                sa_cliff_swallow.set_flock_size(flock_size)
71:        #
72:        ###### Display Sighting Data ##############
```

```
73:        print()
74:        print("Thank you.")
75:        print("The following data will be forwarded to")
76:        print("the Great Backyard Bird Count.")
77:        print("www.birdsource.org/gbbc")
78:        #
79:        print()
80:        print("Sighting on \t", datetime.date.today())
81:        if species == '1':
82:            print("Species: \t European/Barn Swallow")
83:            print("Flock Size: \t", barn_swallow.get_flock_size())
84:            print("Sex: \t\t", barn_swallow.get_sex())
85:        else:
86:            print("Species: \t South Africa Cliff Swallow")
87:            print("Flock Size: \t", sa_cliff_swallow.get_flock_size())
88:            print("Sex: \t\t", sa_cliff_swallow.get_sex())
89: #
90: #########################################################
91: #
92: ########## Call the main function #####################
93: main()
94: pi@raspberrypi ~
```

Notice in lines 84 and 88 of Listing 15.8 that the accessor methods are used to obtain the bird's sex. .get_sex is an accessor method set in the Bird base class (refer to Listing 15.1, lines 26 and 27). Both the subclasses BarnSwallow and SouthAfricaCliffSwallow inherited methods from Bird. Thus, they are able to access the data by using the inherited .get_sex accessor method. This is called *polymorphism*.

DID YOU KNOW

## Polymorphism

*Polymorphism* is the ability of subclasses to have methods with the same name as methods in their base class. This is sometimes called *overriding* a method. You can still access each class's method, by using either base_class.method or subclass.method.

Listing 15.9 shows Python interpreting script1502.py. When the script is run, the subclasses inherit data attributes and methods with no problems.

**LISTING 15.9**   script1502.py Interpreted

```
pi@raspberrypi ~ $ python3 /home/pi/py3prog/script1502.py

The following characteristics are listed
in order to help you determine what swallow
```

```
you have sighted.

A barn swallow is a bird that is migratory.
A cliff swallow is a bird that is not migratory.

Which did you see?
European/Barn Swallow    - 1
African Cliff Swallow    - 2
Type number & press Enter: 1

Approximately, how many did you see? 7

Thank you.
The following data will be forwarded to
the Great Backyard Bird Count.
www.birdsource.org/gbbc

Sighting on       2013-11-06
Species:       European/Barn Swallow
Flock Size:       7
Sex:           unknown
pi@raspberrypi ~ $
```

Again, for simplicity, the data gathered here is overly simplified. However, to aid in your understanding of inheritance, subclasses, and object module files, you are going to improve it!

## TRY IT YOURSELF ▼

### Explore Python Inheritance and Subclasses

In the following steps, you will explore Python inheritance and subclasses by improving the bird-sighting information script, `script1502.py`. Follow these steps, to modify the script and create a new base class and subclass:

1. If you have not already done so, power up your Raspberry Pi and log in to the system.

2. If you do not have the LXDE GUI started automatically at boot, start it now by typing `startx` and pressing Enter.

3. Open up the LXTerminal by double-clicking the LXTerminal icon.

4. At the command-line prompt, type `nano py3prog/script1503.py` and press Enter. This command puts you into the nano text editor and creates the file `py3prog/script1503.py`.

5. Type all the information from `script1502.py` in Listing 15.6 into the nano editor window, pressing Enter at the end of each line. Be sure to take your time here and avoid any typographical errors. You can make corrections by using the Delete key and the up- and down-arrow keys.

BY THE WAY

**Make It Easy**

Instead of doing all this typing, you can download `script1502.py` from informit.com/register. After downloading the script, simply use it instead of creating the new `script1503.py`.

6. Make sure you have entered the code into the nano text editor window, as shown in Listing 15.6. Make any corrections needed.

7. Write out the information from the text editor to the script by pressing Ctrl+O. The script file name shows along with the prompt `File name to write`. Press Enter to write out the contents to the `script1503.py` script.

8. Exit the nano text editor by pressing Ctrl+X.

9. At the command-line prompt, type `nano py3prog/birds.py` and press Enter. This command puts you into the nano text editor and creates the file `py3prog/birds.py`.

10. Type all the information from `birds.py` in Listing 15.4 into the nano editor window. You are creating the `Birds` base class object file that is needed for the Python script.

BY THE WAY

**Continue to Make It Easy**

Instead of doing all this typing, you can download all three files—`birds.py`, `barnswallow.py`, and `sacliffswallow.py`—from informit.com/register. After downloading these object module files, you can skip over the steps to create them!

11. Make sure you have entered the statements into the nano text editor window, as shown in Listing 15.4. Make any corrections needed.

12. Write out the information from the text editor to the file by pressing Ctrl+O. The file name shows along with the prompt `File name to write`. Press Enter to write out the contents to the `birds.py` object file.

13. Exit the nano text editor by pressing Ctrl+X.

14. At the command-line prompt, type `nano py3prog/barnswallow.py` and press Enter. The command puts you into the nano text editor and creates the file py3prog/barnswallow.py.

15. Type all the information from `barnswallow.py` in Listing 15.5 into the nano editor window. You are creating the `BarnSwallow` subclass object file that is needed for the Python script.

16. Make sure you have entered the statements into the nano text editor window, as shown in Listing 15.5. Make any corrections needed.

17. Write out the information from the text editor to the file by pressing Ctrl+O. The file name shows along with the prompt `File name to write`. Press Enter to write out the contents to the `barnswallow.py` object file.

18. Exit the nano text editor by pressing Ctrl+X.

19. Hang in there! You are getting close! At the command-line prompt, type `nano py3prog/sacliffswallow.py` and press Enter. The command puts you into the nano text editor and creates the file `py3prog/sacliffswallow.py`.

20. In the nano text editor window, type the information from lines 17–29 in the `birds.py` file in Listing 15.3 (the comment lines).

21. Now type the following into the nano editor window:

    ```
    from birds import Bird    #import Bird base class
    ```

22. Finish the `SouthAfricaCliffSwallow` subclass object file that needed for the Python script by typing in the Python statements from lines 20 through 41 in Listing 15.3.

23. Make sure you have entered the statements into the nano text editor window as shown in Listing 15.3, along with the additional `import` statement. Make any corrections needed.

24. Write out the information from the text editor to the file by pressing Ctrl+O. The file name shows along with the prompt `File name to write`. Press Enter to write out the contents to the `sacliffswallow.py` object file.

25. Exit the nano text editor by pressing Ctrl+X.

26. Before making the improvements, make sure all is well with your code by typing `python3 py3prog/script1503.py`. If you get any errors, double-check the four files for any typos and make corrections as needed. Next you will begin the improvements.

27. Create a base class called `Sighting` and subclass called `BirdSighting`. For simplicity's sake, you can put them both in one module file. Type `nano py3prog/sightings.py` and press Enter. Python puts you into the nano text editor and creates the object module file `py3prog/sightings.py`.

**28.** Type the following code into the nano editor window:

```
# Sightings base class
#
class Sighting:
    #Initialize Sighting class data attributes
    def __init__(self, sight_location, sight_date):
        self.__sight_location = sight_location  #Location of sighting
        self.__sight_date = sight_date           #Date of sighting

    #Mutator methods for Sighting data attributes
    def set_sight_location(self, sight_location):
        self.__sight_location = sight_location

    def set_sight_date(self, sight_date):
        self.__sight_date = sight_date

    #Accessor methods for Sighting data attributes
    def get_sight_location(self):
        return self.__sight_location

    def get_sight_date(self):
        return self.__sight_date
#
# Bird Sighting subclass (base class: Sighting)
#
class BirdSighting(Sighting):

    #Initialize Bird Sighting data attributes & obtain Bird inheritance.
    def __init__(self, sight_location, sight_date,
                 bird_species, flock_size):

        #Obtain base class data attributes & methods (inheritance)
        Sighting.__init__(self, sight_location, sight_date)

        #Initialize subclass data attributes
        self.__bird_species = bird_species #Bird type
        self.__flock_size = flock_size      #How many in flock.


    #Mutator methods for Bird Sighting data attributes
    def set_bird_species(self,bird_species):
        self.__bird_species = bird_species

    def set_flock_size(self,flock_size):
        self.__flock_size = flock_size
```

```
    #Accessor methods for Bird Sighting data attributes
    def get_bird_species(self):
        return self.__bird_species
    def get_flock_size(self):
        return self.__flock_size
```

**29.** Write out the information from the text editor to the file by pressing Ctrl+O. The file name shows along with the prompt `File name to write`. Press Enter to write out the contents to the `sightings.py` object file.

**30.** Exit the nano text editor by pressing Ctrl+X.

**31.** To modify the script to use these two objects, at the command-line prompt, type `nano py3prog/script1503.py` and press Enter.

**32.** For the first change, in the `Import Modules` section of the script, under the import of the `SouthAfricanCliffSwallow` object file, insert the following lines (which import the new object files into the script):

```
# Sightings object file
from sightings import Sighting
#
# Birds sightings object file
from sightings import BirdSighting
```

**33.** For the second change, in the `Create Variables & Object Instances` section of the script, under the creation of both the barn and cliff swallow object instances, insert the following lines, properly indented:

```
location='unknown'     #Location of sighting
date='unknown'         #Date of sighting
#
bird_sighting=BirdSighting(location,date,species,flock_size)
```

**34.** For the third change, delete both the sections `Obtain Flock Size` and `Mutate Sighted Birds' Flock Size`, along with their Python statements:

```
######## Obtain Flock Size ################
    #
    flock_size=int(input("Approximately, how many did you see? "))
    #
    ###### Mutate Sighted Birds' Flock Size ####
    #
    if species == '1':
        barn_swallow.set_flock_size(flock_size)
    else:
        sa_cliff_swallow.set_flock_size(flock_size)
    #
```

**35.** In place of what you just deleted, add the following:

```
######## Obtain Sighting Information ################
    #
    location=input("Where did you see the birds? ")
    print()
    flock_size=int(input("Approximately, how many did you see? "))
    #
    ###### Mutate Sighted Birds' Information ####
    #
    bird_sighting.set_sight_location(location)
    bird_sighting.set_sight_date(datetime.date.today())
    if species == '1':    #CHANGE
        bird_sighting.set_bird_species('barn swallow')
    else:
        bird_sighting.set_bird_species('SA cliff swallow')
    bird_sighting.set_flock_size(flock_size)
    #
```

(Notice that the mutators, such as `.set_sight_date`, are now all from the `bird_sighting` subclass.)

**36.** For the fourth change, in the section `Display Sighting Data`, delete the following Python statements:

```
print("Sighting on \t", datetime.date.today())
    if species == '1':
        print("Species: \t European/Barn Swallow")
        print("Flock Size: \t", barn_swallow.get_flock_size())
        print("Sex: \t\t", barn_swallow.get_sex())
    else:
        print("Species: \t South Africa Cliff Swallow")
        print("Flock Size: \t", sa_cliff_swallow.get_flock_size())
        print("Sex: \t\t", sa_cliff_swallow.get_sex())
```

**37.** In place of what you just deleted, add the following:

```
print("Sighting Date: \t", bird_sighting.get_sight_date())
print("Location: \t", bird_sighting.get_sight_location())
if species == '1':
    print("Species: \t European/Barn Swallow")
else:
    print("Species: \t South Africa Cliff Swallow")
print("Flock Size: \t", bird_sighting.get_flock_size())
```

(Notice that the accessors, such as `.get_flock_size`, are now all from the `bird_sighting` subclass.)

38. Review the four major changes you just made to `script1503.py` to ensure that there are no typos and that the indentation is correct.

39. Write out the information from the text editor to the script by pressing Ctrl+O. The script file name shows along with the prompt `File name to write`. Press Enter to write out the contents to the `script1503.py` script.

40. Exit the nano text editor by pressing Ctrl+X. All your work is about to pay off!

41. To test your modifications to the script, at the command-line prompt, type `python3 /home/pi/py3prog/script1503.py` and press Enter. Answer the script questions as you please. If there are no problems with your script or object definition file, the output will look similar to Listing 15.10.

**LISTING 15.10**   The `script1503.py` Interpreted

```
pi@raspberrypi ~ $ python3 /home/pi/py3prog/script1503.py

The following characteristics are listed
in order to help you determine what swallow
you have sighted.

A barn swallow is a bird that is migratory.
A cliff swallow is a bird that is not migratory.

Which did you see?
European/Barn Swallow    - 1
African Cliff Swallow    - 2
Type number & press Enter: 1

Where did you see the birds? Indianapolis, Indiana, USA

Approximately, how many did you see? 21

Thank you.
The following data will be forwarded to
the Great Backyard Bird Count.
www.birdsource.org/gbbc

Sighting Date:       2013-11-06
Location:            Indianapolis, Indiana, USA
Species:             European/Barn Swallow
Flock Size:          21
pi@raspberrypi ~ $
```

Good job! You've done quite a bit of work here, but if you are like most other script writers, you probably already see several things to improve in this script. For instance, there are no checks on user-input data. The data should be output to a file, not just displayed to the screen. In addition, to make the data useful, the time of day should be recorded, and more bird species should be added. You can make all sorts of changes to this script!

Here is an idea to start. Start with script1503.py and try adding better bird descriptions to the swallow subclass object module files to aid in their species identification. Now that you know how to create subclasses using inheritance, you can get the script's "ducks in a row."

# Summary

In this hour, you read about the class problem, Python subclasses, and the inheritance solution. You learned how to create an object subclass in the same object module file as the base class. Also, you learned how to create a subclass in its own object module file. Finally, you saw some practical examples of using the base classes and subclasses in a few Python scripts. In Hour 16, "Regular Expressions," you will explore regular expressions.

# Q&A

**Q. What is the difference between an "is a" relationship and a "has a" relationship in Python?**

**A.** An "is a" relationship exists between a subclass and its base class. For example, a barn swallow "is a" bird. A "has a" relationship exists between a class (a subclass or base class) and one of its data attributes or methods. For example, looking at the class definition of a bird, you can see from one of the data attribute statements that a bird "has a" feather.

**Q. I added a subclass definition in a base class object file, and when I try to use one of the stated methods, Python tells me the method is not found. Why?**

**A.** Most likely, you do not have the correct indentation in the class object file. Try re-creating the file within the IDLE 3 editor, which will give you some assistance creating the proper indentation. An even better solution would be to put the subclass in its own object file, using the IDLE 3 editor. Just be sure to include the proper import statements of the base class.

**Q. Do I have to use subclasses?**

**A.** No, there are no Python style police out there waiting to force you to use subclasses. However, good form dictates that you should use subclasses to avoid the duplication issues in the class problem.

# Workshop

## Quiz

1. A superclass is the same thing as a derived class. True or false?

2. What is it called when a subclass receives data attributes and methods from a base class?

3. Which Python statement correctly declares the subclass `Ant` of the base class `Insect`?

   a. `class  Insect(Ant):`

   b. `class  Ant(Insect):`

   c. `from  Insect  subclass  Ant():`

## Answers

1. False. A superclass is also called a base class. A subclass, which inherits data attributes and methods of a base class, is also called a derived class because some of its attributes and methods are derived from the base class.

2. Inheritance is the term used when a subclass receives data attributes and methods from a base class.

3. Answer b is correct. To properly create the subclass `Ant` from the base class `Insect`, you use `class  Ant(Insect):`.

# Regular Expressions

**What You'll Learn in This Hour:**

▶ What regular expressions are

▶ Defining regular expression patterns

▶ How to use regular expressions in your scripts

One of the most common functions used in Python scripts is manipulation of string data. One of the things Python is known for is its ability to easily search and modify strings. One of the features in Python that provides support for string parsing is regular expressions. In this hour, you'll see what regular expressions are, how to use them in Python, and how to leverage them in your own Python scripts.

# What Are Regular Expressions?

Many people have a hard time understanding what regular expressions are. The first step to understanding them is defining exactly what they are and what they can do for you. The following sections explain what a regular expression is and describe how Python uses regular expressions to help with your string manipulations.

## Definition of *Regular Expressions*

A *regular expression* is a pattern you create to filter text. A program or script matches the regular expression pattern you create against data as the data flows through the program. If the data matches the pattern, it's accepted for processing. If the data doesn't match the pattern, it's rejected. Figure 16.1 shows how it works.

While are probably familiar with normal text searching, regular expressions provides a lot more than that. The regular expression pattern makes use of wildcard characters to represent one or more characters in the data stream. You can use a number of special characters in a regular expression to define a specific pattern for filtering data. This means you have a lot of flexibility in how you define your string patterns.

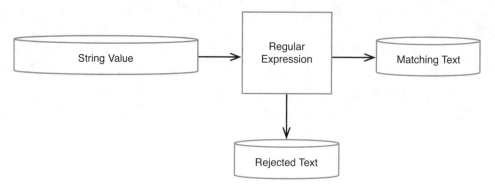

**FIGURE 16.1**
Matching data against a regular expression.

# Types of Regular Expressions

The biggest problem with using regular expressions is that there isn't just one set of them. Different applications use different types of regular expressions. These include such diverse things as programming languages (for example, Java, Perl, Python), Linux utilities (such as the sed editor, the gawk program, and the grep utility), and mainstream applications (such as the MySQL and PostgreSQL database servers).

A regular expression is implemented using a regular expression engine. A regular expression engine is the underlying software that interprets regular expression patterns and uses those patterns to match text.

In the open source software world, there are two popular regular expression engines:

- ▶ The POSIX Basic Regular Expression (BRE) engine
- ▶ The POSIX Extended Regular Expression (ERE) engine

Most open source programs at a minimum conform to the POSIX BRE engine specifications, recognizing all the pattern symbols it defines. Unfortunately, some utilities (such as the sed editor) only conform to a subset of the BRE engine specifications. This is due to speed constraints, as the sed editor attempts to process text in the data stream as quickly as possible.

The POSIX ERE engine is often found in programming languages that rely on regular expressions for text filtering. It provides advanced pattern symbols as well as special symbols for common patterns, such as matching digits, words, and alphanumeric characters. The Python programming language uses the ERE engine to process its regular expression patterns.

# Working with Regular Expressions in Python

Before you can start writing regular expressions to filter data in your Python scripts, you need to know how to use them. The Python language provides the `re` module to support regular expressions. The `re` module is included in the Raspbian Python default installation, so you don't need to do anything special to start using regular expressions in your scripts, other than import the `re` module at the start of a script:

```
import re
```

However, the `re` module provides two different ways to define and use regular expressions. The following sections discuss how to use both methods.

## Regular Expression Functions

The easiest way to use regular expressions in Python is to directly use the regular expression functions provided by the `re` module. Table 16.1 lists the functions that are available.

**TABLE 16.1**    The `re` Module Functions

| Function | Description |
| --- | --- |
| match | Looks for the pattern at the beginning of the string. |
| search | Looks for the pattern anywhere in the string. |
| findall | Finds all substrings that match the pattern and returns them as a list. |
| finditer | Finds all substrings that match the pattern and returns them as an iterator. |

The `re` module functions take two parameters. The first parameter is the regular expression pattern, and the second parameter is the text string to apply the pattern to.

The `match()` and `search()` regular expression functions return either a `True` Boolean value if the text string matches the regular expression pattern or a `False` value if they don't match. This makes them ideal for use in if-then statements.

### The `match()` Function

The `match()` function does what it says: It tries to match the regular expression pattern to a text string. It is a little tricky in that it applies the regular expression string only to the start of the string value. Here's an example:

```
>>> re.match('test', 'testing')
<_sre.SRE_Match object at 0x015F9950>
>>> re.match('ing', 'testing')
>>>
```

The output from the first match indicates that the match was successful. When the match fails, the `match()` function just returns a `False` value, which doesn't show any output in the IDLE interface.

### The `search()` Function

The `search()` function is a lot more versatile than `match()`: It applies the regular expression pattern to an entire string and returns a `True` value if it finds the pattern anywhere in the string. Here's an example:

```
>>> re.search('ing', 'testing')
<_sre.SRE_Match object at 0x015F9918>
>>>
```

This output from the `search()` function indicates that it found the pattern inside the string.

### The `findall()` and `finditer()` Functions

Both the `findall()` and `finditer()` functions returns multiple instances of the pattern if it is found in the string. The `findall()` function returns a list of values that match in the string, as you can see here:

```
>>> re.findall('[ch]at', 'The cat wore a hat')
['cat', 'hat']
>>>
```

The `finditer()` function returns an iterable object that you can use in a `for` statement to iterate through the results.

## Compiled Regular Expressions

If you find yourself using the same regular expression often in your code, you can *compile* the expression and store it in a variable. You can then use the variable everywhere in your code that you want to perform the regular expression pattern match.

To compile an expression, you use the `compile()` function, specifying the regular expression as the parameter and storing the result in a variable, like this:

```
>>> pattern = re.compile('[ch]at')
```

After you store the compiled regular expression, you can use it directly in a `match()` or `search()` function, as shown here:

```
>>> pattern.search('This is a cat')
<_sre.SRE_Match object at 0x015F9988>
>>> pattern.search('He wore a hat')
<_sre.SRE_Match object at 0x015F9918>
>>> pattern.search('He sat in a chair')
>>>
```

The additional benefit of using compiled regular expression patterns is that you can also specify flags to control special features of the regular expression match. Table 16.2 shows these flags and what they control.

**TABLE 16.2** Compiled Flags

| Flag | Shortcut | Description |
|------|----------|-------------|
| DEBUG | | Displays debug information |
| IGNORECASE | I | Performs case-insensitive matching |
| LOCALE | L | Supports characters from the local character set |
| MULTILINE | M | Matches at the beginning and end of each separate line |
| DOTALL | S | Allows the period to match newline characters |
| UNICODE | U | Uses Unicode characters |
| VERBOSE | X | Allows whitespace within the pattern |

For example, by default, regular expression matches are case sensitive. To make a check case insensitive, you just compile the regular expression with the re.I flag, as shown here:

```
>>> pattern = re.compile('[ch]at', re.I)
>>> pattern.search('Cat')
<_sre.SRE_Match object at 0x015F9988>
>>> pattern.search('Hat')
<_sre.SRE_Match object at 0x015F9918>
>>>
```

The search() function can now match the text in either uppercase or lowercase, anywhere in the text.

# Defining Basic Patterns

Defining regular expression patterns falls somewhere between a science and an art form. Entire books have been written about how to create regular expressions for matching different types of data (such as email addresses, phone numbers, or Social Security numbers). Instead of just showing a list of different regular expression patterns, the purpose of this section is to provide the basics of how to use them in daily text searches.

## Plain Text

The simplest pattern for searching for text is to use the text that you want to find in its entirety, as in this example:

```
>>> re.search('test', 'This is a test')
<_sre.SRE_Match object at 0x015F99C0>
>>> re.search('test', 'This is not going to work')
>>>
```

With the `search()` function, the regular expression doesn't care where in the data the pattern occurs. It also doesn't matter how many times the pattern occurs. When the regular expression can match the pattern anywhere in the text string, it returns a `True` value.

The key is matching the regular expression pattern to the text. It's important to remember that regular expressions are extremely picky about matching patterns. Remember that, by default, regular expression patterns are case sensitive. This means they'll match patterns only for the proper case of characters, as shown here:

```
>>> re.search('this', 'This is a test')
>>>
>>> re.search('This', 'This is a test')
<_sre.SRE_Match object at 0x015F9988>
>>>
```

The first attempt here fails to match because the word `this` doesn't appear in all lowercase in the text string; the second attempt, using the uppercase letter in the pattern, works just fine.

You don't have to limit yourself to whole words in a regular expression. If the defined text appears anywhere in the data stream, the regular expression will match, as shown here:

```
>>> re.search('book', 'The books are expensive')
<_sre.SRE_Match object at 0x015F99C0>
>>>
```

Even though the text in the data stream is books, the data in the stream contains the regular expression `book`, so the regular expression pattern matches the data. Of course, if you try the opposite, the regular expression fails, as shown in this example:

```
>>> re.search('books', 'The book is expensive')
>>>
```

You also don't have to limit yourself to single text words in a regular expression. You can include spaces and numbers in your text string as well, as shown here:

```
>>> re.search('This is line number 1', 'This is line number 1')
<_sre.SRE_Match object at 0x015F9988>
>>> re.search('ber 1', 'This is line number 1')
<_sre.SRE_Match object at 0x015F99F8>
>>> re.search('ber 1', 'This is line number1')
>>>
```

If you define a space in a regular expression, it must appear in the data stream. You can even create a regular expression pattern that matches multiple contiguous spaces, like this:

```
>>> re.search('  ', 'This line has  too many spaces')
<_sre.SRE_Match object at 0x015F9988>
>>>
```

The line with two spaces between words matches the regular expression pattern. This is a great way to catch spacing problems in text files!

# Special Characters

As you use text strings in your regular expression patterns, there's something you need to be aware of: There are a few exceptions when defining text characters in a regular expression. Regular expression patterns assign a special meaning to a few characters. If you try to use these characters in your text pattern, you won't get the results you were expecting.

Regular expressions recognize these special characters:

. * [ ] ^ $ { } \ + ? | ( )

As you work your way through this hour, you'll find out what these special characters do in a regular expression. For now, though, just remember that you can't use these characters by themselves in your text pattern.

If you want to use one of the special characters as a text character, you need to escape it. To escape a special character, you add another character in front of it to indicate to the regular expression engine to interpret the next character as a normal text character. The special character that does this is the backslash characters (\).

In Python, as you've learned, backslashes also have special meaning in string values. To get around this, if you want to use the backslash character with a special character, you can create a raw string value, using the r nomenclature:

```
r'textstring'
```

For example, if you want to search for a dollar sign in your text, just precede it with a backslash character, like this:

```
>>> re.search(r'\$', 'The cost is $4.00')
<_sre.SRE_Match object at 0x015F9918>
>>>
```

You can use raw text strings for your regular expressions, even if they don't contain any backslashes. Some coders just get in the habit of always using the raw text strings.

# Anchor Characters

As shown in the "Plain Text" section a little earlier this hour, by default when you specify a regular expression pattern, the pattern can appear anywhere in the data stream and be a match. There are two special characters you can use to anchor a pattern to either the beginning or the end of lines in the data stream: ^ and $.

## Starting at the Beginning

The caret character (^) defines a pattern that starts at the beginning of a line of text in the data stream. If the pattern is located anyplace other than the start of the line of text, the regular expression pattern fails.

To use the caret character, you must place it before the pattern specified in the regular expression, like this:

```
>>> re.search('^book', 'The book store')
>>> re.search('^Book', 'Books are great')
<_sre.SRE_Match object at 0x015F9988>
>>>
```

The caret anchor character checks for the pattern at the beginning of each string, not each line. If you need to match the beginning of each line of text, you need to use the MULTILINE feature of the compiled regular expression, as in this example:

```
>>> re.search('^test', 'This is a\ntest of a new line')
>>>
>>> pattern = re.compile('^test', re.MULTILINE)
>>> pattern.search('This is a\ntest of a new line')
<_sre.SRE_Match object at 0x015F9988>
>>>
```

In the first example, the pattern doesn't match the word test at the start of the second line in the text. In the second example, using the MULTILINE feature, it does.

---

BY THE WAY

### Caret Versus match()

You'll notice that the caret special character does the same thing as the match() function. They're interchangeable when you're working with scripts.

---

## Looking for the Ending

The opposite of looking for a pattern at the start of a line is looking for a pattern at the end of a line. The dollar sign ($) special character defines the end anchor. You can add this special character after a text pattern to indicate that the line of data must end with the text pattern, as in this example:

```
>>> re.search('book$', 'This is a good book')
<_sre.SRE_Match object at 0x015F99F8>
>>> re.search('book$', 'This book is good')
>>>
```

The problem with an ending text pattern is that you must be careful of what you're looking for, as shown here:

```
>>> re.search('book$', 'There are a lot of good books')
>>>
```

Because the book word is plural at the end of the line, it no longer matches the regular expression pattern, even though book is in the data stream. The text pattern must be the very last thing on the line in order for the pattern to match.

## Combining Anchors

There are a couple common situations when you can combine the start and end anchors on the same line. In the first situation, suppose you want to look for a line of data that contains only a specific text pattern, as in this example:

```
>>> re.search('^this is a test$', 'this is a test')
<_sre.SRE_Match object at 0x015F9918>
>>> re.search('^this is a test$', 'I said this is a test')
>>>
```

The second situation may seem a little odd at first, but it is extremely useful. By combining both anchors together in a pattern with no text, you can filter empty strings. Look at this example:

```
>>> re.search('^$', 'This is a test string')
>>> re.search('^$', "")
<_sre.SRE_Match object at 0x015F99F8>
>>>
```

The defined regular expression pattern looks for text that has nothing between the start and end of the line. Because blank lines contain no text between the two newline characters, they match the regular expression pattern. This is an effective way to remove blank lines from documents.

# The Dot Character

The dot special character is used to match any single character except a newline character. The dot character must match some character, though; if there's no character in the place of the dot, the pattern will fail.

Let's take a look at a few examples of using the dot character in a regular expression pattern:

```
>>> re.search('.at', 'The cat is sleeping')
<_sre.SRE_Match object at 0x015F9988>
>>> re.search('.at', 'That is heavy')
<_sre.SRE_Match object at 0x015F99F8>
>>> re.search('.at', 'He is at the store')
<_sre.SRE_Match object at 0x015F9988>
>>> re.search('.at', 'at the top of the hour')
>>>
```

The third test here is a little tricky. Notice that you match the at, but there's no character in front to match the dot character. Ah, but there is! In regular expressions, spaces count as characters, so the space in front of the at matches the pattern. The last test proves this by putting the at in the front of the line and failing to match the pattern.

## Character Classes

The dot special character is great for matching a character position against any character, but what if you want to limit what characters to match? This is called a *character class* in regular expressions.

You can define a class of characters that would match a position in a text pattern. If one of the characters from the character class is in the data stream, it matches the pattern.

To define a character class, you use square brackets. The brackets contain any character you want to include in the class. You then use the entire class within a pattern, just as you would any other wildcard character. This takes a little getting used to, but once you catch on, you see that you can use it to create some pretty amazing results.

Here's an example of creating a character class:

```
>>> re.search('[ch]at', 'The cat is sleeping')
<_sre.SRE_Match object at 0x015F9918>
>>> re.search('[ch]at', 'That is a very nice hat')
<_sre.SRE_Match object at 0x015F99F8>
>>> re.search('[ch]at', 'He is at the store')
>>>
```

This time, the regular expression pattern matches only strings that have a c or h in front of the at pattern.

You can use more than one character class in a single expression, as in these examples:

```
>>> re.search('[Yy][Ee][Ss]', 'Yes')
<_sre.SRE_Match object at 0x015F9988>
>>> re.search('[Yy][Ee][Ss]', 'yEs')
<_sre.SRE_Match object at 0x015F99F8>
```

```
>>> re.search('[Yy][Ee][Ss]', 'yeS')
<_sre.SRE_Match object at 0x015F9988>
>>>
```

The regular expression uses three character classes to cover both lowercase and uppercase for all three character positions.

Character classes don't have to be just letters. You can use numbers in them as well, as shown here:

```
>>> re.search('[012]', 'This has 1 number')
<_sre.SRE_Match object at 0x015F99F8>
>>> re.search('[012]', 'This has the number 2')
<_sre.SRE_Match object at 0x015F9988>
>>> re.search('[012]', 'This has the number 4')
>>>
```

The regular expression pattern matches any lines that contain the numbers 0, 1, or 2. Any other numbers are ignored, as are lines without numbers in them.

This is a great feature for checking for properly formatted numbers, such as phone numbers and zip codes. However, remember that the regular expression pattern can be found anywhere in the text of the data stream. There might be additional characters besides the matching pattern characters.

For example, if you want to match against a five-digit zip code, you can ensure that you only match against five numbers by using the start- and end-of-the-line characters:

```
>>> re.search('^[0123456789][0123456789][0123456789][0123456789][0123456789]$'
, '12345')
<_sre.SRE_Match object at 0x0154FC28>
>>> re.search('^[0123456789][0123456789][0123456789][0123456789][0123456789]$'
, '123456')
>>> re.search('^[0123456789][0123456789][0123456789][0123456789][0123456789]$',
'1234')
>>>
```

If there are fewer than five or more than five numbers in a zip code, the regular expression pattern returns `False`.

# Negating Character Classes

In regular expression patterns, you can reverse the effect of a character class. Instead of looking for a character contained in a class, you can look for any character that's not in the class. To do this, you place a caret character at the beginning of the character class range, as shown here:

```
>>> re.search('[^ch]at', 'The cat is sleeping')
>>> re.search('[^ch]at', 'He is at home')
<_sre.SRE_Match object at 0x015F9988>
>>> re.search('[^ch]at', 'at the top of the hour')
>>>
```

By negating the character class, the regular expression pattern matches any character that's nei-ther a c nor an h, along with the text pattern. Because the space character fits this category, it passes the pattern match. However, even with the negation, the character class must still match a character, so the line with the at in the start of the line still doesn't match the pattern.

## Using Ranges

You may have noticed in the zip code example that it is rather awkward having to list all the possible digits in each character class. Fortunately, you can use a shortcut to avoid having to do that.

You can use a range of characters within a character class by using the dash symbol. You just specify the first character in the range, a dash, and then the last character in the range. The regular expression includes any character that's within the specified character range, depending on the character set you defined when you set up your Raspberry Pi system.

Now you can simplify the zip code example by specifying a range of digits:

```
>>> re.search('^[0-9][0-9][0-9][0-9][0-9]$', '12345')
<_sre.SRE_Match object at 0x01570C98>
>>> re.search('^[0-9][0-9][0-9][0-9][0-9]$', '1234')
>>> re.search('^[0-9][0-9][0-9][0-9][0-9]$', '123456')
>>>
```

This saves a lot of typing! Each character class matches any digit from 0 to 9. The same tech-nique also works with letters:

```
>>> re.search('[c-h]at', 'The cat is sleeping')
<_sre.SRE_Match object at 0x0154FC28>
>>> re.search('[c-h]at', "I'm getting too fat")
<_sre.SRE_Match object at 0x01570C98>
>>> re.search('[c-h]at', 'He hit the ball with the bat')
>>>
```

The new pattern, [c-h]at, only matches words where the first letter is between the letter c and the letter h. In this case, the line with only the word at fails to match the pattern.

You can also specify multiple noncontinuous ranges in a single character class:

```
>>> re.search('[a-ch-m]at', 'The cat is sleeping')
<_sre.SRE_Match object at 0x0154FC28>
>>> re.search('[a-ch-m]at', 'He hit the ball with the bat')
```

```
<_sre.SRE_Match object at 0x01570CD0>
>>> re.search('[a-ch-m]at', "I'm getting too fat")
>>>
```

The character class allows the ranges a through c and h through m to appear before the at text. This range rejects any letters between d and g.

## The Asterisk

Placing an asterisk after a character signifies that the character may appear zero or more times in the text to match the pattern, as shown in this example:

```
>>> re.search('ie*k', 'ik')
<_sre.SRE_Match object at 0x0154FC28>
>>> re.search('ie*k', 'iek')
<_sre.SRE_Match object at 0x01570CD0>
>>> re.search('ie*k', 'ieek')
<_sre.SRE_Match object at 0x0154FC28>
>>> re.search('ie*k', 'ieeek')
<_sre.SRE_Match object at 0x01570CD0>
>>>
```

This pattern symbol is commonly used for handling words that have a common misspelling or variations in language spellings. For example, if you need to write a script that may be used by people speaking either American or British English, you could write this:

```
>>> re.search('colou*r', 'I bought a new color TV')
<_sre.SRE_Match object at 0x0154FC28>
>>> re.search('colou*r', 'I bought a new colour TV')
<_sre.SRE_Match object at 0x01570C98>
>>>
```

The u* in the pattern indicates that the letter u may or may not appear in the text to match the pattern.

Another handy feature is combining the dot special character with the asterisk special character. This combination provides a pattern to match any number of any characters. It's often used between two strings that may or may not appear next to each other in the text:

```
>>> re.search('regular.*expression', 'This is a regular pattern expression')
<_sre.SRE_Match object at 0x0154FC28>
>>>
```

By using this pattern, you can easily search for multiple words that may appear anywhere in the text.

# Using Advanced Regular Expressions Features

Because Python supports extended regular expressions, you have a few more tools available to you. The following sections show what they are.

## The Question Mark

The question mark is similar to the asterisk, but with a slight twist. The question mark indicates that the preceding character can appear zero times or once, but that's all. It doesn't match repeating occurrences of the character. In this example, if the e character doesn't appear in the text, or as long as it appears only once in the text, the pattern matches:

```
>>> re.search('be?t', 'bt')
<_sre.SRE_Match object at 0x01570CD0>
>>> re.search('be?t', 'bet')
<_sre.SRE_Match object at 0x0154FC28>
>>> re.search('be?t', 'beet')
>>>
```

## The Plus Sign

The plus sign is another pattern symbol that's similar to the asterisk, but with a different twist than the question mark. The plus sign indicates that the preceding character can appear one or more times, but it must be present at least once. The pattern doesn't match if the character is not present. In the following example, if the e character is not present, the pattern match fails:

```
>>> re.search('be+t', 'bt')
>>> re.search('be+t', 'bet')
<_sre.SRE_Match object at 0x01570C98>
>>> re.search('be+t', 'beet')
<_sre.SRE_Match object at 0x0154FC28>
>>> re.search('be+t', 'beeet')
<_sre.SRE_Match object at 0x01570C98>
>>>
```

## Using Braces

By using curly braces in Python regular expressions, you can specify a limit on a repeatable regular expression. This is often referred to as an *interval*. You can express the interval in two formats:

- ▶ {m}—The regular expression appears exactly m times.
- ▶ {m,n}—The regular expression appears at least m times but no more than n times.

This feature allows you to fine-tune how many times you allow a character (or character class) to appear in a pattern. In this example, the e character can appear once or twice for the pattern match to pass; otherwise, the pattern match fails:

```
>>> re.search('be{1,2}t', 'bt')
>>> re.search('be{1,2}t', 'bet')
<_sre.SRE_Match object at 0x0154FC28>
>>> re.search('be{1,2}t', 'beet')
<_sre.SRE_Match object at 0x01570C98>
>>> re.search('be{1,2}t', 'beeet')
>>>
```

## The Pipe Symbol

The pipe symbol allows you to specify two or more patterns that the regular expression engine uses in a logical OR formula when examining the data stream. If any of the patterns match the data stream text, the text passes. If none of the patterns match, the data stream text fails.

This is the syntax for using the pipe symbol:

*expr1|expr2|...*

Here's an example of this:

```
>>> re.search('cat|dog', 'The cat is sleeping')
<_sre.SRE_Match object at 0x0154FC28>
>>> re.search('cat|dog', 'The dog is sleeping')
<_sre.SRE_Match object at 0x01570C98>
>>> re.search('cat|dog', 'The horse is sleeping')
>>>
```

This example looks for the regular expression cat or dog in the data stream.

You can't place any spaces within the regular expressions and the pipe symbol, or they'll be added to the regular expression pattern.

## Grouping Expressions

Regular expression patterns can be grouped using parenthesis. When you group a regular expression pattern, the group is treated like a standard character. You can apply a special character to the group just as you would to a regular character. Here's an example:

```
>>> re.search('Sat(urday)?', 'Sat')
<_sre.SRE_Match object at 0x00B07960>
>>> re.search('Sat(urday)?', 'Saturday')
<_sre.SRE_Match object at 0x015567E0>
>>>
```

The grouping of the day ending along with the question mark allows the pattern to match either the full day name or the abbreviated name.

It's common to use grouping along with the pipe symbol to create groups of possible pattern matches, as shown here:

```
>>> re.search('(c|b)a(b|t)', 'cab')
<_sre.SRE_Match object at 0x015493C8>
>>> re.search('(c|b)a(b|t)', 'cat')
<_sre.SRE_Match object at 0x0157CCC8>
>>> re.search('(c|b)a(b|t)', 'bat')
<_sre.SRE_Match object at 0x015493C8>
>>> re.search('(c|b)a(b|t)', 'tab')
>>>
```

The pattern (c|b)a(b|t) matches any combination of the letters in the first group along with any combination of the letters in the second group.

# Working with Regular Expressions in Your Python Scripts

It helps to actually see regular expressions in use to get a feel for how to use them in your own Python scripts. Just looking at the quirky formats doesn't help much; seeing some examples of how regular expressions can match real data can help clear things up!

▼ TRY IT YOURSELF

### Use a Regular Expression

Follow these steps to implement a simple phone number validator script by using regular expressions:

1. Determine what regular expression pattern would match the data you're trying to look for. For phone numbers in the United States, there are four common ways to display a phone number:

   ▶ (123)456-7890

   ▶ (123) 456-7890

   ▶ 123-456-7890

   ▶ 123.456.7890

   This leaves four possibilities for how a customer can enter a phone number in a form. The regular expression must be robust enough to be able to handle any situation.

When building a regular expression, it's best to start on the left side and build the pattern to match the characters you might run into. In this example, there may or may not be a left parenthesis in the phone number. You can match this by using the following pattern:

```
^\(?
```

The caret indicates the beginning of the data. Since the left parenthesis is a special character, you must escape it to search for it as the character itself. The question mark indicates that the left parenthesis may or may not appear in the data to match.

Next comes the three-digit area code. In the United States, area codes start with the number 2 through 9. (No area codes start with the digits 0 or 1.) To match the area code, you use this pattern:

```
[2-9][0-9]{2}
```

This requires that the first character be a digit between 2 and 9, followed by any two digits. After the area code, the ending right parenthesis may or may not be there:

```
\)?
```

After the area code there can be a space, no space, a dash, or a dot. You can group these by using a character group along with the pipe symbol:

```
(| |-|\.)
```

The very first pipe symbol appears immediately after the left parenthesis to match the no-space condition. You must use the escape character for the dot; otherwise, it will take on its special meaning and match any character.

Next comes the three-digit phone exchange number, which doesn't require anything special:

```
[0-9]{3}
```

After the phone exchange number, you must again match either a space, a dash, or a dot:

```
( |-|\.)
```

Then to finish things off, you must match the four-digit local phone extension at the end of the string:

```
[0-9]{4}$
```

Putting the entire pattern together results in this:

```
^\(?[2-9][0-9]{2}\)?(| |-|\.)[0-9]{3}( |-|\.)[0-9]{4}$
```

2. Now that you have a regular expression, plug it into your code by opening a text editor and entering this code:

```
#!/usr/bin/python3

import re
pattern = re.compile(r'^\(?[2-9][0-9]{2}\)?( |-|\.)[0-9]{3}( |-|\.)[0-9]{4}$')

while(True):
    phone = input('Enter a phone number:')
    if (phone == 'exit'):
        break
    if (pattern.search(phone)):
        print('That is a valid phone number')
    else:
        print('Sorry, that is not a valid phone number')
print('Thanks for trying our program')
```

3. Save the file and exit the text editor.

4. Run the file from the Raspberry Pi command prompt or the LXTerminal program in your window:

```
pi@raspberrypi ~ $ python3 script1601.py
Enter a phone number:(555)555-1234
That is a valid phone number
Enter a phone number:333.123.4567
That is a valid phone number
Enter a phone number:1234567890
Sorry, that is not a valid phone number
Enter a phone number:exit
Thanks for trying our program
pi@raspberrypi ~ $
```

This is all there is to it! The script matches the input value against the regular expression pattern and displays the appropriate message.

# Summary

If you manipulate data in Python scripts, you need to become familiar with regular expressions. A regular expression defines a pattern template that's used to filter text in a string value. The pattern consists of a combination of standard text characters and special characters. The regular expression engine uses the special characters to match a series of one or more characters. Python uses the re module to provide a platform for using regular expressions in Python scripts.

You can use the `match()`, `search()`, `findall()`, and `finditer()` functions to filter text from string values in your Python scripts using regular expression patterns.

In the next hour, we'll take a look at how to use exceptions in your Python code. With exceptions, you can add code to your program to handle if things go wrong while the program is running!

# Q&A

**Q. Do regular expressions work in all language characters?**

**A.** Yes, because Python uses Unicode strings, you can use characters from any language in your regular expression patterns.

**Q. Is there a source for common regular expressions?**

**A.** The www.regular-expressions.info website contains lots of different expressions for matching all sorts of data!

**Q. Can I save a regular expression test to use in other programs?**

**A.** Yes, you can create a function (see Hour 12, "Creating Functions") that checks text using your regular expression. You can then copy the function into a module and use that in any program where you need to validate that type of data!

# Workshop

## Quiz

1. What regular expression character matches text at the end of a string?

   **a.** the caret (^)

   **b.** the dollar sign ($)

   **c.** the dot (.)

   **d.** the question mark (?)

2. The caret special character performs the same function in a regular expression as the `match()` Python function. True or false?

3. What regular expression pattern should you use to match both the words `Charlie` and `Charles`?

## Answers

**1.** b. The dollar sign ($) anchors the expression at the end of the string.

**2.** True. You may find it easier to use the `match()` Python function; however, there are plenty of standard regular expressions that use the caret. You can use either format to accomplish the same thing!

**3.** `'Charl[ie]+[es]+'` This regular expression will match if either the `"ie"` or `"es"` characters are at the end of the `"Charl"` string.

# Exception Handling

---

**What You'll Learn in This Hour:**

► What exceptions are
► How to handle exceptions
► How to handle multiple exceptions

In this hour, you will learn about exceptions and how to properly handle them in your Python scripts. To understand exceptions, you will look at the different types that can occur and the tools Python provides to manage them. Properly handling exceptions is a mark of an excellent Python script builder.

# Understanding Exceptions

An *exception* is an error that occurs when a Python script is being run or a command is issued within the Python interactive shell. You might hear people talk about "throwing an exception" or "an exception being raised." They're talking about Python issuing exceptions. There are two primary categories of error exceptions: syntactical and runtime.

## Syntactical Error Exceptions

You learned in Hour 3 that *syntax* refers to the Python commands, their proper order in a Python statement, and additional characters, such as quotation marks ("), that are needed to make a Python statement work properly. Before Python interprets a Python script, the interpreter checks that the syntax of each Python statement is correct. Whenever a Python statement has incorrect syntax, the interpreter generates a syntax error (that is, raises an exception). The exception is appropriately called `SyntaxError`.

In Listing 17.1, a Python `print` statement is missing its ending double quotes. This causes the Python interpreter to raise an exception.

**LISTING 17.1**   A print Function Syntax Error

```
>>> print ("I love my Raspberry Pi!)
  File "<stdin>", line 1
    print ("I love my Raspberry Pi!)
                                    ^
SyntaxError: EOL while scanning string literal
>>>
```

Notice the last line in the Listing 17.1. The word SyntaxError is used along with the helpful message EOL while scanning string literal. This message helps you determine what is wrong in the Python statement's syntax. EOL stands for "end of line." In other words, in this case, the Python interpreter found the end of the line and did not find the closing double quote for the print function.

The error generated in Listing 17.1 came from issuing a single syntactically incorrect statement in the Python interactive shell. A syntax error message looks slightly different when it is generated by a Python statement within a script. Listing 17.2 shows an example.

**LISTING 17.2**   A Syntax Error in a Script

```
pi@raspberrypi ~ $ python3 py3prog/my_errors.py
  File "py3prog/my_errors.py", line 17
    print ("I love my Raspberry Pi!)
                                    ^
SyntaxError: EOL while scanning string literal
pi@raspberrypi ~ $
```

The SyntaxError line of Listing 17.2 is identical to the SyntaxError line of Listing 17.1. However, the first line of the error message denotes the name of the script that has the syntax error as well as the line number in the script where it occurred. Having the line number is very helpful when you're tracking down syntax errors in your scripts!

BY THE WAY

**File Name**

When a syntax error occurs in the Python interactive shell, the interpreter tells you that the file is <stdin> and the line is line 1. This is because a script file is not being used. Instead, the shell is getting Python statements from you interactively.

There is a little trick you can use to help find a raised syntax error quickly in your script. Use the Linux shell cat -n command to display your script, as shown in Listing 17.3.

## LISTING 17.3   A Trick to Display Script Line Numbers

```
pi@raspberrypi ~ $ cat -n py3prog/my_errors.py
     1  # my_errors.py - Demonstrates various Python errors
     2  # Written by Blum and Bresnahan
     3  #
     4  ######################################################
     5  #
     6  ############# Initialize Variables #################
     7  #
     8  my_error = 0
     9  num_1 = 3
    10  num_2 = 4
    11  zero = 0
    12  #
    13  ############# Error Functions ######################
    14  #
    15  def missing_quote ():
    16      print ()
    17      print ("I love my Raspberry Pi!)
    18  #
...
```

Using the `cat -n` command causes Python to display line numbers on the screen for each line in your script. In Listing 17.3, you can easily find line 17, where the syntax error occurred. You can see that the `print` statement on that line does not have the closing double quote!

### DID YOU KNOW

## Jump to It

You can quickly jump to a line with a syntactical error by using the nano text editor. You use the syntax `nano +line_number file_name`. nano opens up the script and goes directly to the line number you specified. For example, to go to directly to line number 17 in the file `py3prog/my_errors.py`, enter `nano +17 py3prog/my_errors.py`.

In the IDLE 3 editor, you can quickly jump to a line with a syntactical error by pressing the key combination Alt+G. A little window opens, asking which line number to go to. Simply type in the line number and press the Enter key.

As you have learned, the Python interpreter finds syntactical errors when it reads each Python statement within a script and checks its syntax. If no errors are found, the Python interpreter translates the statements into something called *bytecode*. When the translation is complete, the bytecode is handed off to the Python virtual machine to run. At this point, a new type of error exception can be raised.

# Runtime Error Exceptions

A *runtime* error is raised when the Python script is running and an error occurs. Often such an error causes the script to halt immediately and produce a traceback. A *traceback* is an error message that, as its name says, traces back to the original runtime error.

---

BY THE WAY

### Illogical Runtime Errors

Runtime errors can come in a couple different flavors. One flavor is a *logic error*. A logic error may cause a script to halt, or it may just produce undesirable results. Logic errors are called *logic errors* because they are attributed to illogical thinking on the part of the script writer.

---

A classic runtime error is trying to divide a number by zero, which is mathematically incorrect. (You math wizards know that dividing by zero is better described as *undefined*.) In Listing 17.4, there are no problems in the Python interactive shell, when the number 3 is divided by the number 4 on lines 2 through 7.

**LISTING 17.4**    Divide-by-Zero Example

```
1: >>>
2: >>> num1=3
3: >>> num2=4
4: >>> zero=0
5: >>> result=num1/num2
6: >>> print (result)
7: 0.75
8: >>> result=num1/zero
9: Traceback (most recent call last):
10:    File "<stdin>", line 1, in <module>
11: ZeroDivisionError: division by zero
12: >>>
```

However, on line 8, when Python attempts to divide 3 by 0, it throws a runtime error and issues a traceback message. The traceback message on line 10 in Listing 17.4 shows that the error occurs in `File "<stdin>"`, which means it is happening in the Python interactive shell. The message on line 11 displays what causes the error exception to occur: `ZeroDivisionError`. As with the `SyntaxError` message, Python gives a little more help by also displaying the message `division by zero`.

As you would expect, a traceback message is a little different when it comes to a runtime error in a script. Listing 17.5 includes code that attempts to divide by zero.

**LISTING 17.5**   **A Runtime Error in a Script**

```
1: pi@raspberrypi ~ $ python3 py3prog/my_errors.py
2:
3: The Classic "Divide by Zero" error.
4:
5: Traceback (most recent call last):
6:   File "py3prog/my_errors.py", line 33, in <module>
7:     main()
8:   File "py3prog/my_errors.py", line 29, in main
9:     divide_zero ()
10:   File "py3prog/my_errors.py", line 23, in divide_zero
11:     my_error = num_1 / zero
12: ZeroDivisionError: division by zero
13: pi@raspberrypi ~ $
```

Remember that a traceback message literally traces back to the original runtime error. Thus, you can see on lines 6 through 11 of Listing 17.5 that the messages starts at the mainline function, main (lines 6 and 7 of Listing 17.5). It then progresses to the divide_zero function (lines 8 and 9 of Listing 17.5), which is called on line 29 of the script. Finally, it zeros in on the problem (lines 10 through 12 of Listing 17.5). As you can see, the culprit is on line 23 of the script, and it is a "division by zero" error.

The my_errors.py script is displayed in Listing 17.6. Sure enough, on line 23, you can see the code that incorrectly divides a number by zero.

**LISTING 17.6**   **The my_errors.py Script**

```
pi@raspberrypi ~ $ cat py3prog/my_errors.py
1: # my_errors.py - Demonstrates various Python errors
2: # Written by Blum and Bresnahan
3: #
4: ######################################################
5: #
6: ############# Initialize Variables #################
7: #
8: my_error = 0
9: num_1 = 3
10: num_2 = 4
11: zero = 0
12: #
13: ############# Error Functions #######################
14: #
15: #def missing_quote ():
16: #     print ()
17: #     print ("I love my Raspberry Pi!)
18: #
```

```
19: def divide_zero ():
20:     print ()
21:     print ("The Classic \"Divide by Zero\" error.")
22:     print ()
23:     my_error = num_1 / zero
24: #
25: ############## Mainline ############################
...
pi@raspberrypi ~ $
```

Understanding error exceptions is important for handling them within a Python script. Now that you have an understanding of exceptions, the next step is to learn how to properly handle them.

# Handling Exceptions

As you have seen, when runtime errors are encountered in a script, the script stops, throws an exception, and produces a traceback message. This is pretty sloppy and could certainly terrify an unsuspecting script user. However, you can handle runtime errors with good form and keep your script users happy. Python provides an exception handler via the `try except` statement.

The basic syntax of a `try except` statement is as follows:

```
try:
    python statements
except exception_name:
    python statements to handle exception
```

Notice that indentation is used with the `try except` statement to indicate which Python statements belong together. The Python statements within the `try` statement area are part of the `try` statement block. Python statements within the `except` statement area are part of the `except` statement block. Together, the two blocks are called the `try except` statement block.

Using the example of the divide–by-zero exception, Listing 17.7 shows a script that could raise this exception, along with a `try except` statement to properly handle it. The `try except` statement block starts on line 15 and ends on line 22. Any Python statement that may raise an exception should be put in a `try` statement block (Listing 17.7 lines 15 through 17) in order to properly handle that exception. This is done in `script1701.py`.

**LISTING 17.7**   A `try except` Statement Block

```
1: pi@raspberrypi ~ $ cat py3prog/script1701.py
2: # script1701 - Properly handle Divide by Zero Exception
3: # Written by Blum and Bresnahan
4: #
5: ##################################################
```

```
6: #
7: ################ Functions #######################
8: #
9: def divide_it ():
10:        print ()
11:        number=int(input("Please enter number to divide: "))
12:        print ()
13:        divisor=int(input("Please enter the divisor: "))
14:        print ()
15:        try:
16:            result = number / divisor
17:            print ("The result is:", result)
18: #
19:        except ZeroDivisionError:
20:            print ("You cannot divide a number by zero.")
21:            print ("Script terminating....")
22:            print ()
23:#
24:############# Mainline ############################
25:#
26:def main ():
27:     divide_it ()
28:#
29:########### Call the Main Function #################
30:#
31:main()
32: pi@raspberrypi ~ $
```

Notice that the script in Listing 17.7 has `input` statements on lines 11 and 13. The script asks the script user for the number to be divided and its divisor. Because a script user could enter the number 0 here for the divisor, the math statement using the input is included in the `try` statement block.

The idea is to allow the script to continue working as normal as long as no exceptions are raised. If an exception is raised, Python statements in the `except` statement block handle it. In Listing 17.8, on lines 1 through 7, the script runs fine. The script user has input appropriate data, and thus no exceptions are raised.

**LISTING 17.8**    Execution of Script `script1701.py`

```
1: pi@raspberrypi ~ $ python3 py3prog/script1701.py
2:
3: Please enter number to divide: 3
4:
5: Please enter the divisor: 4
6:
```

```
7: The result is: 0.75
8: pi@raspberrypi ~ $
9: pi@raspberrypi ~ $ python3 py3prog/script1701.py
10:
11: Please enter number to divide: 3
12:
13: Please enter the divisor: 0
14:
15: You cannot divide a number by zero.
16: Script terminating....
17:
18: pi@raspberrypi ~ $
```

On the second run of the script in Listing 17.8, the user puts 0 as the divisor on line 13. This raises an exception. However, the script is not abruptly halted, nor is a scary traceback message displayed. Instead, because the exception occurs within the `try except` statement block, the exception is "caught," and the Python statements within the `except` statement block run, as shown on lines 15 and 16. This is considered good form for handling raised exceptions within scripts.

# Handling Multiple Exceptions

Often you need to be able to "catch" more than one type of exception for a group of Python statements. For example, using the `script1701.py` script, if the user enters a word instead of a number, Listing 17.9 shows what happens.

**LISTING 17.9**   Additional Exceptions Not Handled

```
pi@raspberrypi ~ $ python3 py3prog/script1701.py

Please enter number to divide: 3

Please enter the divisor: four
Traceback (most recent call last):
  File "py3prog/script1701.py", line 30, in <module>
    main()
  File "py3prog/script1701.py", line 26, in main
    divide_it ()
  File "py3prog/script1701.py", line 12, in divide_it
    divisor=int(input("Please enter the divisor: "))
ValueError: invalid literal for int() with base 10: 'four'
```

Even though the script can handle a ZeroDivisionError exception, it has not been written to handle a ValueError exception. Multiple exceptions can be handled within a `try except` statement block for cases such as this. `script1701.py` was modified to include such a block and was renamed `script1702.py`. The new script is shown in Listing 17.10.

## LISTING 17.10    Handling Multiple Exceptions

```
1: pi@raspberrypi ~ $ cat py3prog/script1702.py
2: # script1702 - Properly handle Division Errors
3: # Written by Blum and Bresnahan
4: #
5: ####################################################
6: #
7: ################# Functions #######################
8: #
9: def divide_it ():
10:     print ()
11: #
12:     try:
13:         # Get numbers to divide
14:         number=int(input("Please enter number to divide: "))
15:         print ()
16:         divisor=int(input("Please enter the divisor: "))
17:         print ()
18:         #
19:         # Divide the numbers
20:         result = number / divisor
21:         print ("The result is:", result)
22:#
23:     except ZeroDivisionError:
24:         print ("You cannot divide a number by zero.")
25:         print ("Script terminating....")
26:         print ()
27:#
28:     except ValueError:
29:         print ("Numbers entered must be digits.")
30:         print ("Script terminating....")
31:         print ()
32:#
33:############# Mainline ###########################
34:#
35:def main ():
36:     divide_it ()
37:#
38:########### Call the Main Function ################
39:#
40: main()
41: pi@raspberrypi ~ $
```

In Listing 17.10, notice that the `input` statements were moved from outside the `try except` statement block to inside it on lines 14 and 16. This is done because the `ValueError` exception can occur when the user is entering input to these statements. The Python statements that can throw an exception must be within the `try` statement block in order for the exceptions to be handled by the `try except` block. Now both `ValueError` and `ZeroDivisionError` exceptions raised by Python statements within the `try` statement block can be properly handled.

Listing 17.11 shows a user attempting input three different times on `script1702.py`.

**LISTING 17.11**    Execution of `script1702.py`

```
 1: pi@raspberrypi ~ $ python3 py3prog/script1702.py
 2:
 3: Please enter number to divide: 3
 4:
 5: Please enter the divisor: 4
 6:
 7: The result is: 0.75
 8: pi@raspberrypi ~ $
 9: pi@raspberrypi ~ $ python3 py3prog/script1702.py
10:
11: Please enter number to divide: 3
12:
13: Please enter the divisor: four
14: Numbers entered must be digits.
15: Script terminating....
16:
17: pi@raspberrypi ~ $
18: pi@raspberrypi ~ $ python3 py3prog/script1702.py
19:
20: Please enter number to divide: 3
21:
22: Please enter the divisor: 0
23:
24: You cannot divide a number by zero.
25: Script terminating....
26:
27: pi@raspberrypi ~ $
```

In Listing 17.11, the script user makes an attempt with no problems on lines 1 through 7. Next, the script user accidently enters the word `four` instead of the number 4 on line 13. The script captures the raised exception and produces a user-friendly message instead of an ugly traceback. Also, notice on lines 18 through 25 in Listing 17.11 that the `ZeroDivisionError` exception is still properly handled.

# Creating Multiple `try` `except` **Statement Blocks**

You can use multiple `try` `except` statement blocks throughout your Python scripts. In fact, it is good form to put the blocks specifically around the Python statements that need them.

For example, looking back at the script `script1702.py` in Listing 17.10, you can see that the `ZeroDivisionError` exception was potentially raised by the statement `result = number / divisor`. The `ValueError` exception could be raised by the `input` statements. Therefore, good form dictates that those statements should be in their own `try` `except` statement blocks.

`script1702.py` has been revised and is now called `script1703.py`, as shown in Listing 17.12. This revised script properly divides the `try` `except` statement blocks for the Python statements.

**LISTING 17.12**   Multiple `try` `except` Statement Blocks

```
1: pi@raspberrypi ~ $ cat py3prog/script1703.py
2: # script1703 - User Determined Division
3: # Written by Blum and Bresnahan
4: #
5: ####################################################
6: #
7: ################# Functions #######################
8: #
9: def divide_it ():
10:      print ()
11: #
12:      try:
13:          # Get numbers to divide
14:          number=int(input("Please enter number to divide: "))
15:          print ()
16:          divisor=int(input("Please enter the divisor: "))
17:          print ()
18:          #
19:      except ValueError:
20:          print ("Numbers entered must be digits.")
21:          print ("Script terminating....")
22:          print ()
23:          exit ()
24:          #
25:      except KeyboardInterrupt:
26:          print ()
27:          print ("Script terminating....")
28:          print ()
29:          exit ()
30: #
31:      try:
32:          # Divide the numbers
33:          result = number / divisor
```

```
34:            print ("The result is:", result)
35: #
36:        except ZeroDivisionError:
37:            print ("You cannot divide a number by zero.")
38:            print ("Script terminating....")
39:            print ()
40:            exit ()
41:            #
42: #
43: ############# Mainline ###########################
44: #
45: ...
46: pi@raspberrypi ~ $
```

Notice in Listing 17.12 that an additional exception has been added to the try except statement block for the Python input statements in lines 25 through 29. This additional exception
has been added for catching keyboard interrupts, such as the user pressing Ctrl+C during data
input.

---

DID YOU KNOW

---

### Exception Groups

Exceptions belong to named exception groups. These exception group names can be used in
the except statement. For example, both ZeroDivisionError and FloatingPointError
exceptions belong to the ArithmeticError base group. An except statement block could use
ArithmeticError as its named exception, like this:

```
except ArithmeticError:
```

However, when an exception is raised, Python looks at the except statements within the try
except statement block in the order in which they are listed in the block. If you have named
except statement blocks for both ArithmeticError and ZeroDivisonError, and a
ZeroDivisionError is raised, the block that comes first is executed.

---

Notice that exit function statements are added in lines 23, 29, and 40 of the script in Listing
17.12 . These statements are needed because the script does not stop executing when a raised
exception is caught by a try except statement block. Specifically, lines 23 and 29 need exit
statements because if the data input is interrupted by a Ctrl+C, the raised exception is caught,
but all the data needed by the division statement on line 33 will not have been entered.

---

BY THE WAY

## Any Python Statements

You can put just about anything within exception statement blocks. For example, instead of issuing an `exit` statement as shown in the preceding examples, you could issue a `return` statement to leave the current function but not exit the Python script.

---

In Listing 17.13, four tests are conducted on `script1703.py` to see if it can properly handle data and a few potential exceptions. The first test simply makes sure the script can properly handle good data.

**LISTING 17.13**   Execution of `script1703.py`

```
pi@raspberrypi ~ $ python3 py3prog/script1703.py

Please enter number to divide: 3

Please enter the divisor: 4

The result is: 0.75
pi@raspberrypi ~ $ python3 py3prog/script1703.py

Please enter number to divide: 3

Please enter the divisor: four
Numbers entered must be digits.
Script terminating....

pi@raspberrypi ~ $ python3 py3prog/script1703.py

Please enter number to divide: 3

Please enter the divisor: 0

You cannot divide a number by zero.
Script terminating....

pi@raspberrypi ~ $ python3 py3prog/script1703.py

Please enter number to divide: 3

Please enter the divisor: ^C
Script terminating....

pi@raspberrypi ~ $
```

The final three tests on `script1704.py` in Listing 17.13 test the newly separated `try except` statement blocks of the script. Notice that the last test uses Ctrl+C. The keyboard interrupt exception is handled gracefully. If this exception were not trapped, it would produce a very long and ugly traceback message.

# Handling Generic Exceptions

So far, you have seen anticipated exceptions being handled. However, few people can determine all the possible error exceptions that may be raised. Fortunately, Python allows a generic exception for unanticipated events.

The syntax for generic exceptions is not too different from that of regular exceptions. You simply leave off the exception name from the `except` statement, as shown in Listing 17.14.

**LISTING 17.14**   A Generic Execution Statement

```
pi@raspberrypi ~ $ cat py3prog/script1703.py
...
    except:
        print ()
        print ("An error has occurred.")
        print ("Script terminating...")
        print ()
        exit ()
...
```

In Listing 17.14, the `script1703.py` has its first `try except` statement block modified to include a generic exception statement. Notice that the only syntax difference between the generic exception statement and the others is that no exception name is listed after `except`. Now if any unforeseen exceptions are raised, the script can handle them in good form.

# Understanding `try except` Statement Options

Several optional items you can use within your `try except` statement blocks provide a great deal of flexibility. These are the three primary options:

- ▶ `else` statement block
- ▶ `finally` statement block
- ▶ `as` variable statement

An optional else statement block follows an except statement block and also contains Python statements. However, these Python statements are executed only if no exceptions were raised by Python statements within a try statement block. The following is an example of an else statement block:

```
else:
    print ()
    print ("Data entered successfully")
```

The optional finally statement is located at the very end of a try except statement block. It also must follow any included else statement blocks. The Python statements within the finally statement block are executed whether an exception was raised or not. The following is an example of a finally statement block:

```
finally:
    print ()
    print ("Script completed.")
```

The as variable statement is not a statement block like the others. Instead, it allows you to capture the exact error message that is issued, when the exception is raised. The beauty here is that no matter which exception is raised, you have the error message. Often script writers write the message to a log file for later review and display a very user-friendly style message to the script user instead. The following is an example of using an as variable statement:

```
except ValueError as input_error:
    print ("Numbers entered must be digits.")
    print ("Script terminating....")
    print ()
    error_log_file = open ('/home/pi/data/error.log', 'a')
    error_log_file.write(input_error)
    error_log_file.close()
    exit ()
```

The variable name here is input_error. If a ValueError exception is raised in this try except statement block, the exact error message is loaded into the input_error variable. A Python statement within the block, error_log_file.write(input_error), then writes that error message out to an error log file.

These three options give a great deal of flexibility in handling exceptions. Now it is time to stop reading about handling exceptions and try to handle a few yourself.

▼ TRY IT YOURSELF

**Explore Python** `try except` **Statement Blocks**

In the following steps, you will explore Python `try except` statement blocks by creating a script that opens a file. At first, the script will produce a traceback message. You will add `try except` statements to handle the raised exceptions using good form. Here's what you do:

1. If you have not already done so, power up your Raspberry Pi and log in to the system.

2. If you do not have the LXDE GUI started automatically at boot, start it now by typing `startx` and pressing Enter.

3. Open a script editor, either nano or the IDLE 3 text editor, to create the script `py3prog/script1704.py`.

4. Type all the information from `script1704.py` shown below. Take your time and avoid any typographical errors:

```
# script1704 - Open a File
# Written by
#
######################################################
#
################# Functions #######################
#
def get_file_name ():    #Get file name
    print ()
#
    try:
        file_name=input("Please enter file to open: ")
        print ()
        return file_name
        #
    except KeyboardInterrupt:
        print ()
        print ("Script terminating....")
        print ()
        exit ()
        #
    except:
        print ()
        print ("An error has occurred.")
        print ("Script terminating...")
        print ()
        exit ()
        #
    #
```

```
def open_it (file_name):        #Open file name

    my_file=open(file_name,'r')
    print ("File", file_name, "opened successfully!")
    my_file.close()
#
############## Mainline #############################
#
def main ():
    file_name = get_file_name ()
    open_it (file_name)
#
############# Call the Main Function ###################
#
main()
```

Notice that the only `try except` statement block is for entering information into the script in the `get_file_name` function.

**5.** Save the editor contents and exit the editor.

**6.** Before you test the script, you need to create a little file for the script to open, so at the command-line prompt, type `echo "I love my Raspberry Pi" >> testfile` and press Enter.

**7.** Test your script by typing `python3 py3prog/script1704.py` and pressing Enter. At the `Please enter file to open:` prompt, type `testfile` and press Enter. This should complete successfully, and you should receive the message `File testfile opened successfully!`, as shown below.

```
pi@raspberrypi ~ $ python3 py3prog/script1704.py

Please enter file to open: testfile

File testfile opened successfully!
pi@raspberrypi ~ $
```

**8.** Now test your script again by typing `python3 py3prog/script1704.py` and pressing Enter. At the `Please enter file to open:` prompt, type `nofile` and press Enter. This should cause the script to abruptly halt (assuming that you do not have a file called `nofile`), and you should get an ugly traceback message similar to what is shown below:

```
pi@raspberrypi ~ $ python3 py3prog/script1704.py

Plvease enter file to open: nofile

Traceback (most recent call last):
  File "py3prog/script1704.py", line 46, in <module>
```

```
    main()
  File "py3prog/script1704.py", line 42, in main
    open_it (file_name)
  File "py3prog/script1704.py", line 32, in open_it
    my_file=open(file_name,'r')
IOError: [Errno 2] No such file or directory: 'nofile'
pi@raspberrypi ~ $
```

Obviously, you need to do more to get script1704.py into shape.

9. Open `script1704.py` in a script editor again. Modify the `open_it` function so that it looks as shown below:

```
def open_it (file_name):          #Open file name
#
    try:
        my_file=open(file_name,'r')
        print ("File", file_name, "opened successfully!")
        my_file.close()
        #
    except Exception as open_error:
        print ("An error exception has been raised.")
        print ("The error message is:")
        print (open_error)
        print ()
        return ()
#
```

10. Notice that the `except` statement uses `Exception` as the name of the exception. This is the overall base group for exceptions. In order to catch the exception using the `as` variable statement into the variable `open_error`, an exception must be named. Using the overall base group `Exception` means that all exceptions will be caught. Save the editor contents to a file and exit the editor.

11. Test your modified script by typing `python3 py3prog/script1704.py` and pressing Enter. At the `Please enter file to open:` prompt, type `nofile` and press Enter. As shown below, your new except statement block should catch the raised exception and display the information desired:

```
pi@raspberrypi ~ $ python3 py3prog/script1704.py

Please enter file to open: nofile

An error exception has been raised.
The error message is:
[Errno 2] No such file or directory: 'nofile'
```

```
pi@raspberrypi ~ $
```

**12.** To clean up `script1704.py` a little more (and give you more experience!), open the script in your favorite script editor again. Modify the `open_it` function so that it looks as shown below:

```
def open_it (file_name):          #Open file name

    try:
        my_file=open(file_name,'r')
        print ("File", file_name, "opened successfully!")
        my_file.close()
#
    except IOError:
        print ("File", file_name, "not found")
        print ("Script terminating...")
        return
        #
    except Exception as open_error:
        print ("An error exception has been raised.")
        print ("The error message is:")
        print (open_error)
        print ()
        return
        #
```

**13.** Notice that the change you made was to add an additional `except` statement block to the function. The additional statement specifically catches any `IOError` exceptions. Save the editor contents to a file and exit the editor.

**14.** Test your modified script by typing `python3 py3prog/script1704.py` and pressing Enter. At the `Please enter file to open:` prompt, type `nofile` and press Enter. The `except` statement block should catch the raised exception and display the information desired, as shown below:

```
pi@raspberrypi ~ $ python3 py3prog/script1704.py

Please enter file to open: nofile

File nofile not found
Script terminating...
pi@raspberrypi ~ $
```

**15.** Open the `script1704.py` script in a script editor again. This time you will be adding the optional `else` and `finally` statements. Modify the `open_it` function so that it looks as shown below:

```
def open_it (file_name):        #Open file name

    try:
        my_file=open(file_name,'r')
#
    except IOError:
        print ("File", file_name, "not found")
        print ("Script terminating...")
        #
    except Exception as open_error:
        print ("An error exception has been raised.")
        print ("The error message is:")
        print (open_error)
        print ()
        #
    else:
        print ("File", file_name, "opened successfully!")
        my_file.close()
        #
    finally:
        return
#
```

**16.** Remember that the `else` statement block is executed only if no exceptions are raised. The `finally` statement block is run whether an exception is raised or not. Save the editor contents to a file and exit the editor.

**17.** Now test your modified script by typing `python3 py3prog/script1704.py` and pressing Enter. At the `Please enter file to open:` prompt, type `nofile` and press Enter. You should see a message similar to what is shown below:

```
pi@raspberrypi ~ $ python3 py3prog/script1704py

Please enter file to open: nofile

File nofile not found
Script terminating...
pi@raspberrypi ~ $
```

Good job! Hopefully you can see the benefits of properly and gracefully handling exceptions.

If you want a little more hands-on experience, go back through the previous hours and look at traceback messages that are displayed in the various listings. What kind of `try except` statement blocks would you write for each one?

# Summary

In this hour, you explored how to handle error exceptions with class. By using `try except` statement blocks, you learned how to eliminate ugly traceback messages and provide a script user with clean and user-friendly error messages. You got to play around with `try except` statement blocks and some of their options in a script.

In Hour 18, "GUI Programming," you will take a major step forward in your Python adventure: You will be learning about GUI programming!

# Q&A

**Q.** I don't know the exact name of an exception that may be raised in my script. Where can I get help?

**A.** If you have a general idea about what kind of error exception may be raised from a Python statement but don't have its exact name to use in a `try except` statement, you can go to docs.python.org/3/library/exceptions.html for a list of Python exception names and a brief description of each one. Also, exception groupings are shown at that site.

**Q.** Can you wrap your mainline function within a `try except` statement?

**A.** Yes, but that would not be considered good form. It's best to keep only the statements that may raise a particular exception with each `try except` statement block.

**Q.** I've heard I can raise my own exceptions. Is that true?

**A.** Yes, it is true. By using the `raise` Python statement, you can raise exceptions to change the flow of a script. These exceptions can be built in or custom made. See docs.python. org/3/tutorial/errors.html for more information on this topic.

# Workshop

## Quiz

**1.** Multiple exceptions can be handled within a `try except` statement block. True or false?

**2.** A syntax error generates a `SyntaxError` message. A runtime error raises a(n) _____ and produces a(n) _____ message.

3. A(n) _____ statement block is executed when exceptions are raised and when they are not raised, and a(n) _____ statement block is executed only when no exceptions are raised.

   **a.** `try; exempt`

   **b.** `finally; else`

   **c.** `else; finally`

## Answers

1. True. Multiple exceptions can be handled within a `try except` statement block, though it is often desirable for each exception to have its own `except` statement block.

2. exception; traceback

3. Answer b is correct. A `finally` statement block is executed when exceptions are raised and when they are not raised, and an `else` statement block is executed only when no exceptions are raised.

# PART IV

# Graphical Programming

# HOUR 18
# GUI Programming

## What You'll Learn in This Hour:

▶ The basics of GUI programming

▶ GUI Python libraries

▶ Exploring the `tkinter` package

▶ How to use GUI programming in Python

When you hear *Python scripting*, the first thing that probably comes to mind is boring command-line scripts. This doesn't have to be the case, though, if you plan on running your Python scripts in a graphical environment—such as the Raspberry Pi. There are plenty of ways to interact with your Python script other than the `input` and `print` statements! In this hour, you'll learn how to add graphical interfaces to your Python scripts to make them look more like Windows programs.

## Programming for a GUI Environment

These days, just about every operating system incorporates some type of graphical user interface (GUI) to allow users to input data and view results. This is true of Linux, which, you'll remember, is the operating system on the Raspberry Pi. While there are several different graphical desktop environments in the Linux world, the Raspbian distribution used on the Raspberry Pi uses the LXDE desktop package to provide a graphical desktop interface for users.

You can leverage the graphical desktop environment of the LXDE package with your Python scripts to create a fancy window-oriented interface for your programs that will help give your scripts a more professional look and feel.

Before we dive into the coding, though, it's a good idea to first go through all the terms used in GUI programming. If you're brand new to the GUI programming world, there may be some things that you've seen and used but never knew actually have names. The following sections walk through some of the terminology and features of a GUI environment that you'll need to become familiar with when coding your Python scripts.

## The Window Interface

When you make the move to GUI programming, you need to learn a new set of terms. For starters, the main area in a window is called the *frame*. The frame contains all the objects the program uses to interact with the user, and it is the central point in a GUI program.

The frame is composed of objects called *widgets* (short for *window gadgets*) that display and retrieve information. Most graphical programming languages provide a library of widgets for you to use in your programs. While not an official standard, there's a common set of widgets available in just about every graphical programming environment. Table 18.1 lists the widgets you'll run into in your Python GUI programming.

**TABLE 18.1**  Window Widgets

| Widget | Description |
| --- | --- |
| Frame | Provides the overall widow area for placing other widgets in the window. |
| Label | Places text in the window area. |
| Button | Triggers an event in the window when clicked. |
| Checkbutton | Allows the user to select or deselect an item. |
| Entry | Provides an area to enter or display a single line of text. |
| Listbox | Displays multiple values to select from. |
| Menu | Creates the menu toolbar at the top of the window. |
| Progressbar | Indicates that something is happening in the background. |
| Radiobutton | Allows the user to select one item from a group of options. |
| Scrollbar | Controls the view in a list box or frame. |
| Separator | Places a horizontal or vertical bar in the window. |
| Spinbox | Allows the user to select a value from a range of numbers. |
| Text | Provides an area to enter or display multiple lines of text. |

Each widget has its own set of properties that define how it appears in the program window and how to handle any data or actions that occur while the user interacts with the window.

## Event-Driven Programming

Programming for a GUI environment is a bit different from command-line programming in the way that Python handles the program code. In a command-line program, the order of the program code controls what happens next. For example, the program prompts the user for input, processes the input, and then displays the results on the command line, based on the input. The program user can only respond to input requests from the program.

In contrast, a GUI program displays an entire set of interaction widgets all at once, all in the same window. The program user gets to decide which widget gets processed next. Because code doesn't know which widget the user will activate at any given time, it has to use a feature called *event-driven programming* to process code. In event-driven programming, Python calls different methods within the program, based on what event (or action) happens in the GUI window. There isn't a set flow to the program code; it's just a bunch of methods that individually run in response to an event.

For example, your user can enter data into a text widget, but nothing happens until the user presses a button in the program window to submit the text. The button triggers an event, and your program code must detect that event and then run the code method to read the text in the text field and process it.

The key to event-driven programming is linking widgets in the window to events and then linking the events to the code modules in the program. An *event handler* is in charge of this process.

For the program to work, you must create separate modules that Python calls when it receives an event from each widget. The event handlers do the bulk of the work in GUI programs. They retrieve the data from the widgets, process the data, and then display the results in the window, using other widgets. This may seem a bit cumbersome at first, but once you get used to coding with event handles, you'll see how easy it is to work in a GUI environment.

# Examining Python GUI Packages

A lot of people have worked hard to simplify GUI programming in the Python environment. Standard library packages help you create GUI widgets from your Python scripts and build your graphical programs. Table 18.2 describes the most popular GUI packages used in the Linux world.

**TABLE 18.2  Popular Linux GUI Packages**

| Package | Description |
| --- | --- |
| tkinter | Uses the TK graphical library and is the default GUI library that is included as part of the standard Python library suite. |
| PyGTK | Uses the GTK+ graphical library, which is used in the GNOME desktop environment. |
| PyQT | Uses the QT graphical library, which is used in the KDE desktop environment. |
| wxPython | Uses the wxWidgets library, which is a multiplatform graphical environment. |

The `tkinter` package is one of the older graphical packages used in Python, and it's therefore one of the most popular packages. Since Python includes the `tkinter` package by default, it's commonly used to create graphical Python programs on the Raspberry Pi, and we use it in this hour.

# Using the `tkinter` Package

Since the Raspberry Pi Python libraries include the `tkinter` package by default, we use it to demonstrate creating GUI programs in Python scripts. Once you become familiar with how one graphical library package works, it's not too difficult to use any of the others.

You need to follow three basic steps to create a GUI application using the `tkinter` package:

1. Create a window.
2. Add widgets to the window.
3. Define the event handlers for the widgets.

The following sections walk through each of these steps to show how you would build a GUI application using `tkinter` in your Python scripts.

## Creating a Window

In a GUI environment, everything revolves around a window. The first step to creating a GUI program is to create the main window for your application, called the *root window*.

You do that by creating a Tk object, which controls all the aspects of your window. To create a Tk object, you first need to import the `tkinter` library, and then you instantiate a Tk object, like this:

```
from tkinter import *
root = Tk()
```

This creates a main window object and assigns it to the variable named `root`. However, this default window does not have any size, title, or features.

You next need to run a few default Tk object methods for the window to set up some of the window features. Two common methods are the `title()` method to set a title for the window, which will appear in the title bar at the top of the window, and the `geometry()` method, which sets the size of the window. Here's how you use them:

```
root.title('This is a test window')
root.geometry('300x100')
```

After you set these methods, you need to use the `mainloop()` method, which puts the window into a loop, waiting for a window widget to trigger an event. As events occur in the window,

Python intercepts them and passes them to your program code. For example, if you click the X at the top-right corner of the window, Python captures that event and knows to close out the window. (Later on, you'll code your own events to add to the window.) Listing 18.1 shows the `tkinter` window code to create a simple window.

**LISTING 18.1**  The `script1801.py` Code

```
#!/usr/bin/python3
from tkinter import *
root= Tk()
root.title('This is a test window')
root.geometry('300x100')
root.mainloop()
```

To run the `script1801.py` code, you need to be in the LXDE graphical desktop on the Raspberry Pi. Once you're in the desktop, you can start the script from the command line by opening the LXTerminal utility and then running the code from the command prompt, like this:

```
pi@raspberrypi ~$ python3 script1801.py
```

You don't see anything happen at the command prompt, but you should see a simple window object appear on your desktop, as shown in Figure 18.1.

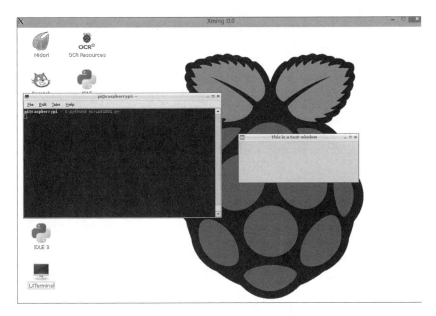

**FIGURE 18.1**
A default `Tk` window with no widgets.

Congratulations! You've just written a Python GUI program! There aren't any widgets to interact with, so to close out the window, you have to click the X in the upper-right corner of the window. Next you will start adding some widgets to your window to make things happen.

# Adding Widgets to the Window

After you've created the root window, you're ready to start working on the widgets for your interface. There are three steps involved with adding widgets to a window:

**1.** Create a frame template in the root window.

**2.** Define a positioning method to use for placing widgets in the frame.

**3.** Place the widgets in the frame, using the positioning method you've chosen.

The following sections walk through these steps.

### Creating a Frame Template

The first step in the process of adding widgets to your window is to create a template for the window widget layout. The `tkinter` package uses the `Frame` object to create an area for you to place widgets in the window. However, you don't use the `Frame` object directly in your window code; instead, you must create a child class to define all the window methods and attributes, based on the `Frame` class. (For more info on child classes, see Hour 15, "Employing Inheritance"). While you can call your `Frame` object child class anything you want, the most popular name for this class is `Application`, as shown here:

```
class Application(Frame):
```

After you create the child class, you need to create a constructor for it. Remember from Hour 14, "Exploring the World of Object-Oriented Programming," that you define a constructor by using the `__init__()` method. This method uses the keyword `self` as the first parameter, and it takes the root `Tk` window object that you created as the second parameter. This is what links the `Frame` object to the window.

You now have a basic template that you can use to create a window `Frame` class:

```
class Application(Frame):
    """My window application"""

    def __init__(self, master):
        super(Application, self).__init__(master)
        self.grid()
```

The class definition to create the window and frame isn't very long, but it is somewhat complicated. The constructor that you built for the `Application` class contains two statements.

The super() statement imports the constructor method from the parent Frame class for the Application class, passing the root window object. The last statement in the constructor defines the positioning method used for the frame. This example uses the tkinter grid() method. (You'll learn more about that feature in the next section.)

Now that you have your Application class template, you can use it to create a window. Listing 18.2 shows the script1802.py file, which is a basic code template that you'll use to create a window.

**LISTING 18.2    The script1802.py File**

```
#!/usr/bin/python3

from tkinter import *

class Application(Frame):
    """Build the basic window frame template"""

    def __init__(self, master):
        super(Application, self).__init__(master)
        self.grid()

root = Tk()
root.title('Test Application window')
root.geometry('300x100')
app = Application(root)
app.mainloop()
```

When you run the script1802.py program, you might notice that it looks just like the window you created using the bare Tk object, shown in Figure 18.1. The difference is that now the window has a frame, so you can start adding widgets to the Application object to fill in the window. The script1802.py code shows the basic template that you'll use for most of your Python GUI programs.

## Positioning Widgets

The key to a user-friendly GUI application is the placement of the widgets in the window area. Too many widgets grouped together can make the user interface confusing.

In the example in the previous section, you used the grid() method to position widgets in the frame. The tkinter package provides three ways to position widgets in the window:

▶ Using a grid system

▶ Packing widgets into available places

▶ Using positional values

The last method, using positional values, requires that you define the precise location of each widget, using X and Y coordinates within the window. While this provides the most accurate control over where your widgets appear, it can be somewhat difficult to work with when you're first starting out.

The packing method pretty much does what it says: It attempts to pack widgets into a window as best it can in the space available. When you choose this method, Python places the widgets in the window for you, starting at the top left and moving along to the next available space, either to the right or below the previous widget. The packing method works fine for small windows with just a few widgets, but if you have a larger window, things can quickly get cluttered and out of alignment.

The compromise between the positional method and the packing method is the grid method. The grid method creates a grid system in the window, using rows and columns, somewhat like a spreadsheet. You place each widget in the window at a specific row and column location. You can define a widget to span multiple rows or columns, so you have some flexibility in how the widgets appear.

The `grid()` method defines three parameters for placing the widget in the window:

```
object.grid(row = x, column = y, sticky = n)
```

The `row` and `column` values refer to the cell location in the layout, starting with row 0 and column 0 as the top-left cell in the window. The `sticky` parameter tells Python how to align the widget inside the cell. There are nine possible sticky values:

- ▶ `N`—Places the widget at the top of the cell.
- ▶ `S`—Places the widget at the bottom of the cell.
- ▶ `E`—Right-aligns the widget in the cell.
- ▶ `W`—Left-aligns the widget in the cell.
- ▶ `NE`—Places the widget at the top-right corner of the cell.
- ▶ `NW`—Places the widget at the top-left corner of the cell.
- ▶ `SE`—Places the widget at the bottom-right corner of the cell.
- ▶ `SW`—Places the widget at the bottom-left corner of the cell.
- ▶ `CENTER`—Centers the widget in the cell.

In this hour, you'll use the grid method of positioning the widgets.

## Defining Widgets

Now that you have a `Frame` object and a positioning method, you're ready to start placing some widgets inside your window. You can define widgets directly in the class constructor for the `Application` class, but it's become somewhat standard in Python circles to create a special method called `create_widgets()` and then place the statements to create the widgets inside that method. You can just call the `create_widgets()` method from inside the class constructor.

When you add the `create_widgets()` method to the `Application` class, the constructor looks like this:

```
def __init__(self, master):
    super(Application, self).__init__(master)
    self.grid()
    self.create_widgets()
```

The `create_widgets()` method contains all the statements to build the widget objects that you want to appear in your window. Listing 18.3 shows the `script1803.py` program, which demonstrates a simple example of this.

**LISTING 18.3** The `script1803.py` File

```
1: #!/usr/bin/python3
2: from tkinter import *
3:
4: class Application(Frame):
5:     """Build the basic window frame template"""
6:
7:     def __init__(self, master):
8:         super(Application, self).__init__(master)
9:         self.grid()
10:         self.create_widgets()
11:
12:     def create_widgets(self):
13:         self.label1 = Label(self, text='Welcome to my window!')
14:         self.label1.grid(row=0, column=0, sticky= W)
15:
16: root = Tk()
17: root.title('Test Application window with Label')
18: root.geometry('300x100')
19: app = Application(root)
20: app.mainloop()
```

The `create_widgets()` method contains two lines of code to define a `Label` widget object for the window. Line 13 defines the actual `Label` object, and line 14 applies the `grid()` method to position the `Label` widget in the window. (Yes, that's correct: You need to specify the placement method for both the `Frame` object and the individual widget objects inside the frame.)

When you run the `script1803.py` file, you see a window like the one shown in Figure 18.2.

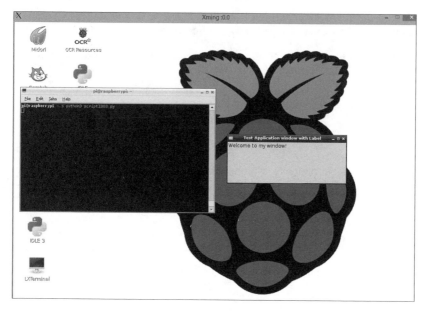

**FIGURE 18.2**
Displaying the simple test window.

The window contains the `Label` object that you defined in the `create_widgets()` method, and the label text appears in the window frame.

## Defining Event Handlers

The next step in building a GUI application is to define the events that the window uses. Widgets that can generate events (such as when the application user clicks a button) use the `command` parameter to define the name of a method Python calls when it detects the event.

For example, to link a button to an event method, you write code like this:

```
def create_widgets(self):
    self.button1 = Button(self, text="Submit", command = self.display)
    self.button1.grid(row=1, column=0, sticky = W)

def display(self):
    print("The button was clicked in the window")
```

The `create_widgets()` method creates a single button to display in the window area. The `Button` class constructor sets the command parameter to `self.display`, which points to the `display()` method in the class.

For now, the test `display()` method just uses a `print()` statement to display a message back in the command line, where you started the program. Now things are starting to look more like a GUI program! Listing 18.4 shows the `script1804.py` code, which creates the window with the button and event defined.

**LISTING 18.4**  The `script1804.py` Code File

```
#!/usr/bin/python3
from tkinter import *
class Application(Frame):
    """Build the basic window frame template"""

    def __init__(self, master):
        super(Application, self).__init__(master)
        self.grid()
        self.create_widgets()

    def create_widgets(self):
        self.label1 = Label(self, text='Welcome to my window!')
        self.label1.grid(row=0, column=0, sticky=W)
        self.button1 = Button(self, text='Click me!', command=self.display)
        self.button1.grid(row=1, column=0, sticky=W)

    def display(self):
        """Event handler for the button"""
        print('The button in the window was clicked!')

root = Tk()
root.title('Test Button events')
root.geometry('300x100')
app = Application(root)
app.mainloop()
```

When you run the program from the LXTerminal, the window appears separate on the desktop. However, when you click the Click me! button, the text from the `print()` method still appears in the LXTerminal window, as shown in Figure 18.3.

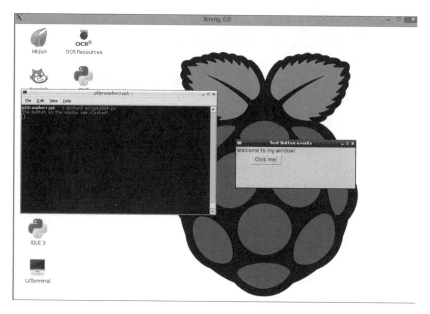

**FIGURE 18.3**
Demonstrating the `Button` event method.

A typical GUI program contains lots of different event handlers, one for each widget that can trigger an event. Sometimes trying to keep track of all the event handles can be challenging. This is where using the `docstring` feature can come in handy. It's always a good idea to place a one-line `docstring` value in each event handler method to help describe what it does, as well as what widget triggers it, as shown here:

```
def display(self):
    """Event handler for the button to display text in the command line"""
    print('The button was clicked!')
```

You don't have to get too fancy with the `docstring`. Just place enough information in it to help link up the event handler to the appropriate widget.

# Exploring the `tkinter` Widgets

Now that you've seen the basics of how widgets interact in a GUI program, you're ready to see the different types of widgets available for you to use. Each widget contains attributes and methods that you can use to customize the widget in your window. The following sections show some of the most popular widgets used in Python GUI programs and how to use them.

# Using the `Label` Widget

The `Label` widget allows you to place text inside the window. This widget is often used to identify other widgets, such as `Entry` or `Textbox` areas, to help your program users know how to interact with the widgets.

To add a `Label` widget to your window, you define the text to display with the `text` parameter, as shown here:

```
self.label1 = Label(self, text='This is a test label')
```

There's not too much that you have to worry about with labels. The hardest part is positioning them inside the frame area in the window.

# Adding the `Button` Widget

Buttons provide a way for application users to trigger event handlers in an application, such as to let it know when there's data in a form that needs to be read. This is the basic format for creating a `Button` widget:

```
self.button1 = Button(self, text='Submit', command=self.calculate)
```

You must assign the `Button` widget to a unique variable name within the `Application` class. With the `Button` widget, you should make sure to point the command parameter to the associated event handler method in the `Application` class. If you don't specify the command parameter, the button won't do anything when it's clicked. Also, you need to use the `grid()` method to position the button where you want it in the window frame area.

# Working with the `Checkbutton` Widget

The `Checkbutton` widget provides an on-or-off type of interface. If the `Checkbutton` widget is checked, it returns a 1 value, and if the widget is not checked, it returns a 0 value. The `Checkbutton` widget is most commonly used to make selections of one or more items from a list (such as selecting the toppings on a pizza).

Working with the `Checkbutton` widget is a bit tricky. You can't directly access the `Checkbutton` widget to find out whether it has been selected. Instead, you need to create a special variable that can hold a value that represents the check box status. This is called a *control variable*.

You create a control variable by using one of four special methods:

- ▸ `BooleanVar()`—For Boolean 0 and 1 values
- ▸ `DoubleVar()`—For floating-point values
- ▸ `IntVar()`—For integer values
- ▸ `StringVar()`—For text values

Because the `Checkbutton` widget returns a Boolean value, you should use the `BooleanVar()` control variable method. You must define this control variable as an attribute of the class object so that you can reference it in the event handler method. You most often do this in the `__init__()` method, as shown here:

```
self.varCheck1 = BooleanVar()
```

Then to link the control variable to the `Checkbutton` widget, you use the `variable` parameter when you define the `Checkbutton` widget:

```
self.check1 = Checkbutton(self, text='Option1', variable=self.varCheck1)
```

So with the `Checkbutton` widget, instead of using an event handler, you link the widget to a control variable. The `text` parameter in the `Checkbutton` object defines the text that appears next to the check box in the window.

To retrieve the status of the `Checkbutton` widget in your code, you need to use the `get()` method for the control variable, like this:

```
option1 = self.varCheck1.get()
if (option1):
    print('The checkbutton was selected')
else:
    print('The checkbutton was not selected')
```

Listing 18.5 shows the `script1805.py` program, which demonstrates how to use a `Checkbutton` object in a program.

## LISTING 18.5    The `script1805.py` Code

```
#!/usr/bin/python3
from tkinter import *
class Application(Frame):
    """Build the basic window frame template"""

    def __init__(self, master):
        super(Application, self).__init__(master)
        self.grid()
        self.varSausage = IntVar()
        self.varPepp = IntVar()
        self.create_widgets()

    def create_widgets(self):
        self.label1 = Label(self, text='What do you want on your pizza?')
        self.label1.grid(row=0)
        self.check1 = Checkbutton(self, text='Sausage', variable =
    self.varSausage)
        self.check2 = Checkbutton(self, text='Pepperoni', variable =
```

```
   self.varPepp)
      self.check1.grid(row=1)
      self.check2.grid(row=2)
      self.button1 = Button(self, text='Order', command=self.display)
      self.button1.grid(row=3)

   def display(self):
      """Event handler for the button, displays selections"""
      if (self.varSausage.get()):
         print('You want sausage')
      if (self.varPepp.get()):
         print('You want pepperoni')
      if ( not self.varSausage.get() and not self.varPepp.get()):
         print("You don't want anything on your pizza?")
      print('----------')

root = Tk()
root.title('Test Checkbutton events')
root.geometry('300x100')
app = Application(root)
app.mainloop()
```

The code in the `script1805.py` file should look pretty familiar to you by now. It creates the `Application` child class for the `Frame` object, defines the constructor, and defines the widgets to place in the frame, including the two `Checkbutton` widgets. The button uses the `display()` method for its event handler. The `display()` method retrieves the two control variable values used for the `Checkbutton` widgets and displays a message in the command line, based on which `Checkbutton` widget is selected.

## Using the `Entry` Widget

The `Entry` widget is one of the most versatile widgets you'll use in your applications. It creates a single-line form field. A program user can use this field to enter text to submit to the program, or your program can use it to display text dynamically in the window.

Creating an `Entry` widget isn't very complicated:

```
self.entry1 = Entry(self)
```

The `Entry` widget doesn't itself call an event handler. Normally, you link another widget, such as a button, to an event handler that then retrieves the text that's in the `Entry` widget or displays new text in the `Entry` widget. To do that, you need to use the `Entry` widget's `get()` method to retrieve the text in the form field or the `insert()` method to display text in the form field.

Listing 18.6 shows the `script1806.py` program, which shows how to use the `Entry` widget both for input and output for Python window scripts.

**LISTING 18.6**    The `script1806.py` Code

```python
#!/usr/bin/python3
from tkinter import *

class Application(Frame):
    """Build the basic window frame template"""

    def __init__(self, master):
        super(Application, self).__init__(master)
        self.grid()
        self.create_widgets()

    def create_widgets(self):
        self.label1 = Label(self, text='Please enter some text in lower
case')
        self.label1.grid(row=0)

        self.text1 = Entry(self)
        self.text1.grid(row=2)

        self.button1 = Button(self, text='Convert text',
command=self.convert)
        self.button1.grid(row=6, column=0)
        self.button2 = Button(self, text='Clear result',
command=self.clear)
        self.button2.grid(row=6, column=1)
        self.text1.focus_set()

    def convert(self):
        """Retrieve the text and convert to upper case"""
        varText = self.text1.get()
        varReplaced = varText.upper()
        self.text1.delete(0, END)
        self.text1.insert(END, varReplaced)

    def clear(self):
        """Clear the Entry form"""
        self.text1.delete(0,END)
        self.text1.focus_set()

root = Tk()
root.title('Testing and Entry widget')
root.geometry('500x200')
app = Application(root)
app.mainloop()
```

The `button1` widget links to the `convert()` method, which uses the `get()` method to retrieve the text in the `Entry` widget, converts it to uppercase, and then uses the `insert()` method to place the converted text back in the `Entry` widget to display it. Before it can do that, though, it uses the `delete()` method to remove the original text from the `Entry` widget. The `focus_set()` method is a handy tool: It allows you to tell the window which widget should get control of the cursor, preventing our window user from having to click in the widget first.

## Adding a `Text` Widget

For entering large amounts of text, you can use the `Text` widget. It provides for multiline text entry or displaying multiple lines of text. The `Text` widget has the following syntax:

```
self.text1 = Text(self, options)
```

You can use quite a few options to control the size of the `Text` widget in the window and how it formats the text contained within the display area. The most commonly used options are `width` and `height`, which set the size of the `Text` widget area in the window. (`width` is defined in characters and `height` in lines.)

As with to the `Entry` widget, you retrieve the text in a `Text` widget by using the `get()` method, you remove text from the widget by using the `delete()` method, and you add text to the widget by using the `insert()` method. However, there's a bit of a twist to these methods in the `Text` widget. Because the widget works with multiple lines of text, the index values you specify for the `get()`, `delete()`, and `insert()` methods is not a single numeric value. It's actually a text value that has two parts:

```
"x.y"
```

In this case, `x` is the row location (starting at 1), and `y` is the column location (starting at 0). So, to reference the first character in the `Text` widget, you use the index value `"1.0"`.

Listing 18.7 shows the `script1807.py` file, which demonstrates the basic use of the `Text` widget.

### LISTING 18.7   The `script1807.py` File

```
#!/usr/bin/python3

from tkinter import *
class Application(Frame):
    """Build the basic window frame template"""

    def __init__(self, master):
        super(Application, self).__init__(master)
        self.grid()
        self.create_widgets()
```

```
    def create_widgets(self):
        self.label1 = Label(self, text='Enter the text to convert:')
        self.label1.grid(row=0, column=0, sticky =W)

        self.text1 = Text(self, width=20, height=10)
        self.text1.grid(row=1, column=0)
        self.text1.focus_set()

        self.button1 = Button(self, text='Convert', command=self.convert)
        self.button1.grid(row=2, column=0)
        self.button2 = Button(self, text='Clear', command=self.clear)
        self.button2.grid(row=2, column=1)

    def convert(self):
        varText = self.text1.get("1.0", END)
        varReplaced = varText.upper()
        self.text1.delete("1.0", END)
        self.text1.insert(END, varReplaced)

    def clear(self):
        self.text1.delete("1.0", END)
        self.text1.focus_set()

root = Tk()
root.title = 'Text widget test'
root.geometry('300x250')
app = Application(root)
app.mainloop()
```

When you run the `script1807.py` file, you get a window that shows the `Text` widget and the two buttons. You can then enter larger blocks of text into the `Text` widget, click the Convert button to convert the text to all uppercase, and click the Clear button to delete the text.

## Using a `Listbox` Widget

The `Listbox` widget provides a listing of multiple values for your application user to choose from. When you create the `Listbox` widget, you can specify how the user selects items in the list with the `selectmode` parameter, as shown here:

```
self.listbox1 = Listbox(self, selectmode=SINGLE)
```

These options are available for the `selectmode` parameter:

▶ **SINGLE**—Select only one item at a time.

▶ **BROWSE**—Select only one item but the items can be moved in the list.

▶ **MULTIPLE**—Select multiple items by clicking them one at a time.

▶ **EXTENDED**—Select multiple items by using the Shift and Control keys while clicking items.

After you create the `Listbox` widget, you need to add items to the list. You do that with the `insert()` method, as shown here:

```
self.listbox1.insert(END, 'Item One')
```

The first parameter defines the index location in the list where the new item should be inserted. You can use the keyword `END` to place the new item at the end of the list. If you have a lot of items to add to the `Listbox` widget, you can place them in a list object and use a `for` loop to insert them all at once, as in the following example:

```
items = ['Item One', 'Item Two', 'Item Three']
for item in items:
        self.listbox1.insert(END, item)
```

Retrieving the selected items from the `Listbox` widget is a two-step process. First, you use the `curselection()` method to retrieve a tuple that contains the index of the selected items (starting at 0):

```
items = self.listbox1.curselection()
```

Once you have the tuple that contains the index values, you use the `get()` method to retrieve the text value of the item at that index location:

```
for item in items:
    strItem = self.listbox1.get(item)
```

Listing 18.8 shows the `script1808.py` file, which demonstrates how to use the `Listbox` widget in your programs.

## LISTING 18.8   The `script1808.py` Program Code

```
#!/usr/bin/python3
from tkinter import *

class Application(Frame):
    """Build the basic window frame template"""

    def __init__(self, master):
        super(Application, self).__init__(master)
        self.grid()
        self.create_widgets()
```

```
    def create_widgets(self):
        self.label1 = Label(self, text='Select your items')
        self.label1.grid(row=0)
        self.listbox1 = Listbox(self, selectmode=EXTENDED)
        items = ['Item One', 'Item Two', 'Item Three']
        for item in items:
            self.listbox1.insert(END, item)
        self.listbox1.grid(row=1)
        self.button1 = Button(self, text='Submit', command=self.display)
        self.button1.grid(row=2)

    def display(self):
        """Display the selected items"""
        items = self.listbox1.curselection()
        for item in items:
            strItem = self.listbox1.get(item)
            print(strItem)
        print('----------')

root = Tk()
root.title('Listbox widget test')
root.geometry('300x200')
app = Application(root)
app.mainloop()
```

When you run the `script1808.py` code, anything you select from the list box will appear in the command-line window when you click the Submit button.

## Working with the `Menu` Widget

A staple of GUI programs is the menu bar at the top of the window. The menu bar provides drop-down menus so that program users can quickly make selections. You can create menu bars in your `tkinter` windows by using the `Menu` widget.

To create the main menu bar, you link the `Menu` widget directly to the `Frame` object. Then you use the `add_command()` method to add individual menu entries. Each `add_command()` method specifies a `label` parameter to define what text appears for the menu entry and a `command` parameter to define the method to run when the menu entry is selected. Here's how it looks:

```
menubar = Menu(self)
menubar.add_command(label='Help', command=self.help)
menubar.add_command(label='Exit', command=self.exit)
```

This creates a single menu bar at the top of the window, with two selections: Help and Exit. Finally, you need to link the menu bar to the root `Tk` object by adding this command:

```
root.config(menu=self.menubar)
```

Now when you display your application, it will have a menu bar at the top, with the menu entries that you defined.

You can create drop-down menus by creating additional Menu widgets and linking them to your main menu bar Menu widget. That looks like this:

```
menubar = Menu(self)
filemenu = Menu(menubar)
filemenu.add_command(label='Convert', command=self.convert)
filemenu.add_command(label='Clear', command=self.clear)
menubar.add_cascade(label='File', menu=filemenu)
menubar.add_command(label='Quit', command=root.quit)
root.config(menu=menubar)
```

After you create the drop-down menu, you use the add_cascade() method to add it to the top-level menu bar and assign it a label.

Now that you've learned about the popular widgets that you'll use in your programs, you can work out a real program to test them out!

## TRY IT YOURSELF

### Create a Python GUI Program

In the following steps, you'll create a GUI application that can calculate your bowling average after three games. Just follow these steps to get your program up and running:

1. Create a file called script1809.py in the folder for this hour.

2. Open the script1809.py file and enter the code shown here:

```
#!/usr/bin/python3
from tkinter import *

class Application(Frame):
    """Build the basic window frame template"""

    def __init__(self, master):
        super(Application, self).__init__(master)
        self.grid()
        self.create_widgets()

    def create_widgets(self):
        menubar = Menu(self)
        filemenu = Menu(menubar)
        filemenu.add_command(label='Calculate', command=self.calculate)
```

```
        filemenu.add_command(label='Reset', command=self.clear)
        menubar.add_cascade(label='File', menu=filemenu)
        menubar.add_command(label='Quit', command=root.quit)
        self.label1 = Label(self, text='The Bowling Calculator')
        self.label1.grid(row=0, columnspan=3)
        self.label2 = Label(self, text="Enter score from game 1:")
        self.label3 = Label(self, text='Enter score from game 2:')
        self.label4 = Label(self, text='Enter score from game 3:')
        self.label5 = Label(self, text="Average:")
        self.label2.grid(row=2, column=0)
        self.label3.grid(row=3, column=0)
        self.label4.grid(row=4, column=0)
        self.label5.grid(row=5, column=0)
        self.score1 = Entry(self)
        self.score2 = Entry(self)
        self.score3 = Entry(self)
        self.average = Entry(self)
        self.score1.grid(row=2, column=1)
        self.score2.grid(row=3, column=1)
        self.score3.grid(row=4, column=1)
        self.average.grid(row=5, column=1)
        self.button1 = Button(self, text="Calculate Average",
command=self.calculate)
        self.button1.grid(row=6, column=0)
        self.button2 = Button(self, text='Clear result', command=self.clear)
        self.button2.grid(row=6, column=1)
        self.score1.focus_set()
        root.config(menu=menubar)

    def calculate(self):
        """Calculate and display the average"""
        numScore1 = int(self.score1.get())
        numScore2 = int(self.score2.get())
        numScore3 = int(self.score3.get())
        total = numScore1 + numScore2 + numScore3
        average = total / 3
        strAverage = "{0:.2f}".format(average)
        self.average.insert(0, strAverage)

    def clear(self):
        """Clear the Entry forms"""
        self.score1.delete(0,END)
        self.score2.delete(0,END)
        self.score3.delete(0,END)
        self.average.delete(0,END)
        self.score1.focus_set()
```

```
root = Tk()
root.title('Bowling Average Calculator')
root.geometry('500x200')
app = Application(root)
app.mainloop()
```

3. Save the `script1809.py` file.

4. Run the `script1809.py` program from an LXTerminal session in your desktop. A window like the one shown in Figure 18.4 should appear.

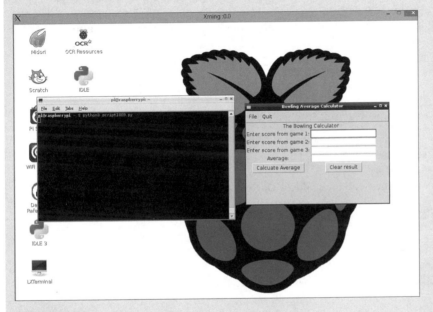

**FIGURE 18.4**
The Bowling Average Calculator program.

You should recognize all the widgets used in the `script1809.py` program. This program uses four `Entry` widgets—three to enter the bowling scores, and one to display the resulting average. One important feature to notice is that the values retrieved from the `Entry` widgets are strings, so you have to convert the values into numeric values for the calculations.

Now you have all the skills you need to start creating fancy GUI programs in your Python scripts.

# Summary

In this hour, you dove into the world of programming GUI programs. There are several different libraries you can use for creating GUI programs, and in this hour, you used the `tkinter` library, which comes installed in the standard Python libraries.

The `tkinter` library allows you to create windows with all the standard features that you're used to seeing in commercial GUI programs—text entry forms, selection boxes, buttons, and menu bars. You just write the Python code to create the window, add the widgets you want to the window, and then link methods to the events generated by the widgets.

In the next hour, we'll take a look at how to create games using Python programming. There's a great tool available to help make creating games a breeze, and we'll walk through just how to use that.

# Q&A

**Q.** Can you link more than one widget to the same method?

**A.** Yes, you can use the same method for multiple events. For example, you can link a `Button` widget to the same method that you link to from a `Menu` item.

**Q.** Can you link more than one method to a single widget?

**A.** No, you can link each widget to only one method. However, you can run a separate method from within the original method's code.

# Workshop

## Quiz

1. What type of widget should you use to display a list of items from which you can select multiple items?

    **a.** `Entry`

    **b.** `Checkbutton`

    **c.** `Listbox`

    **d.** `Button`

2. You can use an `Entry` widget to both retrieve text entered by the user and display text from your program. True or false?

3. The process of linking widgets in a window to specific methods inside the Python code is called what type of programming?

# Answers

1. c. `Listbox`. The `Listbox` widget allows you to display multiple items that you can select from.

2. True. The `Entry` widget provides a textbox area where you can display text from your program, or allow the user to enter data that your program can read.

3. Event-driven programming is what links the widgets that you display in the GUI window to methods inside the Python program code.

# Game Programming

---

**What You'll Learn in This Hour:**

▶ Why you might want to program games
▶ Different Python game interfaces
▶ How to set up and use the `PyGame` library
▶ How to handle action in a game
▶ How to create a simple game script

In this hour, you will learn about creating games using Python. You will learn game scripting basics, how to create a game screen, how to add text, how to add images, and even how to animate those game images. In the process, you'll learn about the `PyGame` library.

## Understanding Game Programming

Why should you learn to program games? The simple answer is that you will become a stronger script writer in Python. Game programming is different from other programming in that it stretches both the programmer and the computer.

Think about computers built specifically for gaming. They tend to have the fastest CPUs, larger memory chips, and the best video cards. This is because games can be large consumers of a computer system's resources.

For a game developer, getting a designed game from paper to Python script can be a big challenge. Game scripts use all the various aspects of a scripting language, such as user input, file input/output, mathematical manipulation, various graphical interfaces, and so on. Also, developing a game forces you to be creative and a good problem solver.

---

**Developer Versus Designer**

In creating a marketable game, a *game developer* is the person who writes the code. A *game designer*, on the other hand, determines the game's appearance, rules, and goals. For the purposes of this hour, you can be both the game designer and the developer.

---

In essence, understanding game development can help make you a well-rounded script writer. Game writing is commonly used to instruct beginners as well as to polish old-timers. As you learn to write Python scripts on a Raspberry Pi, game development will help you solidify Python concepts.

# Learning About Game Frameworks and Libraries

Several game frameworks and libraries are available in Python. Table 19.1 lists a few of them.

**TABLE 19.1**   Python Game Frameworks and Libraries

| Name | Description |
| --- | --- |
| Blender3D | A popular library that has modeling, animating, and 3D rendering capabilities. |
| cocos2d | A framework for building 2D games, demos, and other graphical/interactive applications. |
| fifengine | Also called `fife`, a cross-platform framework for creating games that provides support for different isometric perspectives. With `fifengine`, you can write games in C++ or Python. |
| kivy | A cross-platform framework that includes multimedia support, support for various input devices, multitouch support (including a multitouch mouse simulator), and a very fast graphics engine. |
| Panda3D | A very full-featured framework for game development that includes 3D graphics and a game engine. With `Panda3D`, you can write games in C++ or Python. Used commercially as well as privately. |
| PyGame | A portable and cross-platform Python module library that provides additional features to the SDL library. Allows the creation of games and multimedia programs in Python. |
| Pyglet | A cross-platform multimedia Python library that provides an object-oriented programming interface for game development. Has some nice features, such as handling MP3 music formats. |

| Name | Description |
|------|-------------|
| PySoy | A 3D cloud-based game engine that has an object-oriented API. It is designed for game development and provides multiplatform rapid entertainment. |
| Python-Ogre | A Python interface to several C++ libraries, specifically the Ogre 3D graphics library. Also supports several other graphics and gaming libraries for the purpose of game development. |

As you can see, there are many development tools available for creating games with Python. Having many varieties of game development frameworks and libraries allows you to pick the tools best suited for your game's design.

DID YOU KNOW

## What Is SDL?

You will often see the acronym SDL when reading about game frameworks and libraries. It stands for Simple DirectMedia Layer, which is an open-source cross-platform alternative to the DirectX API. Basically, SDL is a package of multimedia and graphics libraries that provide the necessities of game development.

In this hour, you learn about the PyGame library, and you'll create a simple game. Keep in mind that entire books have been written on game development, including several that focus solely on Python game programming. The game-writing basics you'll learn this hour will give you a good shove in the right direction with game script writing. This hour will also reinforce the Python skills you have learned so far.

BY THE WAY

## Play Python Games

Want to try playing some Python games? Go to the "Specific Games" section of the page wiki.python.org/moin/PythonGames.

# Setting Up the PyGame Library

The PyGame library is a package. You learned in Hour 13, "Working with Modules," that a group of modules can be put together in a package for use in Python scripts. The PyGame package is a collection of modules and objects that will help you create games using Python.

### Preinstalled PyGame Games

You'll find several Python game scripts and their supporting files in the /home/pi/python_games directory. You can play these games if you have the PyGame library installed. These scripts make great learning tools, too.

PyGame for Python is really only for developing simple games, not graphical wonders. (You need Panda3D for that kind of fancy stuff.) The real glory of PyGame is that it will help you learn more script-writing principles while providing instantaneous positive feedback to you as a script writer.

## Checking for PyGame

The PyGame package is typically not installed by default. To check your system, enter the Python interactive shell and try importing PyGame, as shown in Listing 19.1.

**LISTING 19.1    Checking for PyGame**

```
>>>
>>> import pygame
Traceback (most recent call last):
  File "<stdin>", line 1, in <module>
ImportError: No module named pygame
>>>
```

If you receive an ImportError message as shown in Listing 19.1, then the PyGame package is not installed. However, if you do not receive an error, you do have it and can skip the next section, on installing PyGame.

## Installing PyGame

Typically, to install a Python package, you use a variation of the command sudo apt-get install. Unfortunately, at the time of this writing, the PyGame library is not easy to install for Python3 on the Raspberry Pi. The steps in this section guide you through the process. You must follow them very carefully and in the correct order if you want to successfully install PyGame.

### Checking for a Change

Hopefully, by this time there is an easier way to install the PyGame library on your Raspberry Pi. Check elinux.org/RPi_Debian_Python3#PyGame_Module, in the "PyGame Module" section, for potential updates.

# Making Sure Your System Is Up to Date

The first step in installing PyGame is to make sure your system is all up-to-date, with any missing items installed. You do this by logging in to your Raspberry Pi, and typing in at the shell prompt the command sudo apt-get dist-upgrade --fix-missing, and press Enter. (Note the two hyphens in front of fix-missing.) Listing 19.2 shows this process in action.

**LISTING 19.2**    Getting Your System Up to Date

```
pi@raspberrypi ~ $ sudo apt-get dist-upgrade --fix-missing
Reading package lists... Done
Building dependency tree
Reading state information... Done
Calculating upgrade... Done
0 upgraded, 0 newly installed, 0 to remove and 0 not upgraded.
pi@raspberrypi ~ $
```

You can see in Listing 19.2 that no items are missing, and all the packages are up-to-date.

If your system is not up-to-date, the process will ask if you want to install the packages. Answer Y to have the packages installed.

WATCH OUT!

## Check Available Space

You need to install several libraries and packages in order to obtain PyGame. Be sure to keep a close eye on the available space you have on your Raspberry Pi's storage medium. Use the shell command df to see a quick summary of the space currently used.

# Installing the Tools for Building Python Modules

Next, you need to install some tools for building Python modules. To do so, at the shell prompt type sudo apt-get install python3-dev python3-numpy and press Enter. Listing 19.3 shows the successful installation of these tools.

**LISTING 19.3**    Installing Development Tools

```
pi@raspberrypi ~ $ sudo apt-get install python3-dev python3-numpy
Reading package lists... Done
Building dependency tree
Reading state information... Done
The following extra packages will be installed:
    idle3 libexpat1-dev libpython3.2 libssl-dev libssl-doc libssl1.0.0 python3
    python3-minimal python3.2 python3.2-dev python3.2-minimal
```

```
Suggested packages:
...
Setting up python3-dev (3.2.3-6) ...
Setting up python3-numpy (1:1.6.2-1.2) ...
Processing triggers for menu ...
pi@raspberrypi ~ $
```

DID YOU KNOW

### 404 Not Found

If you get errors similar to `404 Not Found` when you try to install the tools for building Python modules, first check and make sure you are connected to the Internet. To do this, open up the Midori web browser and see if you can reach any of your favorite websites. If you can connect to the Internet, next try repairing your system by typing the command `sudo apt-get update` and press Enter. Now try typing `sudo apt-get install python3-dev python3-numpy` and pressing Enter again.

## Obtaining the `PyGame` **Source Code**

To proceed, you need to obtain the PyGame source code and a software package that will allow you to do so. At the command line, type `sudo apt-get install mercurial`, as shown in Listing 19.4, and press Enter.

**LISTING 19.4**    Obtaining the `PyGame` Source Code

```
pi@raspberrypi ~ $ sudo apt-get install mercurial
Reading package lists... Done
Building dependency tree
Reading state information... Done
The following extra packages will be installed:
  mercurial-common
Suggested packages:
  qct vim emacs kdiff3 kdiff3-qt kompare meld xxdiff tkcvs mgdiff
  python-mysqldb python-pygments python-openssl
The following NEW packages will be installed:
  mercurial mercurial-common
0 upgraded, 2 newly installed, 0 to remove and 191 not upgraded.
Need to get 2,404 kB of archives.
After this operation, 8,022 kB of additional disk space will be used.
Do you want to continue [Y/n]? y
Get:1 http://mirrordirector.raspbian.org/raspbian/ wheezy/main mercurial-
common all 2.2.2-3 [2,320 kB]
...
Creating config file /etc/mercurial/hgrc.d/hgext.rc with new version
```

```
pi@raspberrypi ~ $
pi@raspberrypi ~ $ hg clone https://bitbucket.org/pygame/pygame
destination directory: pygame
requesting all changes
...
664 files updated, 0 files merged, 0 files removed, 0 files unresolved
pi@raspberrypi ~ $
pi@raspberrypi ~ $ cd pygame
pi@raspberrypi ~/pygame $ pwd
/home/pi/pygame
pi@raspberrypi ~/pygame $
```

Next, you use the hg command from the newly install mercurial program by typing hg clone
https://bitbucket.org/pygame/pygame and pressing Enter. Finally, you need to change
your directory to the newly copied PyGame directory by typing cd pygame and pressing Enter.
Both of these commands are also shown above in Listing 19.4.

# Installing Additional Software Packages

Now you need to install the following additional software packages:

- ▶ libsdl-dev

- ▶ libsdl-image1.2-dev

- ▶ libsdl-mixer1.2-dev

- ▶ libsdl-ttf2.0-dev

- ▶ libsmpeg-dev

- ▶ libportmidi-dev

- ▶ libavformat-dev

- ▶ libswscale-dev

These software packages, also called *dependencies*, are needed for the PyGame library. To install
these dependencies, type sudo apt-get install and then each software package name. You
can do several at one time, as shown in Listing 19.5.

**LISTING 19.5**   Installing Dependencies

```
pi@raspberrypi ~/pygame $ sudo apt-get install libsdl-dev libsdl-image1.2-dev
libsdl-mixer1.2-dev libsdl-ttf2.0-dev
Reading package lists... Done
Building dependency tree
...
```

```
pi@raspberrypi ~ $ sudo apt-get install libsmpeg-dev libportmidi-dev
libavformat-dev libswscale-dev
Reading package lists... Done
...
Setting up libsmpeg-dev (0.4.5+cvs20030824-5) ...
Setting up libswscale-dev (6:0.8.6-1+rpi1) ...
pi@raspberrypi ~ $
```

There are a lot of packages to install here. Don't worry if you see lots of things going on during their installation. Now would be a good time to go get a cup of coffee.

## Building and Installing PyGame

Now that you have a copy of the PyGame source code, the development tools, and all the necessary dependencies, you can finally build PyGame and install it. To do so, you need to enter two commands.

First, type python3 setup.py build and press Enter. When that is complete, type sudo python3 setup.py install and press Enter.

BY THE WAY

### Be Patient!

When you issue the above commands to build and install PyGame, each command may take a long time to complete. Be patient and wait for each command to fully finish, before moving on in this chapter.

That's it! You are all done installing the PyGame library. You can now move on to more fun tasks, such as creating a game.

# Using PyGame

The PyGame library comes with lots of tools and features and also with a great deal of community support. A great site to visit is the PyGame wiki, at www.pygame.org. There you will find module and object documentation, tutorials, a lovely reference index, and general news about the PyGame library.

Several modules within PyGame help with building games. Table 19.2 shows a list of the various modules. Just reading through this table should get you excited!

**TABLE 19.2**  `PyGame` Modulesv

| Name | Description |
| --- | --- |
| pygame.camera | Provides a camera interface (experimental). |
| pygame.cdrom | Controls audio CD-ROM. |
| pygame.cursors | Provides cursor resources. |
| pygame.display | Provides display screen controls. |
| pygame.draw | Draws shapes. |
| pygame.event | Interacts with events and queues. |
| pygame.examples | Shows examples of programs. |
| pygame.font | Loads and renders fonts. |
| pygame.freetype | Enables enhanced loading and rendering of fonts. |
| pygame.gfxdraw | Draws shapes (experimental). |
| pygame.image | Does image handling. |
| pygame.joystick | Interacts with joysticks, gamepads, and trackballs. |
| pygame.key | Interacts with the keyboard. |
| pygame.locals | Provides PyGame constants. |
| pygame.mask | Provides Image masks. |
| pygame.math | Provides vector classes (experimental). |
| pygame.midi | Interacts with MIDI input and output. |
| pygame.mixer | Loads and plays sounds. |
| pygame.mouse | Interacts with the mouse. |
| pygame.movie | Plays back MPEG video. |
| pygame.music | Controls streamed audio. |
| pygame.pixelcopy | Copies general pixel arrays. |
| pygame.scrap | Provides clipboard support. |
| pygame.sndarray | Accesses sound sample data. |
| pygame.surfarray | Accesses surface pixel data, using array interfaces. |
| pygame.time | Monitors time. |
| pygame.transform | Transforms surfaces. |

BY THE WAY

### Experimental Modules

Notice the modules listed as "experimental." You should not use them in any game you plan to keep for a while. Those modules may change dramatically and could break your game.

The PyGame library also includes object classes that make building a Python game much easier (see Table 19.3).

**TABLE 19.3   PyGame Object Classes**

| Name | Description |
| --- | --- |
| pygame.Color | Used for color representations. |
| pygame.Overlay | Used for video overlay graphics. |
| pygame.PixelArray | Used for direct pixel access of surfaces. |
| pygame.Rect | Used for storing rectangular coordinates. |
| pygame.sprite | Used for basic game object classes. |
| pygame.Surface | Used for representing images on the game screen. |

At this point, you may be a little overwhelmed. That is okay because this is an overwhelming topic. Don't worry. This hour will take you step by step through some of these modules and objects to get you started writing games in Python.

DID YOU KNOW

### Sprites

PyGame documentation often refers to the game pieces or characters in a game as *sprites*. When it does, it is referring to the Sprite object class.

## Loading and Initializing PyGame

To get started using the PyGame library in your Python game script, you need to do three primary things:

1. Import the PyGame library.

2. Import local PyGame constants.

3. Initialize the PyGame modules.

To import the `PyGame` library, you use the `import` command, like this:

```
import pygame
```

This imports all the `PyGame` modules and object classes. However, you can import individual modules and classes, if desired.

---

BY THE WAY

### Speeding It Up

Once you have learned how to write Python game scripts, you can do a few things to speed up your game. One of them is to import only the modules from the `PyGame` library that you actually use in the script.

---

The local `PyGame` constants module, which contains top-level variables, was originally created to make a game script writer's life easier. One `import` statement and you have all the necessary `PyGame` constants needed. To import these constants, you use a variation of the `import` command, as shown here:

```
from pygame.locals import *
```

Finally, you need to initialize the `PyGame` modules. This is how you initialize all the `PyGame` modules you have imported:

```
pygame.init()
```

Once you have all the modules and constants loaded and everything initialized, you can begin to use these various `PyGame` items in your game script.

## Setting Up the Game Screen

In setting up your game screen, you need to determine the following items:

- ▶ Game screen size
- ▶ Screen colors
- ▶ Screen background

To set up the size of your game screen, you need to use the `.display` module. The syntax is as follows:

```
pygame.display.set_mode(width, height)
```

You set the results to a variable name, such as GameScreen, to create a Surface object:

```
GameScreen = pygame.display.set_mode(1000,700)
```

A Surface object is a PyGame object that allows the representation of images on the computer's display. Think of it as a way to create a "playing surface" on which your game will be played.

Another nice thing about creating the game surface is that it will default to the best graphics mode on your current hardware.

## BY THE WAY

### Where You're Located

To keep your bearings when you're beginning to create a game, it's a good idea to make the game screen smaller than your computer's display. This will allow you to see the GUI underneath and provide a visual reference point. For example, while developing the game, keep the game screen size to 600 pixels wide and 400 pixels high. Once you have the game script working perfectly, you can change it to the full computer display size.

You can use any colors on your screen that your computer display can handle. To set up your colors, use the following syntax to add variables to your script:

```
color_variable = Red,Green,Blue
```

Red,Green,Blue handles standard RGB color settings. For example, the color red is represented by the RGB numbers 255,0,0, and the color blue is represented by the RGB numbers 0, 0, 255. A few color examples and their RGB settings are shown here:

```
black=0,0,0
white=255,255,255
blue=0,0,255
red=255,0,0
green=0,255,0
```

Once you have the screen size set up and the color variables created, you can fill the screen's background with the color of your choice. To do so, you need to use the Surface object, which was created to represent the game screen, and use the fill module of that object, as in the following example:

```
GameScreen.fill(blue)
```

In this example, the defined GameScreen Surface object would have the screen's background filled with the color blue. You can make your screen's background a picture, if you want to. However, it's best to start simply as you design your game script and keep the screen's background a plain color.

# Putting Text on the Game Screen

Putting text on your game screen can be tricky. First, you must determine what font you want to use and whether that font exists on your system. `PyGame` provides a default game font that you can use. The module to use is `pygame.font`. Within the font module, you can create a `Font` object by using the syntax *game_font_variable* = `pygame.font.Font(`*font*, *size*`)`, like this:

```
GameFont=pygame.font.Font(None, 60)
```

Notice in this example that the font used is `None`. This is how to set the font equal to the `PyGame` default game font.

To create a text image, you need to use the `Font` object, which was created to represent the font, and use the `render` module of that object.

```
GameTextGraphic=GameFont.render("Hello",True,white)
```

In this example, the word `"Hello"` is the text to be displayed. It is displayed in the `PyGame` default font with a color of white. (Remember that white's color definition, defined earlier this hour, equals `255,255,255`.)

The second argument in the example, `True`, is set to make the displayed text characters have smooth edges. It is a Boolean argument, so if you set it to `False`, the characters do not have smooth edges.

You've done a lot of work, but you still haven't displayed the text to the screen yet! To put the text on the screen, you need to use the `Surface` object created to represent the game's screen. The module of the object to use is the `.blit` module. Earlier, you created the `Surface` object `GameScreen` to represent the game screen. The Python statements below are used to display the text:

```
GameScreen.blit(GameTextGraphic,(100,100))
pygame.display.update()
```

The text graphic `GameTextGraphic` is the first argument to the statement; it tells the `.blit` module what is to be displayed on the screen. The second argument, `(100,100)`, is the location on the game screen where the text should be displayed. Finally, the `pygame.display.update()` function displays the game screen and the graphics it contains on the screen.

Reading about all this can be rather confusing. Trying it yourself will help you understand these concepts. Remember that one of the great features of game programming is that you get quick feedback. Therefore, in the following Try It Yourself, you are going to build a game screen and display text on it.

▼ TRY IT YOURSELF

### Create a Game Screen and Display Text Using PyGame

In the following steps, you will import and initialize the PyGame library, set up a game screen, and display a simple test message on that screen. You will do all this via a Python game script you create, which will be used as a basis for the other Try It Yourself section in this hour. Follow these steps:

1. If you have not already done so, power up your Raspberry Pi and log in to the system.

2. If you do not have the LXDE GUI started automatically at boot, start it now by typing startx and pressing Enter.

3. Open a script editor, such as nano or the IDLE 3 text editor, and create the script py3prog/script1901.py.

4. Type all the information for script1901.py shown below. Take your time and avoid any typographical errors:

```
#script1901.py - Simple Game Screen & Text
#Written by <Insert your Name>
#
###########################################
#
##### Import Modules & Variables ######
import pygame                 #Import PyGame library
#
from pygame.locals import * #Load PyGame constants
#
pygame.init()                #Initialize PyGame
#
# Set up the Game Screen ###############
#
ScreenSize=(1000,700)        #Screen size variable
GameScreen=pygame.display.set_mode(ScreenSize)
#
# Set up the Game Colors ###############
#
black = 0,0,0
white = 255,255,255
blue = 0,0,255
red = 255,0,0
green = 0,255,0
#
# Set up the Game Font #################
#
DefaultFont=None             #Default to PyGame font
```

```
GameFont=pygame.font.Font(DefaultFont,60)
#
# Set up the Game Text Graphic #########
#
GameText="Hello"
GameTextGraphic=GameFont.render(GameText,True,white)
#
###### Draw the Game Screen & Add Game Text #####
#
GameScreen.fill(blue)
GameScreen.blit(GameTextGraphic,(100,100))
pygame.display.update()
#
```

5. Test your game script by exiting the editor, typing `python3 py3prog/script1901.py`, and pressing Enter. If you get any syntax errors, fix them. If you don't get any errors, you probably just saw the game screen with its text appear briefly on the screen and then disappear. (You will address this in the next step.)

6. Open `script1901.py` in a script editor. Under the `import pygame` line, add `import time` to import the `time` module, as shown here:

```
##### Import Modules & Variables ######
import pygame                    #Import PyGame library
import time                      #Import Time module
#
```

7. On the very last line of `script1901.py`, add the line `time.sleep(10)`. This causes your Python game script to pause, or "sleep," for 10 seconds after it writes the game screen to the monitor.

8. Now test your modifications by exiting the editor, typing `python3 py3prog/script1901.py`, and pressing Enter. You should now see (at least for 10 seconds) the game screen and your text displayed. (You will learn later this hour how to control the screen display without using the `time` module.)

9. Just for fun, open `script1901.py` in a script editor again. This time, add a new color, called `RazPiRed`, to the game colors, as shown here:

```
# Set up the Game Colors ###############
#
black = 0,0,0
white = 255,255,255
blue = 0,0,255
red = 255,0,0
RazPiRed = 210,40,82
green = 0,255,0
#
```

BY THE WAY

**Colors**

You can create just about any color in a game by using the RGB settings. Several sites show samples of various colors, along with the RGB settings to achieve them. One such site is www.taylored-mktg.com/rgb/.

**10.** Use the new color for the game screen's background by changing `blue` to `RazPiRed` in the `fill` module attribute, as shown here:

```
GameScreen.fill(RazPiRed)
```

**11.** Change the font by setting the variable `DefaultFont` to `FreeSans`, as shown here:

```
DefaultFont='/usr/share/fonts/truetype/freefont/FreeSans.ttf'
```

(There are some fonts installed by default on your Raspberry Pi. One of them is FreeSans. You can use any installed font instead of the `PyGame` default font.)

**12.** To make things more interesting, change that boring text message from `"Hello"` to `"I love my Raspberry Pi!"`, as shown here:

```
GameText="I love my Raspberry Pi!"
```

**13.** Test your latest game modifications by exiting the editor, typing `python3 py3prog/script1901.py`, and pressing Enter. You should now see (at least for 10 seconds) a screen similar to that in Figure 19.1.

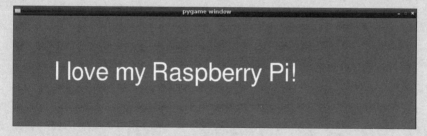

**FIGURE 19.1**
The `script1901.py` game screen.

Good job! You can see the benefits of the instant feedback of writing game scripts. To get more experience, try changing the screen's background color and the location of the text message and see what effects those changes have.

# Learning More About PyGame

Displaying colors and text on a screen is fun, but it doesn't exactly make a game. You need to learn a few more basic concepts before you can write game scripts.

## Staying in the Game

As you saw in the Try It Yourself section, a Python game script displays the game's screen and then exits. So how do you keep the game running? You use a loop construct and the pygame. event module.

You learned about events in Hour 18, "GUI Programming." The PyGame library provides the pygame.event module to monitor these events and event objects to handle them. The following is an example of a typical game loop:

```
1: while True:
2:     for event in pygame.event.get():
3:         if event.type in (QUIT,KEYDOWN):
4:             sys.exit()
```

The main loop is a while loop on line 1, and it will continue to run until the sys.exit() operation exits the loop on line 4. (Note that to use this operation, you need to import the sys module into your game scripts.) In order to reach sys.exit(), the main loop captures any events. If an event occurs, Python handles it in the for loop on line 2 by assigning the event to the variable event. Python checks the event's type in line 3, using the .type method. If the event is either a quit (QUIT) or a key being pressed on the keyboard (KEYDOWN), then sys.exit() runs, and the game exits.

In simple terms, if you press a key on the keyboard while the game is running, the game quits gracefully. Thus, to keep your game running, until a QUIT event occurs, you need to put all the screen drawing and updating within a main game loop. Some of the PyGame event types you can check for include the following:

▶ QUIT

▶ KEYDOWN

▶ KEYUP

▶ MOUSEMOTION

▶ MOUSEBUTTONUP

▶ MOUSEBUTTONDOWN

▶ USEREVENT

# Drawing Images and Shapes

Almost all games include some sort of graphic game pieces. These game pieces can be either imported images or shapes you design yourself.

Creating shapes is easy, thanks to the PyGame module `pygame.draw`. You can use this module to draw circles, squares, hearts, and so on. Table 19.4 shows a few of the methods available in the `pygame.draw` module.

**TABLE 19.4**  A Few `pygame.draw` Module Methods

| Name | Description |
| --- | --- |
| `pygame.draw.arc` | Draws an arc. |
| `pygame.draw.circle` | Draws a circle. |
| `pygame.draw.line` | Draws a single line. |
| `pygame.draw.lines` | Draws several lines. |
| `pygame.draw.polygon` | Draws a polygon. |
| `pygame.draw.rect` | Draws a rectangle. |

BY THE WAY

### Get Help

Don't forget that help is readily available on each of the `pygame.draw` module's methods. You learned about getting help for modules in Hour 13. You can get into the Python interactive shell by typing `python3`, and then you can load PyGame by typing `import pygame`. You can see all the available methods for `pygame.draw` (or any other module) by typing `help(pygame.draw)`. You can get help on an individual method, such as `pygame.draw.circle`, by typing `help(pygame.draw.circle)`. The help shows you a description and all the needed arguments for each method. Remember to press Q to quit out of help when you are done.

You need to do a more work to put an image on your game screen. First, be aware that the PyGame library may not be built to support all image file formats. You can find out more by using the `pygame.image` module and the `.get_extended` method. Listing 19.6 shows an example.

**LISTING 19.6**   Testing `PyGame` for Image Handling

```
>>> import pygame
>>> pygame.image.get_extended()
1
>>>
```

Within the Python interactive shell, if you issue the command `pygame.image.get_extended` and it returns `1` (for true), then you have support for most image file types, including `.png`, `.jpg`, `.gif`, and others. However, if it returns `0` (False), then you can use only uncompressed `.bmp` image files. After you determine what image files your `PyGame` library can handle, you can choose an image to load into a game from the appropriate image file.

To load an image into your Python game script, you need to use the `pygame.image.load` method. Before you do so, it's best to set up a variable name to contain the image file, as shown in this example:

```
# Set up the Game Image Graphics
GameImage="/usr/share/raspberrypi-artwork/raspberry-pi-logo.png"
GameImageGraphic=pygame.image.load(GameImage)
```

This works fine, except that it can be a little slow when loading the image. In fact, it will be slow every time the image has to be redrawn on the screen. To speed it up, you can use the `.convert` method instead on the image, as shown here:

```
# Set up the Game Image Graphics
GameImage="/usr/share/raspberrypi-artwork/raspberry-pi-logo.png"
GameImageGraphic=pygame.image.load(GameImage).convert()
```

However, using `.convert` introduces a new problem: no transparency. In Figure 19.2, you can see the image loaded, but there is a white rectangle around the image.

To make the image transparent and allow the background images to show through, you use the `.convert_alpha()` method instead, as in this example:

```
# Set up the Game Image Graphics
GameImage="/usr/share/raspberrypi-artwork/raspberry-pi-logo.png"
GameImageGraphic=pygame.image.load(GameImage).convert_alpha()
```

This allows the image to have transparency and display the background behind it. In other words, you do not get a white box around the image on the game screen. Figure 19.3 shows the effects of using `.convert_alpha()`.

Once your image is loaded, you simply display it to the screen by using the `Surface` object and the same method you use for text: `.blit`. The following is an example of this:

```
#
GameScreen.blit(GameImageGraphic,(300,0))
```

**FIGURE 19.2**
A loaded game image with no transparency.

**FIGURE 19.3**
A loaded game image with transparency.

Just as with text, the variable representing the image is passed to the `.blit` method, along with where on the `Surface` object you would like the image to be displayed: (`width, height`). However, the game screen still does not show this image! To have the game screen redrawn, remember that you need to use `pygame.display.update()`.

To review, you use `.blit` for the game screen background, you use `.blit` for the text and images (or shapes) on the screen, and then you update the game screen.

## Putting Sound into the Game

Now that you have text and graphics, it's time to learn how to add sound to your Python game script. The `PyGame` library makes adding sounds to a game very easy.

Before you add sound to your game, make sure the sound output is working properly on your Raspberry Pi. If you have your Raspberry Pi hooked up to a television or a computer monitor with built-in speakers through HDMI, then sound will be traveling over the HDMI cable. If you don't have either one of those options, you can hook up a set of computer speakers to the standard 3.5 mm audio-out port. (Review Hour 1, "Setting Up the Raspberry Pi," for more help with this setup.)

To test your sound, you can use one of the sound files from the pre-installed Python games in the `/home/pi/python_games` directory. A nice loud one is the `match1.wav` file. At the command line, type `sudo aplay /home/pi/python_games/match1.wav`. You should hear a sound. If you don't, you may need to adjust your volume or conduct other sound troubleshooting techniques.

To add a sound to your game, you use the `pygame.mixer` module to create a `Sound` object variable, as in this example:

```
# Setup Game Sound ######
ClickSound=pygame.mixer.Sound('/home/pi/python_games/match1.wav')
```

Once you have your `Sound` object created, you can play it by using the `.play` method, as shown here:

```
#
ClickSound.play()
time.sleep(.25)
```

Notice that in the example, after the sound is played, the `time.sleep` method is used, with a sleep time of 1/4 second that adds a little delay to the game. (Without this delay, you might not hear the sound play due to buffering issues.)

A better way to handle the needed delay in a game script, so that a sound can be heard, is to use the `pygame.time.delay` method. This method is superior in that you do not have to load the `time` module (which slows down the game), and you can finely tune the delay time. The `pygame.time.delay` method uses milliseconds as its arguments instead of seconds. Thus, to play a sound, you use code like this:

```
#
ClickSound.play()
pygame.time.delay(300)
```

Notice that this example delays the game by 300 milliseconds. Now your sound will play, and the game will be a little bit faster.

# Dealing with `PyGame` Action

At this point, you have text, an image, and sound in your game script. You're now ready to get things moving.

## Moving Graphics Around the Game Screen

Like a motion picture, graphics moving around your screen are an illusion. What happens "behind the scene" is that the images are redrawn quickly enough to give the appearance of movement. To simulate the appearance of movement in a Python script, you can use the `Surface` object's `.get_rect` method.

When you use the `.get_rect` method on a loaded image, it returns two items: the current coordinates of the image and the image's size. By default, the image's current coordinates on the game surface are 0, 0. However, when you use the method, you can tell it where to place the image on the game surface.

The size that the `.get_rect` method returns is the not the exact same size as the image. The method assumes a rectangular shape around the image to obtain the size and position information. Thus, `.get_rect` stands for "get the rectangle area around the image and return the rectangle's current position on the game screen."

In the following example, the image used previously is used again as a `Surface` object. You can see that a new variable, called `GameImageLocation`, is created, using the `.get_rect` method:

```
GameImage="/usr/share/raspberrypi-artwork/raspberry-pi-logo.png"
GameImageGraphic=pygame.image.load(GameImage).convert_alpha()
#
GameImageLocation=GameImageGraphic.get_rect()
```

To change the location of the image on the screen (that is, make it move), you use the `.move` method. This method uses an *offset* to complete the "move" of the image. In essence, the script moves the image so many pixels down and right on the game screen if the offset numbers are positive. If the offset numbers are negative, the script moves the images up and left.

The following code sets the offset to [10, 10], which means it will move 10 pixels down and 10 pixels to the right:

```
# Set up Graphic Image Movement Speed #####
ImageOffset=[10,10]
#
#Move Image around
GameImageLocation=GameImageLocation.move(ImageOffset)
```

Often the offset is called *speed* because the higher the number, the higher the "speed" across the screen.

Keep in the mind that the .move method does not redraw the image on the screen. You still need to use the .blit method to redraw the image and python.display.update to redraw the screen, as shown here:

```
GameScreen.fill(blue)
GameScreen.blit(GameImageGraphic,GameImageLocation)
pygame.display.update()
```

Notice that the .fill method is used before .blit draws the GameImageGraphic onto the screen. Doing it this way "erases" all the previous images on the game screen and then redraws the game image in its new location. This provides an illusion of the image moving.

## Interacting with Graphics on the Game Screen

There are many ways to creatively interact with the graphics on the screen. Back in the "Staying in the Game" section of this hour, you learned how to use events and event types to exit a game. You can also use events to control the interaction in your game.

One popular method is to use the mouse and a *collide point* (a particular event occurring on top of a Surface object). First, you need to set up the mouse in your game script. Several event types concern the mouse. An easy one is MOUSEBUTTONDOWN, which simply indicates that one of the buttons on the mouse has been pressed. To trap this event, you use the following Python statements:

```
for event in pygame.event.get():
          if event.type == pygame.MOUSEBUTTONDOWN:
```

If event.type matches pygame.MOUSEBUTTONDOWN, then you can test for a collide point. The .collidepoint method helps here. You simply pass the method the position of the other object to test for a collision. If it returns True, then the additional actions can be taken.

The following example tests the current image's location on the game screen and determines whether the mouse pointer has collided with the game image. In other words, it tests whether the mouse pointer clicked the image:

```
for event in pygame.event.get():
        if event.type == pygame.MOUSEBUTTONDOWN:
            if GameImageLocation.collidepoint(pygame.mouse.get_pos()):
                sys.exit()
```

If the mouse clicked the image, then the game is exited. Of course, you can set up any kind of desired action for a collide point. It doesn't have to be "exit the game."

Reading about all this is one thing, but trying it is another. This is especially true when it comes to understanding moving images on your game screen. In the following Try It Yourself section, you'll play a little bit with moving an image around the screen and trying out different "speeds."

▼ TRY IT YOURSELF

**Create Game Images and Move Them About the Screen**

In the following steps, you are going to load an image, move the image about the screen at different speeds, learn how to keep an image on the screen, create a collide point for action, and resize an image. Fasten your safety belt and please keep your hands and legs inside the cart for the duration of this ride:

1. If you have not already done so, power up your Raspberry Pi and log in to the system.

2. If you do not have the LXDE GUI started automatically at boot, start it now by typing startx and pressing Enter.

3. Open a script editor, such as nano or the IDLE 3 text editor, and create the script py3prog/script1902.py.

4. Type all the information for script1902.py shown below. Take your time and avoid any typographical errors:

```
#script1902.py - Move Game Image
#Written by <Insert your Name>
#
##########################################
#
##### Import Modules & Variables ######
import pygame                    #Import PyGame library
import sys                       #Import System module
#
from pygame.locals import *      #Load PyGame constants
#
pygame.init()                    #Initialize PyGame
#
# Set up the Game Screen ###############
#
```

```
ScreenSize=(1000,700)              #Screen size variable
GameScreen=pygame.display.set_mode(ScreenSize)
#
# Set up the Game Color ###############
#
blue = 0,0,255
#
# Set up the Game Image Graphics #######
#
GameImage="/usr/share/raspberrypi-artwork/raspberry-pi-logo.png"
GameImageGraphic=pygame.image.load(GameImage).convert_alpha()
#
GameImageLocation=GameImageGraphic.get_rect()   #Current location
#
ImageOffset=[10,10]  #Moving speed
#
# Set up the Game Sound ###############
#
ClickSound=pygame.mixer.Sound('/home/pi/python_games/match1.wav')
#
#
###### Play the Game ###################################
#
while True:
        for event in pygame.event.get():
            if event.type in (QUIT,MOUSEBUTTONDOWN):
                ClickSound.play()
                pygame.time.delay(300)
                sys.exit()
        #Move game image
        GameImageLocation=GameImageLocation.move(ImageOffset)
        #Draw screen images
        GameScreen.fill(blue)
        GameScreen.blit(GameImageGraphic,GameImageLocation)
        #Update game screen
        pygame.display.update()
    #
```

5. Test your game script by exiting the editor, typing `python3 py3prog/script1902.py`, and pressing Enter. If you get any syntax errors, fix them. If you don't get any errors, watch as the Raspberry Pi image moves...right off the screen! Click anywhere in the game screen with your mouse to play a sound and end the game.

6. To keep the Raspberry Pi image on the game screen, open `script1902.py` in a script editor again. The first change you need to make this little tweak to the `ScreenSize` variable:

```
ScreenSize = ScreenWidth,ScreenHeight = 1000,700
```

By adding `ScreenWidth` and `ScreenHeight` in the middle of the variable assignment, you essentially create two additional variables in one assignment statement! These variables are needed in the next small change to the game script.

7. To keep the image on the screen, change the `ImageOffset` variable at the appropriate time. To do this, under the `.move` method for the Surface object `GameImageLocation`, add the following lines:

```
if GameImageLocation.left < 0 or GameImageLocation.right > ScreenWidth:
    ImageOffset[0] = -ImageOffset[0]
if GameImageLocation.top < 0 or GameImageLocation.bottom > ScreenHeight:
    ImageOffset[1] = -ImageOffset[1]
```

In essence, this little tweak causes the Raspberry Pi image to go the opposite direction whenever it "hits" an edge of the game screen.

8. Test your changes to the game script by exiting the editor, typing `python3 py3prog/script1902.py`, and pressing Enter. You should see the Raspberry Pi image move and appear to bounce off the screen edges. Click anywhere in the game screen with your mouse to play a sound and end the game.

9. Open `script1902.py` in a script editor again. To get a feel for what is meant by changing the image's "speed," change the `ImageOffset` variable as follows:

```
ImageOffset=[50,50]
```

10. Test the speed change to the game script by exiting the editor, typing `python3 py3prog/script1902.py`, and pressing Enter. You should see the Raspberry Pi image move "faster" than it did before. Click anywhere in the game screen with your mouse to play a sound and end the game.

11. Again open `script1902.py` in a script editor and change the `ImageOffset` variable back to its original setting, as follows:

```
ImageOffset=[10,10]
```

12. Add a collide point to force the user to click the Raspberry Pi image in order to end the game. To do this, change the `for` loop construct concerning the event so it looks as follows:

```
for event in pygame.event.get():
    if event.type in (QUIT,MOUSEBUTTONDOWN):
        if GameImageLocation.collidepoint(pygame.mouse.get_pos()):
            ClickSound.play()
            pygame.time.delay(300)
            sys.exit()
```

13. Test the collide point in the game script by exiting the editor, typing `python3 py3prog/script1902.py`, and pressing Enter. Use your mouse to click anywhere in the game screen *except* on the Raspberry Pi image. Nothing should happen. Finally, click somewhere on the Raspberry Pi image to play a sound and end the game. You've created a nice collide point, but the Raspberry Pi image is so large that clicking it is hardly a game.

14. Open `script1902.py` in a script editor and change the size of the Raspberry Pi image. To do this, under where you load the image and convert it using the `.convert_alpha` method, add the following two lines to make the Raspberry Pi image smaller:

```
# Resize image (make smaller)
GameImageGraphic=pygame.transform.scale(GameImageGraphic,(75,75))
```

15. Exit the editor, type `python3 py3prog/script1902.py`, and press Enter to test this change. Does the Raspberry Pi image seem a lot smaller? It should. Chase the image around the screen until you can click somewhere on it to play a sound and end the game. Now you need one more tweak: You are going to make it a little harder to catch that raspberry.

16. Make the Raspberry Pi image move "faster" every time it hits a screen wall. To do this, open `script1902.py` in your favorite script editor. Change the whole "Keep game image on screen" section so that it matches the following:

```
#Keep game image on screen
if GameImageLocation.left < 0 or GameImageLocation.right > ScreenWidth:
    ImageOffset[0] = -ImageOffset[0]
    #Speed it up
    if ImageOffset[1] < 0:
        ImageOffset[1] = ImageOffset[1] - 1
    else:
        ImageOffset[1] = ImageOffset[1] + 1
    #
if GameImageLocation.top < 0 or GameImageLocation.bottom > ScreenHeight:
    ImageOffset[1] = -ImageOffset[1]
    #Speed it up
    if ImageOffset[0] < 0:
```

```
            ImageOffset[0] = ImageOffset[0] - 1
        else:
            ImageOffset[0] = ImageOffset[0] + 1
    #
```

**17.** Exit the editor, type `python3 py3prog/script1902.py`, and press Enter to test the last change. Let the Raspberry Pi image hit the game screen walls several times before you begin chasing it with the mouse pointer. Does it appear to speed up? Careful! Don't wait too long to chase it, or you may never catch it!

That was fun. But this hour is almost over, and you have only touched your little toe into the Python gaming world's ocean. Listing 19.7 is a last gift to help you on your way: the Raspberry Pie game.

**LISTING 19.7**    The Raspberry Pie Game Script

```
#script1903.py - The Raspberry Pie Game
#Written by Blum and Bresnahan
#######################################
#
##### Import Modules & Variables ######
import pygame       #Import PyGame library
import random       #Import Random module
import sys           #Import System module
#
from pygame.locals import *  #Local PyGame constants
#
pygame.init()                    #Initialize PyGame objects
#
##### Set up Functions ##############
#
# Delete a Raspberry
def deleteRaspberry (RaspberryDict, RNumber):
    key1 = 'RasLoc' + str(RNumber)
    key2 = 'RasOff' + str(RNumber)
    #
    #Make a copy of Current Dictionary
    NewRaspberry = dict(RaspberryDict)
    #
    del NewRaspberry[key1]
    del NewRaspberry[key2]
    #
    return NewRaspberry
#
```

```
# Set up the Game Screen #############
#
ScreenSize = ScreenWidth,ScreenHeight = 1000,700
GameScreen=pygame.display.set_mode(ScreenSize)
#
# Set up the Game Color ##############
blue=0,0,255
#
# Set up the Game Image Graphics ######
#
GameImage="/usr/share/raspberrypi-artwork/raspberry-pi-logo.png"
GameImageGraphic=pygame.image.load(GameImage).convert_alpha()
GameImageGraphic=pygame.transform.scale(GameImageGraphic,(75,75))
#
GameImageLocation=GameImageGraphic.get_rect()  #Current location
#
ImageOffset=[10,10]  #Starting Speed
#
# Build the Raspberry Dictionary ######
#
RAmount = 17    #Number of Raspberries on screen
Raspberry = {}  #Initialize the dictionary
#
for RNumber in range(RAmount): #Create the Raspberry dictionary
       Position_x = (ImageOffset[0] + RNumber) * random.randint(9,29)
       Position_y = (ImageOffset[1] + RNumber) * random.randint(8,18)
       RasKey = 'RasLoc' + str(RNumber)
       Location = GameImageLocation.move(Position_x, Position_y)
       Raspberry[RasKey] = Location
       RasKey = 'RasOff' + str(RNumber)
       Raspberry[RasKey] = ImageOffset
#
# Setup Game Sound #################
#
ClickSound=pygame.mixer.Sound('/home/pi/python_games/match1.wav')
#

###### Play the Game ####################################
#
while True:
       for event in pygame.event.get():
             if event.type == pygame.MOUSEBUTTONDOWN:
                   for RNumber in range(RAmount):
                         RasLoc = 'RasLoc' + str(RNumber)
                         RasImageLocation = Raspberry[RasLoc]
                         if RasImageLocation.collidepoint(pygame.mouse.get_pos()):
                               deleteRaspberry(Raspberry,RNumber)
                               RAmount = RAmount - 1
```

```
                        ClickSound.play()
                        pygame.time.delay(50)
                        if RAmount == 0:
                                sys.exit()
#Redraw the Screen Background ##############
GameScreen.fill(blue)
#
#Move the Raspberries around the screen #####
for RNumber in range(RAmount):
        RasLoc = 'RasLoc' + str(RNumber)
        RasImageLocation = Raspberry[RasLoc]
        RasOff = 'RasOff' + str(RNumber)
        RasImageOffset = Raspberry[RasOff]
        #
        NewLocation = RasImageLocation.move(RasImageOffset)
        #
        Raspberry[RasLoc] = NewLocation    #Update location
        #
#Keep Raspberries on screen #################
        if NewLocation.left < 0 or NewLocation.right > ScreenWidth:
            NewOffset = -RasImageOffset[0]
            if NewOffset < 0:
                NewOffset = NewOffset - 1
            else:
                NewOffset = NewOffset + 1
            #
            RasImageOffset = [NewOffset, RasImageOffset[1]]
            Raspberry[RasOff] = RasImageOffset #Update offset
            #
        if NewLocation.top < 0 or NewLocation.bottom > ScreenHeight:
            NewOffset = -RasImageOffset[1]
            if NewOffset < 0:
                NewOffset = NewOffset - 1
            else:
                NewOffset = NewOffset + 1
            #
            RasImageOffset = [RasImageOffset[0],NewOffset]
            Raspberry[RasOff] = RasImageOffset #Update offset
        #
        GameScreen.blit(GameImageGraphic,NewLocation) #Put on Screen
#
    pygame.display.update()
#
```

Often games have multiple images on the game screen. The Raspberry Pie game creates 17 Raspberry Pi images. (Supposedly, it takes about 4×17 raspberries to make a full-sized raspberry pie.) The game uses a dictionary to create the different raspberry images and keep track of their current locations and individual offset settings.

Each image must be clicked with a mouse to be eliminated. The game ends when all the raspberry images have been removed from the game screen. Figure 19.4 shows how the game looks in action.

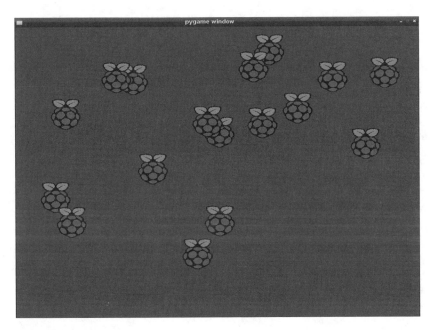

**FIGURE 19.4**
The Raspberry Pie game in action.

You can make a number of changes and improvements to the Raspberry Pie game script. The following are a few suggestions of changes you can try to further your adventure in learning:

▶ Create a "quit" option to leave the game using a keyboard sequence of keys.

▶ Add a game header to the game screen.

▶ Make each raspberry "pop" and disappear when it is clicked.

▶ Draw a hot baked pie image at the end of the game.

▶ Rewrite the Raspberry Pi images as objects instead of dictionary entries.

Game script writing really lets you be creative. Hopefully this small list of suggested changes will get you started writing your own Python game scripts.

# Summary

In this hour, you read about various Python game frameworks. You learned how to load and use the `PyGame` library to create game scripts. You got to write a couple simple games, add sounds to the game, and create player interaction with moving game images. In Hour 20, "Using the Network," you will expand your Python knowledge and learn about networking with Python.

# Q&A

**Q. I enjoyed this hour. How else can I improve my Python game script writing skills?**

**A.** For further challenges, take a look at www.pyweek.org. This site runs game programming contests twice a year to test your new-found Python game script writing skills.

**Q. I want to use a gamepad in my game. Can `PyGame` handle that?**

**A.** Yes, `PyGame` has an entire module, `pygame.joystick`, for interacting with gamepads, trackballs, and joysticks.

**Q. I'd like to have a picture as my game screen instead of a plain color. How can I make that happen?**

**A.** Instead of using the `.fill` method to fill the game screen with a plain color, you first load an image, similarly to the way you loaded the Raspberry Pi image this hour. Next, you make sure that image is as big as the game screen. Then you use the `.blit` method to draw it on the screen first, before you `.blit` the other images. You can create some really cool game backgrounds this way!

# Workshop

## Quiz

1. `Panda3D` is a sprite that you can use in a game script, and it comes pre-created in the `PyGame` library. True or false?

2. What `PyGame` method allows you to add a slight pause (milliseconds) in the game?

3. Which `PyGame` method checks for two images or an image and a mouse pointer being at the same place at the same time?

   a. `.collisionpoint`

   b. `.collidepoint`

   c. `.get_event`

# Answers

1. False. `Panda3D` is a full-feature framework for game development, including 3D graphics and a game engine. You can use it to write games in C++ or Python.

2. The `pygame.time.delay` method allows you to pause a game for a specified number of milliseconds.

3. Answer b is correct. The `.collidepoint` method tests the current image's location on the game screen and determines whether it has collided with another image or events location on the game surface.

# PART V

# Business Programming

# Using the Network

## What You'll Learn in This Hour:

► The Python network modules
► How to interact with email and web servers
► How to create your own client/server applications

These days, it's almost a necessity for programs to be able to interact with networks. Fortunately, there are lots of different modules available to help write network-aware Python applications that can interact with lots of different types of network servers. In this hour, you'll first take a look at the different network modules available for Python, and then you'll explore how to use Python scripts to interact directly with email servers and web servers. Finally, you'll wrap up this hour by creating your own client/server programs using Python.

# Finding the Python Network Modules

The Python v3 language supports lots of different networking features. However, because of the modular approach to Python programming, you often have to find just the right network module to use for your specific networking needs. Table 20.1 shows the different network-related modules that you can use in your Python v3 programs.

**TABLE 20.1** Python Networking Modules

| Module | Description |
|---|---|
| asyncore | Acts as an asynchronous socket handler. |
| asynchat | Provides additional functionality for the asyncore module. |
| cgi | Provides basic CGI scripting support for web servers. |
| cookie | Enables cookie object manipulation for web servers. |
| cookielib | Provides client-side cookie support. |
| email | Supports formatting of email messages (including MIME). |

| Module | Description |
| --- | --- |
| `ftplib` | Acts as an FTP client module. |
| `gopherlib` | Acts as a Gopher client module. |
| `httplib` | Acts as an HTTP client module for retrieving webpages. |
| `imaplib` | Is the IMAP 4 client module for reading mail from email servers. |
| `mailbox` | Reads text in several different Linux mailbox formats. |
| `mailcap` | Provides access to MIME configurations through `mailcap` files. |
| `mhlib` | Provides access to MH-formatted mailboxes. |
| `nntplib` | Is the NNTP client module for reading network news feeds. |
| `poplib` | Is the POP client module for reading mail from email servers. |
| `robotparser` | Provides support for parsing web server robot files. |
| `SimpleXMLRPCServer` | Acts as a simple XML-RPC server. |
| `smtpd` | Acts as an SMTP server module for creating an email server. |
| `smtplib` | Acts as an SMTP client module for sending email messages. |
| `telnetlib` | Acts as a Telnet client module for communicating with servers. |
| `urlparse` | Supports interpreting URLs. |
| `urllib` | Supports reading data from web servers. |
| `xmlrpclib` | Provides client support for the XML-RPC protocol. |

Each network module has its own documentation on how to use it within your Python programs, so you may have to do a bit of digging at the docs.python.org website to find just what you're looking for. The next two sections demonstrate a couple examples of how to use the modules to provide networking features in your Python programs. First you'll see how to incorporate email features into your Python scripts, and then you'll see how to read data from webpages and process it in your Python scripts.

# Working with Email Servers

One popular network feature you may run into with your Python scripts is the ability to send email messages. This can come in handy if you have automated Python scripts that run on their own, and you need to know if they fail, or if you'd just like to get the data results from your script sent to you in a convenient format, without having to log in to the Raspberry Pi to view the data.

The `smtplib` module provides just what you need to interface your Python scripts with the email system on your server. The following sections first explore how email works in the Linux environment and then tackle how to use the `smtplib` module to send email messages from Python scripts.

## Email in the Linux World

Sometimes the hardest part of using email in your Python programs is understanding just how the email system works in Linux. Knowing what software packages perform what particular tasks is crucial for getting emails from your Python scripts into your inbox.

One of the main goals of the Linux operating system was to modularize software. Instead of having one monolithic program handle all the required pieces of a function, Linux developers created smaller programs that each handle a smaller piece of the total functionality of the system.

This philosophy was also used when implementing the email systems used in Linux systems. In Linux, email functions are divided into separate pieces, each assigned to a different program. Figure 20.1 shows how most open-source email software modularizes email functions in the Linux environment.

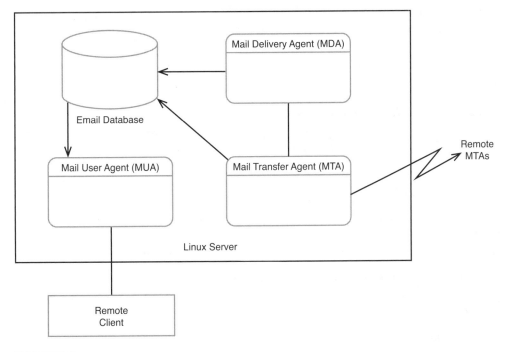

**FIGURE 20.1**
The Linux modular email environment.

As you can see in Figure 20.1, in the Linux environment, the email process is normally divided into three functions:

▶ The Mail Transfer Agent (MTA)

▶ The Mail Delivery Agent (MDA)

▶ The Mail User Agent (UA)

The MTA is the core of the Linux email system. It's responsible for handling both incoming and outgoing mail messages in the system. It maintains mailboxes for each user account on the system and can accept messages for each user. If your Raspberry Pi is directly connected to the Internet, it can also send messages destined for recipients on remote hosts as well.

Closely related to the MTA, the MDA delivers the messages the MTA server receives. The strength of the MDA is that it's highly customizable. This is where you can program out-of-office messages or rules to route incoming messages to different folders in your Inbox.

If you want to support email mailboxes directly on your Raspberry Pi, you have to install at least an MTA package, and you can optionally install a fancier MDA package. By far the two most popular MTA email packages that you'll see in the Linux environment are sendmail and Postfix. As it turns out, both of these packages combine the MTA and MDA functions into one software application, providing a full email server for your system. However, the Raspberry Pi's Raspbian distribution doesn't install either package by default, so you don't have any email capability for the user accounts on the Raspberry Pi.

### BY THE WAY

#### sendmail and Postfix on the Raspberry Pi

While not installed by default, both the sendmail and Postfix programs are available in the Raspbian software repository. You can install either package by using the `apt-get` installation tool. If you want to read your mail messages from the command line, you should also install the `mailx` program.

Setting up and configuring a full-blown email server on the Internet is not an easy task. However, if all you need to do is send email messages from your Python scripts to external email addresses, there's an easier way than having to set up your own mail server.

We haven't talked about the MUA package yet. The job of MUA is to provide a method for users to interface with their existing mailboxes (either on the local system or a remote system) to read and send email messages.

The `smtplib` module in Python provides full MUA capabilities, allowing your scripts to connect to any email server to send email messages—even servers that require encrypted authentication!

And what's even better, the smtplib package has become so popular that it's included in the standard Python library modules, so it's already available on your Raspberry Pi.

# The smtplib **Library**

The smtplib library includes three classes to create an SMTP connection to a remote email server and send out messages:

- ▶ **SMTP**—The SMTP class connects to the remote email server, using either the standard SMTP or the extended ESMTP.

- ▶ **SMTP_SSL**—The SMTP_SSL class allows you to establish an encrypted session to a remote email server.

- ▶ **LMTP**—The LMTP class offers a more advanced method of connecting to ESMTP servers.

For the scripts in this hour, you'll be using the SMTP class, but you'll also see how to encrypt the password transaction inside the SMTP session. This provides the least amount of overhead for the session while still keeping transactions secure.

Inside the SMTP class are several methods that you can use to set up and establish the connection with the email server. Table 20.2 lists these methods.

**TABLE 20.2**   The SMTP Class Methods

| Method | Description |
| --- | --- |
| connect(*host*, *port*) | Connects to the specified email server. |
| helo(*hostname*) | Identifies you to the remote email server, using the standard HELO SMTP message. |
| ehlo(*hostname*) | Identifies you to the remote email server, using the extended EHLO SMTP message. |
| login(*user*, *password*) | Logs in to the remote email server if it requires authentication. |
| startttls(*keyfile*, *certfile*) | Uses TLS security to encrypt the SMTP session. |
| sendmail(*from*, *to*, *message*, *mail options*, *rcpt options*) | Sends a mail message to the designated recipients. |
| quit() | Closes the SMTP session. |

To send an email message, you should follow these steps:

  **1.** Instantiate an SMTP class object, using the remote server connection information.

2. Send an EHLO message to the remote server.

3. Place the connection into TSL security mode.

4. Send the login information for the server.

5. Create the message to send.

6. Send the message.

7. Quit the connection.

The next section walks through each of these steps in creating a simple Python script for sending out email messages.

## Using the `smtplib` Library

The first step in the process of creating a Python script for sending out email messages is to instantiate the SMTP object class. When you do this, you need to provide the host name and port address required to connect to your email server, like this:

```
import smtplib
smtpserver = smtplib.SMTP('smtp.gmail.com', 587)
```

BY THE WAY

### Remote Email Servers

This example shows the host name and port used for the popular Gmail email server. Most email servers have specific host names and ports that you can use to connect to send email messages via email clients such as smart phone apps instead of using the web interface. You should be able to find that information on the Frequently Asked Questions (FAQ) webpages for your email server. If not, you may have to contact the tech support group for your email server to get that information.

After you instantiate the SMTP object class, you can start the login process. For the `login()` function, you must provide the user ID and password that you use to connect to your email server using your normal email client. Usually they're just your standard email address and password. To make the connection secure, you should also use the `starttls()` method. The process looks like this:

```
smtpsrtver.ehlo()
smtpserver.starttls()
smtpserver.ehlo()
smtpserver.login('myuserid', 'mypassword')
```

You may have noticed that the code uses the `ehlo()` method twice. Some email servers require the client to re-introduce itself after switching to encrypted mode. Therefore, it's a good idea to always use the extra `ehlo()` method, which doesn't break anything on servers that don't need it.

After you establish the connection and login, you're ready to compose and send your message. The message must be formatted using the RFC2822 email standard, which requires the message to start out with a To: line to identify the recipients, a From: line to identify the sender, and a Subject: line. Instead of creating one huge text string with all that info, it's easier to create separate variables with that info and then "glue" them all together to make the final message, like this:

```
to = 'person@remotehost.com'
from = 'rich@myhost.com'
subject = 'This is a test'
header = 'To:' + to + '\n'
header = header + 'From:' + from + '\n'
header = header + 'Subject:' + subject + '\n'
body = 'This is a test message from my Python script!'
message = header + body
```

Now the `message` variable contains the required RFC2822 headers, plus the contents of the message to send. It's easy to change any of the individual parts, if needed.

Now you need to send the message and close the connection, as shown here:

```
smtpserver.sendmail(from, to, message)
smtpserver.quit()
```

To send the message, the `sendmail()` method also needs to know the from and to information. The `to` variable can either be a single email address or a list object that contains multiple recipient email addresses.

In the following Try It Yourself, you'll write your own Python script to send email messages.

### TRY IT YOURSELF ▼

## Send Email Messages

In the following steps, you'll create a window application using the `tkinter` library (see Hour 18, "GUI Programming") to send an email message to one or more recipients. Just follow these steps:

**1.** Create the file `script2001.py` in the folder for this hour.

**2.** Enter the code shown here in the `script2001.py` file:

```
1:  #!/usr/bin/python3
2:  from tkinter import *
3:  import smtplib
```

```
 4:
 5:    class Application(Frame):
 6:        """Build the basic window frame template"""
 7:
 8:        def __init__(self, master):
 9:            super(Application, self).__init__(master)
10:            self.grid()
11:            self.create_widgets()
12:
13:        def create_widgets(self):
14:            menubar = Menu(self)
15:            menubar.add_command(label='Send', command=self.send)
16:            menubar.add_command(label='Quit', command=root.quit)
17:            self.label1 = Label(self, text='The Quick E-mailer')
18:            self.label1.grid(row=0, columnspan=3)
19:            self.label2 = Label(self, text="Enter the recipients:")
20:            self.label3 = Label(self, text='Enter the Subject:')
21:            self.label4 = Label(self, text='Enter your message here:')
22:            self.label2.grid(row=2, column=0)
23:            self.label3.grid(row=3, column=0)
24:            self.label4.grid(row=4, column=0)
25:
26:            self.recipients = Entry(self)
27:            self.subj = Entry(self)
28:            self.body = Text(self, width=50, height=10)
29:            self.recipients.grid(row=2, column=1, sticky = W)
30:            self.subj.grid(row=3, column=1, sticky = W)
31:            self.body.grid(row=5, column=0, columnspan=2)
32:
33:            self.button1 = Button(self, text="Send message",
   command=self.send)
34:            self.button1.grid(row=6, column=0, sticky = W)
35:
36:            self.recipients.focus_set()
37:            root.config(menu=menubar)
38:
39:        def send(self):
40:            """Retrieve the text, build the message, and send it"""
41:            server = 'smtp.gmail.com'
42:            port = 587
43:            sender = 'user@gmail.com'
44:            password = 'xxxxxxxx'
45:            to  = self.recipients.get()
46:            tolist = to.split(',')
47:            subject = self.subj.get()
48:            body = self.body.get('1.0', END)
49:            header = 'To:' + to + '\n'
```

```
50:            header = header + 'From:' + sender + '\n'
51:            header = header + 'Subject:' + subject + '\n'
52:            message = header + body
53:
54:            smtpserver = smtplib.SMTP(server, port)
55:            smtpserver.ehlo()
56:            smtpserver.starttls()
57:            smtpserver.ehlo()
58:            smtpserver.login(sender, password)
59:            smtpserver.sendmail(sender, tolist, message)
60:            smtpserver.quit()
61:            self.body.delete('1.0', END)
62:            self.body.insert(END, 'Message sent')
63:
64: root = Tk()
65: root.title('The Quick E-mailer')
66: root.geometry('500x300')
67: app = Application(root)
68: app.mainloop()
```

3. In lines 41–44, replace the `server`, `port`, `sender`, and `password` variable values with the information required for your email server.

4. Save the file.

5. Open the LXTerminal session in your LXDE desktop on the Raspberry Pi.

6. Run the `script2001.py` program from the command line, like this:

```
pi@raspberrypi ~ $ python3 script2001.py
```

You should see the window interface, as shown in Figure 20.2.

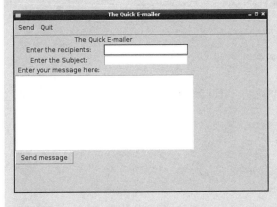

**FIGURE 20.2**
The Quick E-mailer application main window.

To send your message to multiple recipients, you just place a comma after each recipient in the To line. The code uses the `split()` string method to split the comma-delimited string into a list that the `sendmail()` method uses.

---

WATCH OUT!

**Advanced Security in Gmail**

GMail offers an advanced security feature that requires a two-step authentication process. This method won't work with that feature in GMail. It only works with a standard userid/password security.

---

# Working with Web Servers

These days, just about everyone gets information from the Internet. The World Wide Web (WWW) has become a primary source of information for news, weather, sports, and even personal information.

You can leverage this wealth of information on the Internet from your Python scripts. You might be wondering how you can use your Python scripts to extract data from the graphical world of webpages. Fortunately, Python makes it easy.

The Python `urllib` module, which is part of the standard Python library, allows you to interact with a remote website to retrieve information. It retrieves the full HTML code sent from the website and stores it in a variable. The downside is that you then have to parse through the HTML code, looking for the content you need. But fortunately again, Python provides help for doing that!

To summarize, extracting data from websites is basically a two-step process:

1. Connect to the website and retrieve the webpage.

2. Parse the HTML code to find the data you're looking for.

The following sections walk through these two steps to help you retrieve useful information from any website by using a Python script.

## Retrieving Webpages

Retrieving the HTML code for a webpage involves three steps:

1. Connect to the remote web server.

2. Send an HTTP request for the webpage.

3. Read the HTML code that the web server returns.

All these steps are handled with just two simple commands from the `urllib` module (after you import the module):

```
import urllib.request
response = urllib.request.urlopen(url)
html = response.read()
```

The `urlopen()` method attempts to establish the HTTP connection with the remote website specified in the parameter. You need to specify the full `http://` or `https://` format of the address in the URL. The `read()` method then retrieves the HTML code sent from the remote website.

The `read()` method returns the text as binary data instead of as a text string. You can use some of the standard Python tools to convert the HTML code into text (see Hour 10, "Working with Strings") and then use the standard Python searching tools (see Hour 16, "Regular Expressions") to parse through the HTML code, looking for the data you need, in a process called *screen scraping*. However, there's an easier way of extracting, and you'll learn about it next.

## Parsing Webpage Data

While screen scraping is certainly one way to extract data from a webpage, it can be extremely painful. Trying to hunt down individual data elements buried in the HTML code of a webpage can be quite a chore.

If you find the data you want and try to use a positional method of extracting the content (such as looking for the 1,200th character in an HTML document and splicing the next 10 characters), you might be disappointed when, the next time the webpage is updated, the data is at the 1,201st position.

One solution to this problem is to use an HTML parser library. An HTML parser library allows you to parse through the individual HTML elements contained in the document, looking for specific tags and keywords. This makes the job of searching for data much easier, and it can help your program survive simple changes to the webpage.

There are plenty of HTML parser libraries available in Python. The `HTMLParser` module is included in the standard Python library, but it can be somewhat difficult to work with. In the following Try It Yourself, you will use the LXML module, which is fairly easy to use yet robust enough to help you parse through the webpages you need.

▼ TRY IT YOURSELF

## Install the LXML Module

To complete the web parsing project, you need to install the Python v3 version of the LXML module from the Raspbian Linux distribution software repository. Just follow these steps:

1. Open a command prompt, either from the main Raspberry Pi login interface or from the LXTerminal utility in the graphical desktop.

2. Run the `apt-get` command as the root user account to update your library, like this:

   ```
   pi@raspberrypi ~ $ sudo apt-get update
   ```

3. Run the `apt-get` command as the root user account to install the Python v3 version of the LXML module, like this:

   ```
   pi@raspberrypi ~ $ sudo apt-get install python3-lxml
   ```

WATCH OUT!

## The LXML Module

Be careful. The Raspbian Linux distribution software repository includes both the Python v2 and Python v3 versions of the LXML module. Make sure you install the Python v3 version to use with your Python v3 code! The Python v3 version is `python3-lxml`, while the Python v2 version is `python-lxml`.

Now that you have the LXML module installed, you can import it into your program and use its features. There are two specific features that you're interested in:

▶ The `etree` methods, which break an HTML document down into the individual HTML code elements in the document.

▶ The `cssselect` methods, which can parse CSS data embedded in HTML documents.

Let's take a closer look at using each of these features.

## Using the `etree` Methods to Parse HTML

The `etree` methods break an HTML document down into the individual HTML elements. If you're familiar with HTML code, you've seen the HTML elements that are used to define the layout and structure of the webpage. Here's a quick example of the HTML code in a simple webpage:

```
<!DOCTYPE html>
<html>
<head>
```

```
<title>This is a test webpage</title>
</head>
<body>
<h1>This is a test webpage!</h1>
<p>This webpage contains a simple title and two paragraphs of text</p>
<p>This is the second paragraph of text on the webpage</p>
<h2>This is the end of the test webpage</h2>
</body>
</html>
```

The `etree` methods can return each HTML element in the document as a separate object that you can manipulate. Here's the code required to extract the HTML elements from the `html` variable returned from the `urllib` process shown earlier:

```
import lxml.etree
encoding = lxml.etree.HTMLParser(encoding='utf-8')
doctree = lxml.etree.fromstring(html, encoding)
```

First, you need to define the encoding that you want to convert the raw binary HTML data into. The `encoding` variable contains the encoding object to use. This example defines the `utf-8` encoding scheme, which can handle most languages in the world.

The second statement uses the `fromstring()` method to produce a list that contains the string values of all the HTML elements and their values. The `doctree` variable contains a list of the individual HTML elements and their values. You can search the list values, looking for the data, or if you know exactly which position in the element list your data appears, you can jump directly there. That method is a little better than using the regular expression method to search for data, but you can still make things easier!

Most webpages use Cascading Style Sheets (CSS) to differentiate important content on the webpage. The next step is to leverage that information to look for the specific data you want.

## Using CSS to Find Data

Now that you have the webpage data broken down into the separate elements, you can use the `CSSSelector()` method in the `lxml` module to try to parse the data even further, based on CSS information in the webpage.

You may need to do some hunting around through the raw HTML code to figure out just what unique features make the data you're looking for stand out. Most modern webpages use CSS classes to define CSS styles for specific content on the webpage. It looks something like this:

```
<div class="day-temp-current temp-f">79</div>
```

In this example, the current temperature value is surrounded by an HTML `<div>` element that's assigned to a specific CSS class to customize how it appears on the webpage. With the `lxml` method, you can only find the `<div>` element, but with the `CSSSelector()` method, you can search for the specific CSS class to find the exact data, like this:

```
from lxml.etree.cssselect import CSSSelector
div = CSSSelector("div.'day-temp-current temp-f'")
temp = div(doctree)[0]
```

The `CSSSelector()` method specifies the HTML element and the CSS class that you're looking for. This is the syntax:

```
CSSSelector("element.class")
```

In this example, the class name that you're looking for contains a space (which is allowed in CSS), and this complicates things a bit. Because of the space, you need to place quotes around the class name in the parameter, and you also need to place quotes around the entire value.

The `CSSSelector()` method sets up the item you're searching for, and then you just need to feed the result from the `etree` parser into it. The result will be a list of all the elements that match both the HTML element and the CSS class. Hopefully, that only applies to one item in the webpage and gives you the data you want!

▼ TRY IT YOURSELF

### Find the Current Temperature

Plenty of websites can tell you the current temperature in your city. Follow these steps to write a Python script that contacts one of those sites, retrieves the temperature, and then displays it:

1. Find the specific website URL that has the information you're looking for. For this exercise, use the popular Yahoo Weather webpage to look up the current temperature in Chicago, Illinois. If you go to the main weather.yahoo.com webpage, you have to enter the city and state information. After you do that, you are redirected to a different URL that contains the weather data. For Chicago, this is the URL:

   ```
   http://weather.yahoo.com/unitd-states/illinois/chicago-2379574/
   ```

   Make note of this, as it's the URL that you need to use for your `urlopen()` method.

2. Use the View Source feature in your browser to look at the raw HTML code for the webpage. Look for the data you're interested in, and see what HTML elements are around it. For the current temperature on the Yahoo Weather page, you might find this:

   ```
   <div class="day-temp-current temp-f">79&deg;
   ```

   Armed with the URL and the data you're looking for, you're ready to write the Python script.

3. Create the file `script2002.py` in this hour's folder on your Raspberry Pi.

**4.** Open the `script2002.py` file with an editor and enter the code shown here:

```
1:  #!/usr/bin/python3
2:
3:  import urllib.request
4:  import lxml.etree
5:  from lxml.cssselect import CSSSelector
6:
7:  url = 'http://weather.yahoo.com/united- states/illinois/chicago-2379574/'
8:  response = urllib.request.urlopen(url)
9:  html = response.read()
10:
11: parser = lxml.etree.HTMLParser(encoding='utf-8')
12: doctree = lxml.etree.fromstring(html, parser)
13:
14: div = CSSSelector("div.'day-temp-current temp-f'")
15: temp = div(doctree)[0].text[0:-1]
16: print('The current temperature in Chicago is', temp)
```

**5.** Save the file.

**6.** Run the script from a command line, as shown here:

```
pi@raspberrypi ~ $ python3 script2002.py
The current temperature in Chicago is 79
pi@raspberrypi ~ $
```

This example had to add one extra step in processing the data (refer to line 15). Unfortunately, the data returned from the element contained the `&deg;` HTML code to make the fancy degree symbol on the webpage. Depending on the terminal you use to run the program, that code may produce an odd ASCII character in the text output. To avoid that, you use the `text` method to convert the data into a text string, and then you use string splicing to remove the odd character at the end of the string value, like this:

```
temp = div(doctree)[0].text[0:-1]
```

The beauty of this script is that after you extract the temperature data from a webpage, you can do whatever you want with it, such as create a table of temperatures to track historical temperature data. You can then schedule the script to run at regular intervals to track the temperature throughout the day (or even combine it with the email script to automatically email it to yourself)!

### The Volatility of the Internet

The Internet is a dynamic place. Don't be surprised if you spend hours working out the precise location of data on a webpage, only to find that it's moved a couple weeks later, breaking your script. In
fact, it's quite possible that this example won't work by the time you read this book. If you know the
process for extracting data from webpages, as shown in this Try It Yourself, you can then apply that
principle to any situation.

# Linking Programs Using Socket Programming

Besides connecting to other servers, Python also allows you to create your own servers on a network. You can write a server application that listens for connections from client programs and
communicates with the client programs across the network, allowing you to move your application around the network.

In the following sections, you'll first learn how the client/server paradigm works in network
programming, and then you'll see how to create your own server and client programs by using
Python scripts.

## What Is Socket Programming?

Before diving into client/server programming, it's a good idea to have an understanding of how
client and server programs operate. Obviously, the client program and the server program each
have different responsibilities in the connection and transfer of data.

A server program listens to the network for requests coming from clients. A client program initi-
ates a request to the server for a connection. Once the server accepts the connection request, a
two-way communication channel is available for each device to send and receive data. Figure
20.3 shows this process.

As you can see in Figure 20.3, the server must perform two functions before it can communicate
with the client. First, it must set up a specific TCP port to listen for incoming requests. When a
connection request comes in, it must then accept the connection.

The client's responsibility is much simpler. All it must do is attempt to connect to the server on
the specific TCP port on which the server is listening. If the server accepts the connection, the
two-way communication is available, and the data can be sent.

Once a connection is established between the server and the client, the two devices must use
some sort of communication process (or protocol). If both devices attempt to listen for a message
at the same time, they'll deadlock, and nothing will happen. Likewise, if they both attempt to
send a message at the same time, nothing will be accomplished. It's your job as the network programmer to decide what protocol rules your client and server programs must follow in order to
communicate.

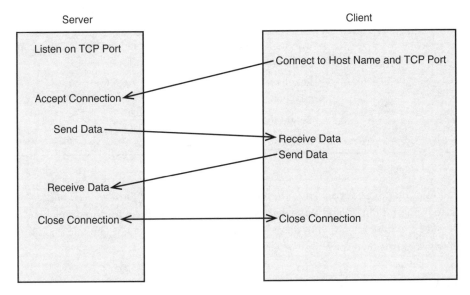

**FIGURE 20.3**
The client/server communication process.

You create client/server programs by using sockets. Sockets are the interface between your program and the physical network connection on the client and server devices. Creating the code to interact with the network connection is called *socket programming*.

# The Python `socket` Module

Python includes the `socket` module in the standard Python libraries to help you write network programs. This module provides all the methods you need for both the server and client sides of the connection. Table 20.3 shows the methods you use to write network programs.

**TABLE 20.3**   Python `socket` Module Methods

| Method | Description |
|---|---|
| `accept` | Accepts an incoming connection. |
| `bind(address)` | Binds the socket to an address and a port. |
| `close()` | Closes an established connection. |
| `connect(address)` | Connects to a server. |
| `gethostbyname(host)` | Returns the IP address of a host name. |
| `gethostname()` | Returns the host name of the local system. |
| `gethostbyaddr(address)` | Returns the host name of an IP address. |

| Method | Description |
| --- | --- |
| listen(*backlog*) | Listens for incoming connections and buffers backlog connections, if necessary. |
| rcv(*bufsize*) | Receives up to *bufsize* bytes of data from the connection. |
| send(*data*) | Sends the data through the connection. |
| socket(*family*, *type*, *proto*) | Creates a network socket interface. |

There are also lots of methods for converting host addresses into different formats used on the Internet. This hour doesn't dig into all those details, as you don't need them for the simple scripts you'll create. If you're interested in them, though, check out the socket module documentation on the docs.python.org website for more information.

## Creating the Server Program

To demonstrate the creation of a client/server program using Python, you're going to set up a simple network application. The server program you'll create will listen for incoming connection requests on TCP port 5150. When a connection request comes in, the server will accept it and then send a welcome message to the client.

The server program will then wait to receive a message from the client. If it receives a message, the server will display that message and then send the same message back to the client. After sending the message, the server will loop back to listen for another message. This loop will continue until the server receives a message that consists of the word exit. When that happens, the server will terminate the session.

To create the server program, you use the socket module to create a socket and listen for incoming connections on TCP port 5150, as shown in this example:

```
import socket
server = socket.socket(socket.AF_INET, socket.SOCK_STREAM)
host = ''
port = 5150
server.bind((host, port))
server.listen(5)
```

The socket() method parameters are somewhat complex, but they're also somewhat standard. The AF_INET parameter tells the system that you're requesting a socket that uses a IPv4 network address, not an IPv6 network address. The SOCK_STREAM parameter tells the system that you want a TCP connection instead of a UDP connection. (You use SOCK_DGRAM for a UDP connection.) TCP connections are more reliable than TCP connections, and you should use them for most network data transfers.

The bind() function establishes the program's link to the socket. It requires a tuple value that represents the host address and port number to listen on. If you bind to an empty host address, the server listens on all IP addresses assigned to the system (such as the localhost address and the assigned network IP address). If you need to listen on only one specific address, you can use the gethostbyaddr() or gethostbyname() method in the socket module to retrieve the system's specific host name or address.

Once you bind to the socket, you use the listen() method to start listening for connections. The program halts at this point, until it receives a connection request. When it detects a connection request on the TCP port, the system passes it to your program. Your program can then accept it by using the accept() method, like this:

```
client, addr = server.accept()
```

The two variables are required because the accept() method returns two values. The first value is a handle that identifies the connection, and the second value is the address of the remote client. Once the connection is established, all interaction with the client is done using the client variable. The two methods you use are send() and recv():

```
client.send(b'Welcome to my server!')
data = client.recv(1024)
```

The send() method specifies the bytes that you want sent to the client. Note that this is in byte format and not text format. You can use the standard Python string methods to convert between the text and byte values. The recv() method specifies the size of the buffer to hold the received data and returns the data received from the client as bytes instead of a text string value.

When you're done with the connection, you must close it, like this, to reset the port on the system:

```
client.close()
```

Now that you've seen the basics, you're going to create a server program to experiment with.

**TRY IT YOURSELF** ▼

### Create a Python Server Program

Follow these steps to create a server program to listen for client connections:

1. Create the script2003.py program in the folder for this hour.

2. Open the script2003.py file with a text editor and enter the code shown here:

```
1:  #!/usr/bin/python3
2:
3:  import socket
```

```
4:  server = socket.socket(socket.AF_INET, socket.SOCK_STREAM)
5:  host = ''
6:  port = 5150
7:  server.bind((host, port))
8:  server.listen(5)
9:  print('Listening for a client...')
10: client, addr = server.accept()
11: print('Accepted connection from:', addr)
12: client.send(str.encode('Welcome to my server!'))
13: while True:
14:     data = client.recv(1024)
15:     if (bytes.decode(data) == 'exit'):
16:         break
17:     else:
18:         print('Received data from client:', bytes.decode(data))
19:         client.send(data)
20: print('Ending the connection')
21: client.send(str.encode('exit'))
22: client.close()
```

   **3.** Save the file.

The `script2003.py` program goes through the steps to bind to a socket on the local system (line 7) and listen for connections on TCP port 5150 (line 8). When it receives a connection from a client (line 10), it prints a message in the command prompt and then sends a welcome message to the client (lines 11 and 12). This example uses the `str.encode()` string method to convert the text into a byte value to send and the `bytes.decode()` method to convert the bytes into text values to display.

After sending the welcoming message, the code goes into an endless loop, listening for data from the client and then returning the same data (lines 13 through 19). If the data is the word `exit`, the code breaks out of the loop and closes the connection.

That's it for the server side! Now you're ready to work on the client side of the application.

## Creating the Client Program

The client side of a network connection is a little simpler than the server side. It simply needs to know the host address and the port number that the server uses to listen for connections, and then it can create the connection and follow the protocol that you created to send and receive data.

Just like a server program, a client program needs to use the `socket()` method to establish a socket that defines the communication type. Unlike the server program, it doesn't need to bind to a specific port; the system will assign one to it automatically to establish the connection. With that, you just need five lines of code to establish the connection to the server:

```
import socket
s = socket.socket(socket.AF_INET, socket.SOCK_STREAM)
host = socket.gethostbyname('localhost')
port = 5150
s.connect((host, port))
```

The client code uses the `gethostbyname()` method to find the connection information to the server, based on the server's host name. Since the server is running on the same system, you use the special `localhost` host name.

The `connect()` method uses a tuple value of the host and port number to request the connection with the server. If the connection fails, it returns an exception that you can catch.

Once the connection is established, you can use the `send()` and `recv()` methods to send and receive byte streams. Just as in the server program, when the connection is terminated, you want to use the `close()` method to properly end the session.

---

### TRY IT YOURSELF ▼

### Create a Python Client Program

Follow these steps to create a Python client program that can communicate with the server program you just created:

1. Create the file `script2004.py` in the folder for this hour.

2. Open the `script2004.py` file in a text editor and enter the code shown here:

```
 1: #!/usr/bin/python3
 2:
 3: import socket
 4: server = socket.socket(socket.AF_INET, socket.SOCK_STREAM)
 5: host = socket.gethostbyname('localhost')
 6: port = 5150
 7:
 8: server.connect((host, port))
 9: data = server.recv(1024)
10: print(bytes.decode(data))
11: while True:
12:     data = input('Enter text to send:')
13:     server.send(str.encode(data))
14:     data = server.recv(1024)
```

```
15:     print('Received from server:', bytes.decode(data))
16:     if (bytes.decode(data) == 'exit'):
17:         break
18: print('Closing connection')
19: server.close()
```

3. Save the file.

The `script2004.py` code runs through the standard methods to create the socket and connect to the remote server (lines 4 through 8). It then listens for the welcome message from the server (line 9) and displays the message when it's received (line 10).

After that, the client goes into an endless `while` loop, requesting a text string from the user and sending it to the server (line 13). It listens for the response from the server (line 14) and prints it on the command line. If the response from the server is `exit`, the client closes the connection.

## Running the Client/Server Demo

To run the client/server demo programs, you need to have two separate command-line sessions open. You can do that in your LXDE graphical desktop by opening two separate LXTerminal windows.

In one window, you start the `script2003.py` server program first:

```
pi@raspberrypi ~ $ python3 script2003.py
Listening for a client...
```

When you see the message that the server is listening for a client, you can start the `script2004.py` client program in the other LXTerminal window:

```
pi@raspberrypi ~ $ python3 script2004.py
Welcome to my server!
Enter text to send:this is a test
```

If all goes well, you see the welcome message sent from the server, along with a prompt for the text to send back. Enter a short message and press Enter.

When you send the message from the client, you should see it appear in the server window:

```
Accepted connection from: ('127.0.0.1', 46043)
Received data from client: this is a test
```

The client window should receive the response back from the server:

```
Received from server: this is a test
Enter text to send:
```

When you enter the word exit at the client prompt, both the server and client connections terminate and stop the programs.

Congratulations! You've just written a complete Python client/server application!

---

WATCH OUT!

### Closing Sockets

If you try to restart the script2003.py server program immediately after it closes, you may get an error message that the socket is still in use. By default, Linux systems allow socket connections to linger open for a while after they've been closed, in case any stray network data comes across. The socket usually fully closes and is ready for reuse within a minute or so. You can monitor this by using the netstat -t command. You should see a connection for TCP port 5150 in a TIME_WAIT status while the system is waiting to close the socket.

---

# Summary

In this hour, you explored the world of network programming with Python scripts. You learned about the various modules available that allow your Python scripts to interact with a myriad of network servers. You got to see two specific examples: using Python to send email messages and using Python to parse data from webpages. After that, you took a look at the client/server programming paradigm and how to use the Python socket module to create your own client/server network programs.

In the next hour, we'll explore the world of database programming. Databases have become popular in just about every type of programming, and Python programming is no different. Fortunately, there are some simple libraries that we can use to help us add database features to our programs.

# Q&A

Q. The socket demo program in this hour allows only one client to connect to the server at a time. Can I write a program that allows hundreds of clients to connect at the same time?

A. Yes, you can, but that gets complicated! You have to use a process called *forking* to create a separate program thread to handle each client connection as it comes in. The server program listens for new connections and then forks each client connection to a new thread to handle the protocol process.

# Workshop

## Quiz

1. What Python module should you use to allow your scripts to send email messages to a remote host?

   **a.** `smtplibd`

   **b.** `smtplib`

   **c.** `urllib`

   **d.** `lxml`

2. You can't retrieve data from webpages with your Python scripts because the data is in a graphical format. True or false?

3. What's the order of socket methods required for a server to listen and accept a client connection?

## Answers

1. b. `smtplib`. The smtplibd module allows you to write your own email server programs to receive messages, but not send them.

2. False. Your Python script can read the raw HTML code text and parse the data.

3. You need to use the `socket()`, `bind()`, `listen()`, and then `accept()` methods for the server side of the connection.

# HOUR 21

# Using Databases in Your Programming

**What You'll Learn in This Hour:**

▶ How to use a MySQL database server in your Python scripts
▶ How to use a PostgreSQL database server in your Python scripts

One of the problems with Python scripts is persistent data. You can store all the information you want in your program variables, but at the end of the program, they just go away. There are times when you'd like for your Python scripts to be able to store data that you can use later. In the old days, storing and retrieving data from a Python script required creating a file, reading data from the file, parsing the data, and saving the data back into the file. Trying to search for data in the file meant having to read every record in the file to look for your data. Today, with databases being all the rage, it's a snap to interface your Python scripts with professional-quality open-source databases.

The two most popular open-source databases used in the Linux world are MySQL and PostgreSQL, and both are supported on the Raspberry Pi! In this hour, you'll see how to get these databases running on your Raspberry Pi system and then spend some time getting used to working with them from the command line. You'll then learn how to interact with each one by using your Python script programs.

## Working with the MySQL Database

By far the most popular database available in the Linux environment is the MySQL database. Its popularity has grown as a part of the Linux-Apache-MySQL-PHP (LAMP) server environment, which many Internet web servers use for hosting online stores, blogs, and applications.

The following sections describe how to install and set up a MySQL database in your Raspberry Pi environment, how to create the necessary database objects to use in your Python scripts, and how to write Python scripts to interact with the database.

## Installing MySQL

While the MySQL database isn't installed by default on the Raspberry Pi, installing it is a simple process. The Raspbian Linux distribution has two packages in the software repository to support the MySQL environment: `mysql-client` and `mysql-server`. As you can probably guess, the `mysql-server` package contains the files necessary to install and run the MySQL database server. You'll also install the `mysql-client` package, which contains a command-line interface to the database server that you can use to create the database objects for your Python programs.

To install the MySQL environment, you just use the `apt-get` program:

```
pi@raspberrypi ~ $ sudo apt-get update
pi@raspberrypi ~ $ sudo apt-get install mysql-client mysql-server
```

WATCH OUT!
_____

### The MySQL root User Account

During the process of installing the MySQL server package, the installation script queries you for a password for the MySQL root user account. The MySQL server maintains its own set of user accounts and passwords, separate from the Linux system user accounts. The root user account in the MySQL server has total control over the entire MySQL server. Be sure to remember the password you assign to the MySQL root user account!

_____

When the installation process finishes, the MySQL database server program automatically starts running in background mode. You're all ready to start setting up your MySQL database for your Python application.

## Setting Up the MySQL Environment

Before you can start writing your Python scripts to interact with a database, you need a few database objects to work with. At a minimum, you'll want to have these:

▶ A unique database to store your application data

▶ A unique user account to access the database from your scripts

▶ One or more data tables to organize your data

You build all these objects by using the `mysql` command-line client program. The `mysql` program interfaces directly with the MySQL server, using SQL commands to create and modify each of the objects.

Most of the interactions you make with the database are performed using SQL statements. The SQL language is an industry standard for communicating with different types of databases. You can send any type of SQL statement to the MySQL server by using the `mysql` program. The following sections walk through the different SQL statements you'll need to build the basic database objects for your shell scripts.

## Creating a Database

The MySQL server organizes data into databases. A database usually holds the data for a single application, separating it from other applications that use the database server. Creating a separate database for each Python application helps eliminate confusion and data mix-ups.

You need to use this SQL statement to create a new database:

```
CREATE DATABASE name;
```

This is pretty simple. Of course, you must have the proper privileges to create new databases on the MySQL server. The easiest way to ensure that you do is to log in as the root user account:

```
pi@raspberrypi ~ $ mysql -u root -p
Enter password:
Welcome to the MySQL monitor.  Commands end with ; or \g.
Your MySQL connection id is 51
Server version: 5.5.31-0+wheezy1 (Debian)

Copyright (c) 2000, 2013, Oracle and/or its affiliates. All rights reserved.

Oracle is a registered trademark of Oracle Corporation and/or its
affiliates. Other names may be trademarks of their respective
owners.

Type 'help;' or '\h' for help. Type '\c' to clear the current input statement.

mysql> CREATE DATABASE pytest;
Query OK, 1 row affected (0.00 sec)

mysql>
```

You can see if the new database was created by using the SHOW command:

```
mysql> SHOW DATABASES;
+--------------------+
| Database           |
+--------------------+
| information_schema |
| mysql              |
| performance_schema |
| pytest             |
| test               |
+--------------------+
5 rows in set (0.01 sec)

mysql>
```

This code shows that it was successfully created. You should now be able to connect to the new database with the USE statement:

```
mysql> USE pytest;
Database changed
mysql> SHOW TABLES;
Empty set (0.00 sec)

mysql>
```

The SHOW TABLES command allows you to see if there are any tables created. The Empty set result indicates that there aren't any tables to work with yet. Before you start creating tables, though, there's one other thing you need to do.

## Creating a User Account

So far you've seen how to connect to the MySQL server by using the root administrator account. This account has total control over all the MySQL server objects—very much as the root Linux account has complete control over the Linux system.

It's extremely dangerous to use the root MySQL account for normal applications. If there were a breach of security in the application and an attacker figured out the password for the root user account, all sorts of bad things could happen to your system (and data). To prevent that, it's wise to create a separate user account in MySQL that has privileges only for the database used in the application. You do this with the GRANT SQL statement, as shown here:

```
mysql> GRANT SELECT,INSERT,DELETE,UPDATE ON pytest.* TO test@localhost
IDENTIFIED by 'test';
Query OK, 0 rows affected (0.00 sec)

mysql>
```

This is quite a long command. Let's walk through the pieces to see what it's doing.

The first section defines the privileges the user account has on various database(s). This statement allows the user account to query the database data (the SELECT privilege), insert new data records, delete existing data records, and update existing data records.

The pytest.* entry defines the database and tables that the privileges apply to. This is specified in the following format:

```
database.table
```

As you can see from this example, you're allowed to use wildcard characters when specifying the database and tables. This format applies the specified privileges to all the tables contained in the database named pytest.

Finally, you specify the user account(s) that the privileges apply to (`test`, in this example). The MySQL server also allows you to restrict the assigned privileges to apply only when the user account connects from a specific location. You restrict the `test` user account to only log in from the `localhost` location, which means only from scripts running on your Raspberry Pi system.

The neat thing about the GRANT statement is that if the user account doesn't exist, it creates it. IDENTIFIED BY allows you to set the password for the new user account.

When you're done working with the MySQL server, use the `exit` command to get back to the standard Linux command prompt:

```
mysql> exit;
pi@raspberrypi ~ $
```

You can test the new user account directly from the `mysql` program, like this:

```
pi@raspberrypi ~ $ mysql pytest -u test -p
Enter password:
Welcome to the MySQL monitor.  Commands end with ; or \g.
Your MySQL connection id is 76
Server version: 5.5.31-0+wheezy1 (Debian)

Copyright (c) 2000, 2013, Oracle and/or its affiliates. All rights reserved.

Oracle is a registered trademark of Oracle Corporation and/or its
affiliates. Other names may be trademarks of their respective
owners.

Type 'help;' or '\h' for help. Type '\c' to clear the current input statement.

mysql>
```

The first parameter specifies the default database to use (`pytest`), and as you've already seen, the `-u` parameter defines the login user account, and `-p` to queries for the password (remember, you set that to "test" earlier). After you enter the password assigned to the `test` user account, you should be connected to the server.

Now that you have a database and a user account, you're ready to create some tables for the data.

## Creating Tables

The MySQL server is a relational database. In a relational database, data is organized by data fields, records, and tables. A data field is a single piece of information, such as an employee's last name or a salary. A record is a collection of related data fields, such as the employee ID number, last name, first name, address, and salary. Each record indicates one set of the data fields.

The table contains all the records that hold the related data. Thus, a table called `employees` holds the data records for each employee.

To create a new table in the database, you need to use the CREATE TABLE SQL command, like this:

```
pi@raspberrypi ~ $ mysql -u root -p
Enter password:
mysql> USE pytest;
Database changed
mysql> CREATE TABLE employees (
    -> empid int not null,
    -> lastname varchar(30),
    -> firstname varchar(30),
    -> salary float,
    -> primary key (empid));
Query OK, 0 rows affected (0.14 sec)

mysql>
```

First of all, notice that to create the new table, you need to log in to MySQL using the root user account, since the `test` user account doesn't have privileges to create a new table. The next item to notice is that you specify the `pytest` database in the USE SQL command to connect to the `pytest` database.

WATCH OUT!

## Creating the Table in a Database

It's extremely important that you make sure you're in the right database before you create the new table. Also, you need to make sure you're logged in using the administrative user account (root for MySQL) to create the tables.

Each data field in the table is defined using a data type. The MySQL database supports lots of different data types. Table 21.1 shows some of the more popular data types you may need.

**TABLE 21.1**  MySQL Data Types

| Data Type | Description |
| --- | --- |
| char | A fixed-length string value |
| varchar | A variable-length string value |
| int | An integer value |
| float | A floating-point value |
| boolean | A Boolean true/false value |

| Data Type | Description |
|-----------|-------------|
| `date` | A date value in YYYY-MM-DD format |
| `time` | A time value in HH:mm:ss format |
| `timestamp` | A date and time value together |
| `text` | A long string value |
| `blob` | A large binary value, such as an image or a video clip |

The `empid` data field definition also specifies a data constraint. A data constraint restricts what type of data you can enter to create a valid record. The `not null` data constraint indicates that every record must have an `empid` value specified.

Finally, the `primary key` line defines a data field that uniquely identifies each individual record. This means that each data record must have a unique `empid` value in the table.

After you create the new table, you can use the `SHOW TABLE` command to ensure that it's created.

## Installing the Python MySQL Module

To get your Python scripts to communicate with the MySQL server, you need to use a Python `MySQL/Connector` module. This is where things get a little interesting.

There are quite a few different Python modules for communicating with MySQL servers, but unfortunately, not very many of them have been ported to the Python v3 world yet. The `MySQL/Connector` module, created by the developers of MySQL, has been ported to the Python v3 world, so you can use that in your Python 3 scripts.

The downside is that at the time of this writing, the `MySQL/Connector` module for Python v3 isn't available in the standard Debian Linux software repository yet, so it's also not available in the Raspbian software repository. However, you can download the package from the Debian Experimental software repository and install it on your Raspberry Pi.

TRY IT YOURSELF ▼

### Install the Python v3 `MySQL/Connector` Module

In the following steps, you'll install the `MySQL/Connector` module for Python v3 on your Raspberry Pi system. Here's what you do:

1. Open a browser window and navigate to this URL:

   http://packages.debian.org/experimental/python3-mysql.connector

**2.** In the "Download python3-mysql.connector" section, click the All link. That should take you to the following URL:

```
://packages.debian.org/experimental/all/python3-mysql.connector/download
```

**3.** Click a link to download the package from a repository close to your location. At the time of this writing, this is the download file name:

```
python3-mysql.connector_1.0.9-1_all.deb
```

**4.** Use the dpkg command to install the Debian package, like this:

```
pi@raspberrypi ~ $ sudo dpkg -i python3-mysql.connector_1.0.9-1_all.deb
```

**5.** Open a Python v3 command-line session and try to import the mysql.connector module, as follows:

```
pi@raspberrypi ~ $ python3
Python 3.2.3 (default, Mar  1 2013, 11:53:50)
[GCC 4.6.3] on linux2
Type "help", "copyright", "credits" or "license" for more information.
>>> import mysql.connector
>>>
```

If all goes well, you shouldn't get any error messages about a missing module when you run the import statement. Now you're ready to start working with the MySQL database!

BY THE WAY

**Downloading Debian Packages**

If you don't have a graphical environment setup on your Raspberry Pi, you can download the MySQL/Connector Debian package on a separate workstation, then use SFTP to copy it over to your Raspberry Pi system.

## Creating Your Python Scripts

Now that you have a MySQL database and all the pieces for your Python script to interact with your MySQL database, you can start some coding. The following sections walk through the main processes that you need to use to create and retrieve data.

### Connecting to the Database

The first step in interacting with the MySQL database is to establish a connection from your Python script to the MySQL server. That's done using the connect() method, as shown here:

```
>>> import mysql.connector
>>> conn = mysql.connector.connect(user='test', password='test',
 database='pytest')
>>>
```

You must first import the `mysql.connector` module, and then you can run the `connect()` method from the library. In the `connect()` method, you need to specify the user account and password to connect to the MySQL server, as well as the database name. When you're done interacting with the database, you should use the `close()` method to close the connection.

**WATCH OUT!**

**Database Script Security**

You may have noticed that for the `connect()` method, you must specify the user account and password directly in your Python script. This can be somewhat of a security issue, so make sure to use the proper permissions on your script to protect it from being read by anyone else on your Linux system.

## Inserting Data

After connecting to the database, you can submit SQL statements to the MySQL server to insert new data records. Inserting data into the table is a three-step process. Listing 21.1 shows the `script2101.py` program, which demonstrates this process.

**LISTING 21.1** The `script2101.py` Program

```
1:  #!/usr/bin/python3
2:
3:  import mysql.connector
4:  conn = mysql.connector.connect(user='test', password='test',
 database='pytest')
5:  cursor = conn.cursor()
6:  new_employee = ('INSERT INTO employees '
7:     '(empid, lastname, firstname, salary) '
8:     'VALUES (%s, %s, %s, %s)')
9:
10: employee1 = ('1', 'Blum', 'Barbara', '45000.00')
11: employee2 = ('2', 'Blum', 'Rich', '30000.00')
12:
13: try:
14:     cursor.execute(new_employee, employee1)
15:     cursor.execute(new_employee, employee2)
16:     conn.commit()
17: except:
```

```
18:     print('Sorry, there was a problem adding the data')
19: else:
20:     print('Data values added!')
21: cursor.close()
22: conn.close()
```

First, you must define a cursor to the table (line 5). The cursor is a pointer object; it keeps track of where in the table the current operation will perform. You must have a valid table cursor to be able to insert new data records. You create a cursor by using the cursor() method after you establish the connection. You need to assign the output of the cursor() method to a variable because you need to reference that later on in your script.

Next, you have to create an INSERT statement template that you use to add a new data record (lines 6 through 8). The template uses placeholders for any data locations in the INSERT statement. That allows you to reuse the same template to insert multiple data records.

The new_employee template defines the data fields for the data, as well as a data value placeholder for each of the data values. It uses the %s format for the placeholders, no matter what data type they really are.

After creating the template, you create the tuples that contain the actual data values (lines 10 and 11). Make sure you list the data values in the same order in which you list the data fields in the INSERT statement.

Now you're ready to apply the data values to the INSERT statement template. You do that with the execute() method (lines 14 and 15).

After submitting the new data record, you must run the commit() method for the connection to commit the changes to the database (line 16).

After you run the script2101.py script, you can use the mysql command-line program to verify that the new data values have been entered, as shown here:

```
pi@raspberrypi ~ $ python3 script2101.py
Data values added!
pi@raspberrypi ~ $ mysql -u test -p
Enter password:

mysql> use pytest;
Reading table information for completion of table and column names
You can turn off this feature to get a quicker startup with -A
```

```
Database changed
mysql> select * from employees;
+-------+----------+-----------+--------+
| empid | lastname | firstname | salary |
+-------+----------+-----------+--------+
|     1 | Blum     | Barbara   |  45000 |
|     2 | Blum     | Rich      |  30000 |
+-------+----------+-----------+--------+
2 rows in set (0.00 sec)

mysql>
```

And there they are! The next step is to write a Python script to retrieve the data that you just stored in the table.

## WATCH OUT!

### Primary Key Data Constraint

Because you defined the empid data field as the primary key for the table, that value must be unique for each data record. If you try to rerun the script2101.py script without changing the data value tuples, the INSERT statements fail due to the duplicate data values.

## Querying Data

The process of querying tables is similar to the process of inserting data records. Your Python script must connect to the MySQL database, create a cursor, and then submit a SELECT SQL statement to retrieve the data using the execute() method.

The difference from the insert process is that with the SELECT query, you need to retrieve data back from the MySQL server. This is where the cursor object comes into play.

The cursor object contains a pointer to the query results. You must iterate through the cursor object using a for loop to extract all the data records returned by the query.

Listing 21.2 shows the script2102.py program code, which demonstrates how to retrieve the data records from the MySQL table.

### LISTING 21.2   The script2102.py Program Code

```
1:   #!/usr/bin/python3
2:
3:   import mysql.connector
4:   conn = mysql.connector.connect(user='test', password='test',
     database='pytest')
5:   cursor = conn.cursor()
```

```
6:
7:   query = ('SELECT empid, lastname, firstname, salary FROM employees')
8:   cursor.execute(query)
9:   for (empid, lastname, firstname, salary) in cursor:
10:    print(empid, lastname, firstname, salary)
11: cursor.close()
12: conn.close()
```

When you write the `for` loop, you must specify a variable for each data field that you return in the SELECT statement. For each iteration, the data fields contain the values for one data record in the query results. When the loop completes, you should have iterated through all the individual data records. When you run the script, you should get a listing of all the data records you stored in the `employees` table:

```
pi@raspberrypi ~ $ python3 script2102.py
1 Blum Barbara 45000.0
2 Blum Rich 30000.0
pi@raspberrypi ~ $
```

Congratulations! You've just written a database program using Python. Now let's take a look at how to do the same thing with the other popular Linux database server, PostgreSQL.

# Using the PostgreSQL Database

The PostgreSQL database started out as an academic project to demonstrate how to incorporate advanced database techniques into a functional database server. Over the years, PostgreSQL has evolved into one of the most advanced open-source database servers available for the Linux environment.

The following sections walk you through getting a PostgreSQL database server installed and running on your Raspberry Pi and then setting up your Python scripts to interact with the PostgreSQL database to store and retrieve data.

## Installing PostgreSQL

For the PostgreSQL server installation on the Raspberry Pi, both the PostgreSQL server and client programs are contained in a single software package. You just install the `postgresql` package by using the `apt-get` utility, as shown here:

```
pi@raspberrypi ~ $ sudo apt-get install postresql
```

After you do this, you have a fully functional PostgreSQL server up and running. The installation process automatically starts the PostgreSQL server, so there's nothing for you to manually run. Next, you need to create the database objects for your Python scripts.

# Setting Up the PostgreSQL Environment

Just as with the MySQL environment, you need to create your database objects in the PostgreSQL environment before you start your Python scripting.

For the Python environment, the command-line program you use to interact with the PostgreSQL server is called `psql`. However, it works a little differently from the `mysql` command-line program in MySQL.

PostgreSQL uses the Linux system user accounts instead of maintaining its own database of user accounts. While this can sometimes be confusing, it does make for a nice, clean way to control user accounts in PostgreSQL. All you need to do is ensure that each PostgreSQL user has a valid account on the Linux system; you don't have to worry about a whole separate set of user accounts.

Another major difference between MySQL and PostgreSQL is that the administrator account in PostgreSQL is called postgres, not root. When you installed the PostgreSQL package on your Raspberry Pi, the installation process created a postgres user account on the system so the PostgreSQL administrative user account can exist.

To interact with the PostgreSQL server, you need to run the `psql` program as the postres user account. It looks like this:

```
pi@raspberrypi ~ $ sudo -u postgres psql
psql (9.1.9)
Type "help" for help.

postgres=#
```

The default `psql` prompt indicates the name of the database you are connected to. The pound sign (#) in the prompt indicates that you're logged in with the administrative user account. To exit the `psql` command prompt, you just enter the `\q` meta-command.

Now you're ready to start entering some commands to interact with the PostgreSQL server.

## Creating a Database

Creating a database in PostgreSQL is the same as in MySQL: All you need to do is submit a CREATE DATABASE statement. Just remember to be logged in as the postgres administrative account to create the new database, as shown here:

```
pi@raspberrypi ~ $ sudo -u postgres psql
psql (9.1.9)
Type "help" for help.

postgres=# CREATE DATABASE pytest;
CREATE DATABASE
postgres=#
```

After you create the database, you can use the `\l` meta-command to see if your new database appears in the database listing and then the `\c` meta-command to connect to it. Here's an example:

```
postgres=# \l
        List of databases
   Name    |   Owner   | Encoding
-----------+-----------+----------
 postgres  | postgres  | UTF8
 template0 | postgres  | UTF8
 template1 | postgres  | UTF8
 pytest    | postgres  | UTF8
(4 rows)

postgres=# \c pytest
You are now connected to database "test" as user "postgres".
pytest=#
```

When you connect to the `pytest` database, the `psql` prompt changes to indicate the new database name. This is a great reminder when you're ready to create your database objects: You can easily tell where you are in the system.

When you're done working with the PostgreSQL server, just enter the `\q` command to return to the Linux command prompt.

### BY THE WAY

#### PostgreSQL Schemas

PostgreSQL adds another layer of control, called the *schema*, to the database. A database can contain multiple schemas, and each schema can contain multiple tables. This allows you to subdivide a database for specific applications or users.

By default, every database contains one schema, called `public`. If you're going to have only one application use the database, you're fine with just using the `public` schema. If you'd like to get fancy, you can create new schemas. The following example just uses the `public` schema for the tables.

After you create the database to use in your Python scripts, you need to create a separate user account that your scripts can use to log in to the database.

## Creating a User Account

After you create a new database, the next step is to create a user account that has access to it for your Python scripts. As you've already seen, user accounts in PostgreSQL are significantly different from those in MySQL.

User accounts in PostgreSQL are called *login roles*. The PostgreSQL server matches login roles to the Linux system user accounts. Because of this, there are two common thoughts about creating login roles to run Python scripts that access the PostgreSQL database:

▸ Create a special Linux account with a matching PostgreSQL login role to run all your Python scripts.

▸ Create a PostgreSQL account for each Linux user account that needs to run Python scripts to access the database.

This example uses the second method: You create a PostgreSQL account that matches the default pi Linux system account on the Raspberry Pi. This way, you can run Python scripts that access the PostgreSQL database directly from the default Raspberry Pi user account.

First, you must create the login role, like this:

```
pytest=# CREATE ROLE pi login;
CREATE ROLE
pytest=#
```

This is simple enough. Without the `login` parameter, the role is not allowed to log in to the PostgreSQL server, but it can be assigned privileges. This type of role is called a *group role*. Group roles are great if you're working in a large environment with lots of users and tables. Instead of having to keep track of which user has which type of privileges for which tables, you just create group roles for specific types of access to tables and then assign the login roles to the proper group role.

For simple Python scripting, you most likely won't need to worry about creating group roles, and you can just assign privileges directly to the login roles. That's what you'll do in this example.

However, PostgreSQL also handles privileges a bit differently than MySQL. It doesn't allow you to grant overall privileges to all objects in a database that filter down to the table level. Instead, you need to grant privileges for each individual table you create. While this is kind of a pain, it certainly helps enforce strict security policies. Because of that, though, you need to hold off on assigning privileges until you've created the table for your application. That's the next step in the process.

## Creating a Table

Just like the MySQL server, the PostgreSQL server is a relational database. That means you need to group your data fields into tables. As you can see here, you use the same CREATE TABLE statement to create the `employees` table in the PostgreSQL `pytest` database:

```
pi@raspberrypi ~ $ sudo -u postgres psql
psql (9.1.9)
Type "help" for help.
```

```
postgres=# \c pytest
You are now connected to database "pytest" as user "postgres".
pytest=# CREATE TABLE employees (
pytest(# empid int not null,
pytest(# lastname varchar(30),
pytest(# firstname varchar(30),
pytest(# salary float,
pytest(# primary key (empid));
NOTICE:  CREATE TABLE / PRIMARY KEY will create implicit index "employees_pkey" for
table "employees"
CREATE TABLE
pytest=#
```

Once you have created the table, you can list the tables by using the \dt meta-command:

```
pytest=# \dt
           List of relations
 Schema |    Name    | Type  |  Owner
--------+------------+-------+----------
 public | employees  | table | postgres
(1 row)

pytest=#
```

Now you're ready to assign privileges for the employees table to the pi login role so that it can access the table. Here's how you do this:

```
pytest=# GRANT SELECT,INSERT,DELETE,UPDATE ON public.employees To pi;
GRANT
pytest=#
```

You can now log in to the PostgreSQL server with the pi user account to connect directly to the pytest database. From the pi user account's command prompt, you enter this command:

```
pi@raspberrypi ~ $ psql pytest
psql (9.1.9)
Type "help" for help.

pytest=> \dt
           List of relations
 Schema |    Name    | Type  |  Owner
--------+------------+-------+----------
 public | employees  | table | postgres
(1 row)

pytest=>
```

When you enter the database name on the psql command line, the psql program takes you directly to that database, and you don't have to use the \c meta-command. Even though the owner of the `employees` table is the `postgres` user account, the `pi` login role has privileges to interact with it.

Now you have your table and user account all set. The next step is to install the PostgreSQL module for Python.

# Installing the Python PostgreSQL Module

Very much as in the MySQL environment, Python has several different modules that provide support for communicating with PostgreSQL databases from your Python scripts. Fortunately, there's a Python v3 module for PostgreSQL already in the Raspbian software repository. The oddly named `psycopg2` module provides full support for interacting with the PostgreSQL database from Python scripts. This is what you'll use in the following examples.

The `psycopg2` module is in the `python3-psycopg2` software package. To install it, you just use the `apt-get` utility, like this:

```
pi@raspberrypi ~ $ sudo apt-get install python3-psycopg2
```

The `psycopg2` module has both Python v2 and Python v3 versions, so make sure you install the `python3-psycopg2` module!

# Coding with `psycopg2`

With the `psycopg2` module installed, you're all set to start coding your Python scripts to access the PostgreSQL database. You'll probably notice that many of the methods used are the same ones you've seen in the `MySQL/Connector` module. For most Python database modules, once you learn how to use one of them, it's not too difficult to pick up how to use any others.

The following sections walk through how to connect to the PostgreSQL database, insert new data records, and retrieve data records.

## Connecting to the Database

Before you can interact with the table, your Python script must connect to the PostgreSQL database you created. You do that with the `connect()` method, as shown here:

```
>>>import psycopg2
>>> conn = psycopg2.connect('dbname=pytest')
>>>
```

You may have noticed that the `connect()` method only specifies the database name. By default, it logs in to the PostgreSQL server using the same user account with which you're logged in to the Linux system. Since you're running the script logged in as the `pi` user account, the `connect()` method uses the `pi` login role automatically.

You can also specify a separate user and password in the `connect()` method, as shown here:

```
>>> conn = psycopg2.connect('dbname=pytest user=pi password=mypass')
```

Notice that the parameters are all part of one string value, not separate strings.

Now your Python script is connected to the `pytest` database, and you're ready to start interacting with the tables.

### Inserting Data

After you've connected to the database, you can insert some new data records into your `employees` table. The psycopg2 module provides a similar approach to what you used with the `mysql.connector` module. Listing 21.3 shows the `script2103.py` program, which demonstrates how to add new data elements to the database.

**LISTING 21.3** The `script2103.py` Program Code

```
 1:  #!/usr/bin/python3
 2:
 3:  import psycopg2
 4:  conn = psycopg2.connect('dbname=pytest')
 5:  cursor = conn.cursor()
 6:  new_employee = 'INSERT INTO employees VALUES (%s, %s, %s, %s)'
 7:  employee1 = ('1', 'Blum', 'Katie Jane', '55000.00')
 8:  employee2 = ('2', 'Blum', 'Jessica', '35000.00')
 9:  try:
10:      cursor.execute(new_employee, employee1)
11:      cursor.execute(new_employee, employee2)
12:      conn.commit()
13:  except:
14:      print('Sorry, there was a problem adding the data')
15:  else:
16:      print('Data values added!')
17:  cursor.close()
18:  conn.close()
```

The `execute()` method submits the `INSERT` statement template along with the data tuple to the PostgreSQL server for processing. The data isn't committed to the database, though, until you issue the `commit()` method from the connection. You can run the `script2103.py` program and then check the `employees` table for the data, as shown here:

```
pi@raspberrypi ~ $ python3 script2103.py
Data values added!
pi@raspberrypi ~ $ psql pytest
psql (9.1.9)
Type "help" for help.
```

```
pytest=> select * from employees;
 empid | lastname | firstname | salary
-------+----------+-----------+--------
     1 | Blum     | Katie Jane | 55000
     2 | Blum     | Jessica    | 35000
(2 rows)

pytest=>
```

The data is there! The next step is to write the code to query the table and retrieve the data values.

## Querying Data

To query data, you submit a SELECT statement by using the execute() method. However, to retrieve the query results, you have to use the fetchall() method for the cursor object. Listing 21.4 shows the script2104.py program, which demonstrates how to do this.

**LISTING 21.4**   The script2104.py Program Code

```
1:  #!/usr/bin/python3
2:
3:  import psycopg2
4:  conn = psycopg2.connect('dbname=pytest')
5:  cursor = conn.cursor()
6:  cursor.execute('SELECT empid, lastname, firstname, salary FROM
 employees')
7:  result = cursor.fetchall()
8:  for data in result:
9:      print(data[0], data[1], data[2], data[3])
10: cursor.close()
11: conn.close()
```

The script2104.py program assigns the output of the fetchall() method to the result variable (line 7), which then contains a list of the data records in the query results. It iterates through the list by using a for loop (line 8). The resulting list uses positional index values to reference each data field in the data record. The order of the values matches the order in which you list the data fields in the SELECT statement.

When you run the script2104.py program, you should see a list of the data records stored in the PostgreSQL employees table, as shown here:

```
pi@raspberrypi ~ $ python3 script2104.py
1 Blum Katie Jane 55000.0
2 Blum Jessica 35000.0
pi@raspberrypi ~ $
```

You can use these methods to handle a database of any size. The beauty of using a database server is that all the data crunching and scaling happens behind the scenes, in the database server. Your Python scripts just need to interface with the database server to submit SQL statements to handle the data.

# Summary

This hour you learned how to incorporate open-source databases in your Python scripts. The Raspberry Pi supports both the MySQL and PostgreSQL open-source database servers, and you can use them for storing and retrieving data in your scripts.

First, you learned how to install the MySQL database server, how to set it up, and how to use the `MySQL/Connector` Python module to interact with the database in your Python scripts. Next, you saw how to install and set up the PostgreSQL database server and how to use the `psycopg2` module to interact with it in your Python scripts. Both systems provide advanced data storage and retrieval methods to add great functionality to your Python scripts.

In the next hour, we'll take a look at another popular aspect of programming—Web programming. The Raspberry Pi provides some modules to help you publish your Python programs on the Web.

# Q&A

**Q.** Is the Raspberry Pi really powerful enough to support a full-blown database server such as MySQL or PostgreSQL?

**A.** Yes, you can run the MySQL and PostgreSQL database servers on the Raspberry Pi just fine. I wouldn't recommend trying to support thousands of application users at the same time, but for a small number of concurrent users, Raspberry Pi will hold out just fine!

**Q.** Which database server is better, MySQL or PostgreSQL?

**A.** This has been a longstanding debate in the open-source database world. The general consensus is that the MySQL database server is usually faster, but the PostgreSQL database supports more advanced database features. Which one you decide to use depends on the database requirements for your specific application.

# Workshop

## Quiz

1. What data storage method allows you to easily store application data and retrieve it later, using different Python scripts?

    **a.** Variables

    **b.** Relational databases

    **c.** Library modules

    **d.** Log files

2. Using standard files to store and retrieve data is just as easy as using a relational database server. True or false?

3. When method from the `psycopg2` module should you use to retrieve the data records from a `SELECT` query?

## Answers

1. a. relational databases. Database servers that use relational databases can quickly store and retrieve data behind the scenes, without you having to do much coding.

2. False. With standard files, you must read the data into your Python scripts yourself—and you must search for the data. With relational databases, the database server can do all that work for you.

3. The `fetchall()` method retrieves the data records that result from a `SELECT` query that you send to the PostgreSQL server.

# HOUR 22
# Web Programming

**What You'll Learn in This Hour:**

▶ Installing a web server on your Raspberry Pi
▶ Using CGI to run your Python programs from the Web
▶ How to generate dynamic webpages using Python
▶ How to retrieve form data in your Python web programs

With the popularity of the World Wide Web, these days it's often a requirement to write applications that are Web aware. While Python wasn't intended to be a Web-based programming language, over the years it has evolved to provide many Web features. In this hour, you'll learn how to move your Python programs into the Web world, using the Apache web server and some Python modules that are part of the standard Python library on your Raspberry Pi.

## Running a Web Server on the Pi

Before you can move your Python applications into the Web world, you need to have a web server to host them. While the Raspberry Pi isn't intended to be a production web server that supports thousands of customers, it works just fine as a host for small intranet applications on your local network.

As with just about everything else in the Linux world, there are a few different web servers that you can choose to install on your Raspberry Pi. Here's a list of the most popular ones:

▶ **Apache**—A full-blown production web server environment that runs on many platforms.

▶ **Nginx**—A lightweight web and email server package.

▶ **Monkey HTTP**—A development web server built for the Linux environment.

▶ **lighttp**—A small web server that focuses on performance.

The Apache web server is by far the most popular web server used on the Linux platform. It runs just fine on the Raspberry Pi, as long as you don't try hosting thousands of concurrent users. You'll use the Apache web server in the examples in this hour. The following sections walk you through setting up the popular Apache web server in your Raspberry Pi environment.

## Installing the Apache Web Server

Installing the Apache web server on the Raspberry Pi is a simple process, thanks to the Raspbian software repository. The complete Apache web server package is contained in a single software package, apache2. You simply use the apt-get utility to install it:

```
pi@raspberypi ~ $ sudo apt-get install apache2
```

The apache2 package installs the web server and all the supporting files required to run the server. Table 22.1 shows some of the most important files and folders that you need to become familiar with as you use the Apache web server.

**TABLE 22.1**   apache2 Files and Folders

| File or Folder | Description |
| --- | --- |
| /var/www | Folder for serving web documents |
| /usr/lib/cgi-bin | Folder for serving scripts |
| /etc/apache2 | Folder for the web server configuration files |
| /var/log/error.log | The Apache web server error log file |

The installation process starts the Apache web server automatically, so there's no need to manually start the server. However, you can start and stop the server from the command prompt at any time, if needed, using the service command. To stop the Apache web server, you use this command:

```
sudo service apache2 stop
```

Likewise, to restart the Apache web server, you use this command:

```
sudo service apache2 start
```

When you have the Apache web server running, you can test it. You can open a browser in the LXDE desktop on your Raspberry Pi, or if you know the IP address of your Raspberry Pi, you can connect to it from another client on the network.

If you're connecting from the Raspberry Pi desktop, you can connect to the special localhost host name:

```
http://localhost/
```

If you're connecting from a remote client on your network, you need to know the IP address of your Raspberry Pi. (The IP address may change if you're using Dynamic Host Configuration Protocol [DHCP] to assign addresses on your network.) To find the current IP address assigned to your Raspberry Pi, you use the `ifconfig` command at the command prompt:

```
pi@raspberrypi ~ $ ifconfig
```

When you know the IP address assigned to your Raspberry Pi, you can connect to it from the remote client by specifying the IP address as the URL. For example, if you found out that the IP address assigned to your Raspberry is 10.0.1.70, you would use the URL:

```
http://10.0.1.70/
```

Either way, you should see the generic test webpage, which is shown in Figure 22.1.

**FIGURE 22.1**
The default Apache web server page for the Raspberry Pi.

Now that the web server is running, you can try making a test webpage of your own, which is what we'll cover next.

## Serving HTML Files

The core function of the Apache web server is to serve HTML documents to clients on the network. By default, the Raspberry Pi Apache web server is configured to only serve files in the

/var/www folder on the system, so you must place your web documents under that folder structure.

However, that folder is owned by the root user account. To be able to save files in that folder, you must use the sudo command when you copy them. The following Try It Yourself shows how to publish a simple webpage to test things out.

▼ TRY IT YOURSELF

### Publishing a Webpage

You can create a webpage document in your home folder, and then when you're ready to publish it, you simply copy it over to the proper folder. Just follow these steps:

**1.** Create the script2201.html file in this hour's working folder and enter the code shown here:

```
<!DOCTYPE html>
<html>
<head>
<title>Test HTML Page</title>
</head>
<body>
<h2>This is a test HTML page for my server</h2>
</body>
</html>
```

**2.** Save the file and then exit the editor.

**3.** Copy the script2201.html file to the /var/www folder, like this:

```
pi@raspberrypi ~ $ sudo cp script2201.html /var/www
pi@raspberrypi ~ $
```

**4.** Change the mode of the new file so that it will be readable by everyone:

```
pi@raspberrypi ~ $ sudo chmod +r /var/www/script2201.html
pi@raspberrypi ~ $
```

**5.** Open a browser and navigate to the new webpage by using the URL http://localhost/script2201.html.

Congratulations! You just published a webpage on your Apache web server! The next step is to publish your Python programs.

# Programming with the Common Gateway Interface

There are a few different ways to publish Python programs on the Apache web server. We'll take a look at the oldest and easiest way of do so: using the Common Gateway Interface (CGI).

The following sections describe the CGI and how to use it with your Python programs to allow remote network clients to run and interact with your programs.

## What Is CGI?

The CGI is a feature built into the Apache web server that allows remote clients to run shell scripts on the host server. Allowing unknown visitors to your website to run scripts can be dangerous. However, with the proper security controls, running scripts provides a new dimension to web applications.

By default, the CGI in the Apache web server restricts shell scripts to a specific folder on the server: /usr/lib/cgi-bin. This is where you must place your Python scripts so that remote clients can run them.

## Running Python Programs

To get a Python program to run as a CGI script, you have to modify it a bit. First, you have to tell the shell that it's a Python program. You do this by using the #! command, pointing to the standard path of the python3 application. For the Raspberry Pi environment, you just add this line to the top of your Python programs:

```
#!/usr/bin/python3
```

When the Linux system runs the shell script, it knows to process the program through the python3 interpreter.

Another issue with running your Python program via the Apache web server is that you won't have access to the command prompt to view the output from your program. Instead, the Apache CGI redirects any output from your Python program directly to the web client. That setup is problematic because you must format the output of your Python program so that the web client thinks it's coming from a webpage document.

To format the output of your Python program for a browser to process, you need to add a Multipurpose Internet Mail Extensions (MIME) heading at the start of your output. You do this with the Content-Type header, as shown here:

```
print('Content-Type: text/html')
print('')
```

After you identify the output as an HTML document, you also need to have a blank line before any other output from the program.

▼ TRY IT YOURSELF

## Creating a Python Web Program

Now you're ready to write a test Python program to run on the web server. Follow along with these steps to get things going:

1. Create a file called `script2202.cgi` in the folder for this hour.

2. Open the `script2202.cgi` file and enter the code shown here:

```
#!/usr/bin/python3

import math
radius = 5
area = math.pi * radius * radius
print('Content-Type: text/html')
print('')
print('The area of a circle with radius', radius, 'is', area)
```

3. Save the file and exit the editor.

4. Test your script from the Python command prompt interpreter, like this:

```
pi@raspberrypi ~ $ python3 script2202.cgi
Content-Type: text/html

The area of a circle with radius 5 is 78.53981633974483
pi@raspberrypi ~ $
```

5. If this works, copy the script file to the `/usr/lib/cgi-bin` folder for publishing and make sure web clients can run it by giving everyone execute permissions on the file, like this:

```
pi@raspberrypi ~ $ sudo cp script2202.cgi /usr/lib/cgi-bin
pi@raspberrypi ~ $ sudo chmod +x /usr/lib/cgi-bin/script2202.cgi
pi@raspberrypi ~ $
```

6. Open your browser and browse to the new program. Because the file is in the `cgi-bin` folder, you need to include that in the URL:

```
http://localhost/cgi-bin/script2202.py
```

You should see the results of your program appear in your web browser, as shown in Figure 22.2!

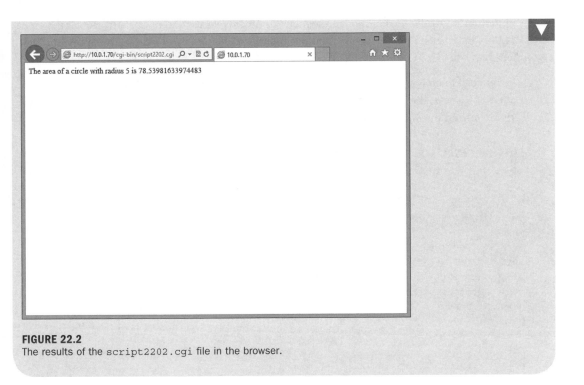

**FIGURE 22.2**
The results of the `script2202.cgi` file in the browser.

Congratulations! You've run your Python program on the Web! However, this is pretty boring, as webpages go. In the next section, you'll take a look at how to spice up your Python webpages a bit.

# Expanding Your Python Webpages

Now that you've seen the basics of how to get your Python programs to run on the Web, you can dive a little deeper into the process. In the following sections, you'll first learn how to format your program code so that it appears more like a real webpage instead of just program output. Then you'll look at how to make your Python webpages more dynamic by enabling them to access database data and display it on the webpage. Finally, you'll add some debugging features to your code in case things go wrong.

## Formatting Output

You may have noticed from the example in Figure 22.2 that just displaying the output from your Python code directly to the web browser isn't all that exciting. Browsers were created to display formatted text, using Hypertext Markup Language (HTML). HTML enables you to use plan-text commands to identify formatting features such as layouts, fonts, and colors. Because all the HTML coding is done in text, you can output that from your Python programs and pass it to the client browser.

All you need to do is add some HTML code to your Python output to help liven things up a bit. Listing 22.1 shows the `script2203.cgi` program, which embeds HTML code inside the Python script to format the output of the Python program.

**LISTING 22.1**  Using HTML in the Python Program Output

```
#!/usr/bin/python3

import math
print('Content-Type: text/html')
print('')
print('<!DOCTYTPE html>')
print('<html>')
print('<head>')
print('<title>The Area of a Circle</title>')
print('</head>')
print('<body>')
print('<h2>Calculating the area of a circle:</h2>')
print('<table>')
print('<tr><th>Radius</th><th>Area</th></tr>')
for radius in range(1,11):
    area = math.pi * radius * radius
    print('<tr><td>', radius, '</td><td>', area, '</td></tr>')
print('</table>')
print('</body>')
print('</html>')
```

After you copy the `script2203.cgi` file to the `/usr/lib/cgi-bin` folder, you can view it in your browser to see the results. Figure 22.3 shows what you should see.

It's amazing what just a little bit of HTML code can do to help with the output of your Python program!

# Working with Dynamic Webpages

Python scripting allows you to create dynamic webpages. Dynamic webpages have the ability to change webpage content, based on some external event, such as updating data in a database.

In Hour 21, "Using Databases in Your Programming," you learned how to store and retrieve data from your Python scripts by using both the MySQL and PostgreSQL database servers running on your Raspberry Pi. You can now combine that knowledge with your CGI knowledge to create dynamic webpages to publish database data directly on your network.

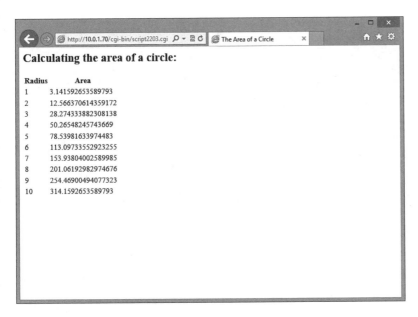

**FIGURE 22.3**
The `script2203.cgi` program output.

In the following Try It Yourself, you'll write a script that can read the `employees` table you cre-ated in Hour 21 and display the information on a webpage.

---

TRY IT YOURSELF ▼

## Publishing Database Data on the Web

The key to dynamic webpages is the ability to work with a behind-the-scenes database to store and manipulate data. In the following steps, you'll use the MySQL database server and the `pytest` database created in Hour 21 to provide dynamic data for your Python web application. Just follow these steps:

1. Create the file `script2204.cgi` in this hour's working folder.

2. Open the `script2204.cgi` file and enter the code shown here:

```
1:  #!/usr/bin/python3
2:
3:  import mysql.connector
4:  print('''Content-Type: text/html
5:
```

```
 6:  <!DOCTYPE html>
 7:  <html>
 8:  <head>
 9:  <title>Dynamic Python Webpage Test</title>
10:  </head>
11:  <body>
12:  <h2>Employee Table</h2>
13:  <table border=1>
14:  <tr><th>EmpID</th><th>Last Name</th><th>First
 Name</th><th>Salary</th></tr>''')
15:
16:  conn = mysql.connector.connect(user='test', password='test',
 database='pytest')
17:  cursor = conn.cursor()
18:
19:  query = ('SELECT empid, lastname, firstname, salary FROM employees')
20:  cursor.execute(query)
21:  for (empid, lastname, firstname, salary) in cursor:
22:      print('<tr><td>', empid, '</td>')
23:      print('<td>', lastname, '</td>')
24:      print('<td>', firstname, '</td>')
25:      print('<td>',  salary, '</td></tr>')
26:  print('</table>')
27:  print('</body>')
28:  print('</html>')
29:  cursor.close()
30:  conn.close()
```

3. Save the file and then exit the editor.

4. Copy the file to the `/usr/lib/cgi-bin` folder and assign it permissions to run, like this:

```
pi@raspberrypi ~ $ sudo cp script2204.cgi /usr/lib/cgi-bin
pi@raspberrypi ~ $ sudo chmod +x /usr/lib/cgi-bin/script2204.cgi
```

5. View the `script2204.cgi` file in your web browser client. You should see the results from the data you entered into the `employees` table in Hour 21 (see Figure 22.4).

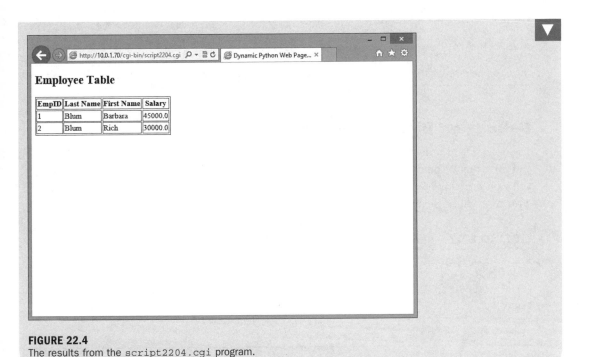

**FIGURE 22.4**
The results from the `script2204.cgi` program.

The `script2204.cgi` file uses a slightly different method for adding the HTML code required to display the output. Instead of using a separate `print()` method for each HTML element in the document, in the `script2204.cgi` file, you used the triple-quote method of creating one long string value (lines 4 through 14). This helps cut down on some of the typing, and makes it a little easier to follow the HTML code embedded in the Python code.

WATCH OUT!

## Web Database Security

The script2204.cgi script embeds the user ID and password for your MySQL database into a file that everyone on the server can read. For testing on a personal Raspberry Pi system, this is not a problem, but on a real system shared by others, doing this isn't such a great idea. One solution is to restrict access to the file to only the Apache web server user account. For the Raspberry Pi, the Apache web server runs as the `www-data` user account. You can use the `chown` command to change the group owner of your files to the `www-data` user account:

```
sudo chown www-data /usr/lib/cgi-bin/script2204.cgi
```

Then you change the permissions on the file so only the www-data user can access it:

```
sudo chmod 700 /usr/lib/cgi-bin/script2204.cgi
```

Now no one else on the system can read the file, but it'll work just fine on the Apache web server.

# Debugging Python Programs

The downside to running your Python programs using CGI is that you don't get any feedback if you have any Python scripting errors in your code. If anything goes wrong in the Python script, you won't get any output in the client browser. Listing 22.2 shows a Python script with a math error that will cause problems.

**LISTING 22.2    Running a Python Program That Has an Error**

```
#!/usr/bin/python3

print('Content-Type: text/html')
print('')
result = 1 / 0
print('This is a test of a bad Python program')
```

You need to copy this code into the /usr/lib/cgi-bin folder as the file script2205.cgi, change the permissions on the file, and then try to run it in your web browser. The mathematical equation in line 5 attempts to divide by 0, which causes an exception in the Python code. However, when you run this in your web browser, you don't see any Python error messages. In fact, you don't see any output in the browser window!

Fortunately, there's an easier way to troubleshoot Python code in your webpages. The Python cgitb module provides simple debugging output for your Python scripts. By referencing the cgitb module, you can run the enable() method to enable debugging in the output. Listing 22.3 shows the script2206.cgi program, which adds the enable() method to the bad Python program code.

**LISTING 22.3    Displaying Errors from a Python Web Program**

```
#!/usr/bin/python3

import cgitb
cgitb.enable()
print('Content-Type: text/html')
print('')
result = 1 / 0
print('This is a test of a bad Python program')
```

The `script2206.cgi` code still has the same division error as the `script2205.cgi` program, but now you have added the `cgitb.enable()` method to enable the debugging feature in the Python script. The `cgitb` debugging feature displays full error messages and code when a Python error occurs in the program.

Now when you run this program from your web browser, you should see a webpage similar to the one shown in Figure 22.5.

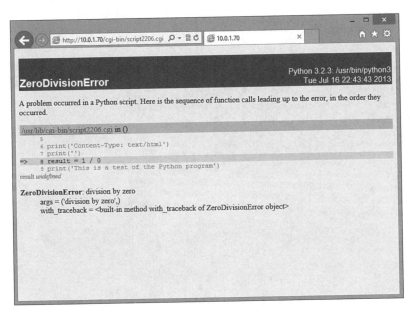

**FIGURE 22.5**
The debugging output from the `script2206.cgi` program.

The error message not only tells you what went wrong but also displays the Python code and what line has the error. Now you can get a better idea of what's going wrong with your Python code, so you can get things working more quickly.

While the `cgitb.enable()` method can be very helpful when you're debugging a Python web application, it can also help out any attackers trying to gain insight into your Python code. Another option you have is to redirect error messages to a log file instead of display them on the webpage. To do that, you need to add a couple parameters to the `enable()` method, as shown here:

```
cgitb.enable(display=0, logdir='path')
```

The `display` parameter determines whether the error message appears on the webpage. (You can set the value to 1 if you want the error to appear both on the webpage and in the log file.) The `logdir` parameter specifies the folder path where you want the log file to be created. It's important to remember that the Apache web server's Linux account (`www-data` on the Raspberry Pi) must have write permissions to that folder. You can use the `/tmp` folder, as shown here, if you don't mind others on your Raspberry Pi system seeing the log files that are generated:

```
cgitb.enable(display=0, logdir='/tmp')
```

Figure 22.6 shows what the webpage shows after you add this line to the `script2206.cgi` program.

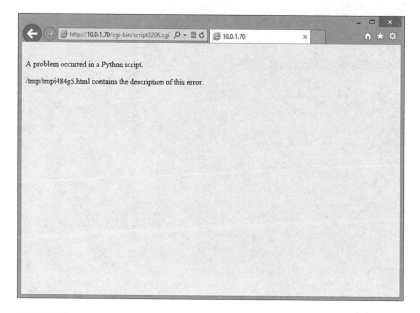

**FIGURE 22.6**
The `cgitb.enable()` output on the webpage.

The log file that is generated contains the HTML code of the error webpage that would have been displayed in the browser.

# Processing Forms

Web applications allow you to easily collect and process data from site visitors. Web forms provide an excellent way to interact with your program users to retrieve dynamic data for processing or storage in databases.

The Python CGI environment provides an easy way for your Python scripts to retrieve and use form data in your web applications. The follow sections walk through how you use it.

# Creating Web Forms

The HTML standard provides elements you can use to create forms. Your site visitors can then fill out these forms to submit data to your web applications. Table 22.2 lists the HTML form elements.

**TABLE 22.2    HTML Form Elements**

| Element | Description |
|---|---|
| checkbox | A box to select or deselect an item |
| fileupload | A single-line text box with a Browse button for finding and specifying file names |
| radio | A button for selecting one of a group of options |
| password | A single-line text box that hides entered characters |
| submit | A button that submits the form data when selected |
| text | A single-line text box |
| textarea | A multiline text box |

To build the form in your webpage, you must use the HTML `<form>` element, which defines the action the browser takes with the form data when the site visitor clicks the Submit button for the form. You use this element to tell your Python script where to pass the form data. Listing 22.4 shows the `script2207.html` file, which creates a simple web form you can use to use to test the `<form>` element.

**LISTING 22.4    Creating a Simple Web Form**

```
1:  <!DOCTYPE html>
2:  <html>
3:  <head>
4:  <title>Web Form Test</title>
5:  </head>
6:  <body>
7:  <h2>Please enter your information</h2>
8:  <br />
9:  <form action='/cgi-bin/script2208.cgi' method='post'>
10: <label>Last Name:</label><input type="text" name="lname" size="30" />
11: <br />
12: <label>First Name:</label><input type="text" name="fname" size="30" />
13: <br />
```

```
14: <label>Age range:</label><br />
15: <input type="radio" name="age" value="20-30" /> 20-30<br />
16: <input type="radio" name="age" value="31-40" /> 31-40<br />
17: <input type="radio" name="age" value="41-50" /> 41-50<br />
18: <input type="radio" name="age" value="51+" /> 51+<br />
19: <br />
20: <label>Select all that apply:</label><br />
21: <input type="checkbox" name="hobbies" value="fishing" /> Fishing<br />
22: <input type="checkbox" name="hobbies" value="golf" /> Golf<br />
23: <input type="checkbox" name="hobbies" value="baseball" /> Baseball<br />
24: <input type="checkbox" name="hobbies" value="football" /> Football<br />
25: <br />
26: <label>Enter your comment:</label><br />
27: <textarea name="comment" rows="10" cols="20"></textarea>
28: <br />
29: <input type="submit" value="Submit your comment" />
30: </form>
31: </body>
32: </html>
```

The `<form>` element in line 9 specifies the location of the Python script that you want to process the form data. The form contains two text boxes, a series of radio buttons to specify one age range as a single return value, a series of check boxes to select one or more hobbies, and a text area for entering comments.

The web form is just HTML code, and you need to place it in the standard /var/www folder to serve it to web clients:

```
sudo cp script2207.html /var/www
sudo chmod +x /var/www/script2207.html
```

You can then open your browser to view the form by using this URL:

```
http://localhost/script2207.html
```

You should see a form like the one in Figure 22.7.

Now you're ready to write the Python script to retrieve and process the form data.

## The `cgi` Module

When the webpage passes the form data back to the Apache web server, it groups the data in key/value pairs. The key is the HTML element name for the form field in the webpage, and the value is the data value that was entered in the form field.

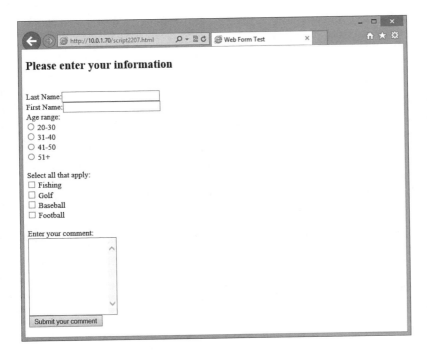

**FIGURE 22.7**
The basic web form used for the Python script.

For example, this HTML code associates the key name lname with the value entered into that form field:

```
<input type="text" name="lname" />
```

The Apache server passes the data into the shell environment for the script so that the Python script can retrieve it.

The Python cgi module provides the necessary elements for your Python script to retrieve the shell environment data so you can process it in your Python script. The FieldStorage class in the cgi module provides the methods for you to access the data. To retrieve the form data, you need to create an instance of the FieldStorage class in your Python code, as shown here:

```
import cgi
formdata = cgi.FieldStorage()
```

After you create the FieldStorage instance, you need to use two methods to retrieve the form data:

- ▶ getfirst()

- ▶ getlist()

The getfirst() method retrieves only the first occurrence of a key name as a string value. You use this to retrieve text box, radio button, and text area form values.

---

WATCH OUT!

### Numeric Form Fields

The getfirst() method returns the form data as string values, even if the string value is a number. You need to use the Python type conversion methods to convert the data into numeric values, if necessary.

---

The getlist() method retrieves multiple values as a list value. You use it to retrieve the check box values. The web form returns the values of any selected check boxes in the list object.

Now you're ready to write the script2208.cgi file to process the form data. Listing 22.5 shows the code you use.

**LISTING 22.5** Processing Form Data in a Python Script

```
1:  #!/usr/bin/python3
2:
3:  import cgi
4:  formdata = cgi.FieldStorage()
5:  lname = formdata.getfirst('lname', '')
6:  fname = formdata.getfirst('fname', '')
7:  age = formdata.getfirst('age', '')
8:  comment = formdata.getfirst('comment', '')
9:
10: print('Content-Type: text/html')
11: print('')
12: print('''<!DOCTYPE html>
13: <html>
14: <head>
15: <title>Form Results</title>
16: </head>
17: <body>
18: <h2>Here are the results from your survey</h2>
19: <br />
20: <table border=1>''')
21:
22: print('<tr><th>Name</th><td>',fname, lname, '</td></tr>')
23: print('<tr><th>Age range</th><td>', age, '</td></tr>')
24: print('<tr><th>Hobbies</th><td>')
25: for item in formdata.getlist('hobbies'):
26:     print(item)v
27: print('</td></tr>')
```

```
28: print('<tr><th>Comments</th><td>', comment, '</td></tr>')
29: print('</table>')
30: print('</body>')
31: print('</html>')
```

In the `script2208.cgi` code, lines 5 through 8 use the getfirst() method to retrieve the single-value form data values: fname, lname, age, and comment. Line 25 uses the getlist() method to retrieve the multivalue hobbies value from the check box options. Because you don't know how many (if any) check boxes were selected, you can use the for statement to iterate through the list to retrieve whatever is there.

When you submit the form from the `script2207.html` file, you should see the output from the `script2208.cgi` script program, as shown in Figure 22.8.

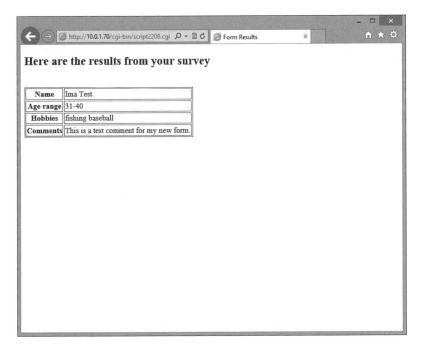

**FIGURE 22.8**
The results of the `script208.cgi` script processing the form data.

After you retrieve the form data in your Python script, you can perform any type of processing on it, including storing it in a database (see Hour 21). Now you're well on your way to writing fully dynamic web applications using Python on your Raspberry Pi!

# Summary

In this hour, you learned how to use Python to create dynamic web applications. Although it was not originally intended for use on the web, Python provides many Web features that you can use in your applications. In this hour, you saw how to use the CGI to run simple Python scripts from a browser, using the Apache web server. Next, you learned how to incorporate HTML code inside the Python output to format an application to display in a browser environment. Finally, you walked through using the `cgi` module to retrieve data from web forms and process it in your Python programs.

In the next hour, we'll take a look at how to create some applications using Python on your Raspberry Pi. The Raspberry Pi is known for its support for high-definition images and video, as well as support for audio. We'll walk through writing some applications that can leverage those features!

# Q&A

**Q.** Are there other ways to run Python scripts from the web server, besides using the CGI?

**A.** Yes, the `mod_python` and the `mod_wsgi` Apache plug-in modules provide direct support for running Python scripts without using the CGI.

**Q.** What are Python web frameworks?

**A.** The web frameworks are modules that provide built-in classes for handling many low-level data retrieval, formatting, and storage features for you. When you use these modules, you can concentrate more on your web application than on the low-level Python coding. Module packages such as Django and Pylons are popular with professional Python web developers.

# Workshop

## Quiz

**1.** Where should you place your Python scripts so they can be viewed from the Apache web server?

   **a.** `/var/log/apache2`

   **b.** `/usr/lib/cgi-bin`

   **c.** `/home/pi`

   **d.** `/etc/apache2`

2. In Python CGI scripts, you can only run modules from the standard Python library. True or false?

3. What cgi module method should you use to retrieve data from a `textarea` form element?

## Answers

1. b. You must place all web Python scripts in the `/usr/lib/cgi-bin` folder for security reasons.

2. False. You can run any module installed in the Raspberry Pi from Python CGI scripts.

3. The `getfirst()` module retrieves the data passed from a `textarea` form element.

# PART VI

# Raspberry Pi Python Projects

# Creating Basic Pi/Python Projects

---

**What You'll Learn in This Hour:**

▸ How to display HD images via Python

▸ How to use Python to play music

▸ How to create a special presentation

In this hour, you will learn how to create some basic projects on your Raspberry Pi using Python. You will learn how to make your very own high-definition image presentation, how to use Python to play a list of music on your Raspberry Pi, and how to create a special presentation using Python.

# Thinking About Basic Pi/Python Projects

The sky is the limit when it comes to creating projects using your Raspberry Pi and Python! The projects in this hour and the next will help you improve and solidify your Python script-writing skills. Also, you will learn to take advantage of some of the nice features on your Raspberry Pi. And, hopefully, by working on these projects, you will be inspired for additional ventures!

This hour covers a few simple projects. You'll create something useful, without spending any additional money. You already have all you need for this hour's basic projects: a Raspberry Pi and Python.

# Displaying HD Images via Python

One of the Raspberry Pi's greatest features is its small size. Carrying around a Pi is even easier than carrying a tablet computer. Another great feature is the Raspberry Pi's HDMI port. The HDMI port allows you to display high-definition images from your Pi.

These two features together make the Raspberry Pi a perfect platform for many uses. You can take your Pi over to a friend's house, hook it up to his or her television, and show your vacation pictures. For a business person, the small size of a Pi makes it ideal for travel and making business presentations. For a student, imagine how impressed your teacher will be to not only see your presentation from the Pi, but to learn that you wrote the script that runs it.

# Understanding High Definition

There can be a lot of confusion concerning HD. Therefore, before you begin to build the scripts this hour, you need to learn about—or review—what is meant by a *high-definition* (*HD*) image.

Other terms for the dimensions of an image are *canvas*, *size*, and *resolution*. This alone can cause confusion! Basically, the dimensions of a picture are the image's width times its height. It is measured in pixels and is often written in the format *width × height*. For example, a picture may have the dimensions 1280×720 pixels.

The dimensions of a picture determine whether it is HD. Larger dimension numbers mean a higher resolution. A higher resolution provides a clearer picture. Thus, if you have a new beautiful product to sell your client, an HD picture of it will be worthwhile. Table 23.1 shows the resolutions for each current definition.

**TABLE 23.1**  Picture Quality Definitions

| Definition Name | Picture Resolution |
| --- | --- |
| Standard definition (SD) | 640×480 pixels |
| High definition (HD) | 1280×720 pixels (minimum) |
| Full high definition (full HD) | 1920×1080 pixels (minimum) |
| Ultra high definition (ultra HD) | 3840×2160 (minimum) |

You may have noticed that dots per inch (dpi) is not mentioned in the definitions in Table 23.1. This is because dpi has nothing to do with the quality of a picture. dpi is actually an old term that has to do with printing pictures on a computer printer.

BY-THE-WAY

### Horizontal and Megapixel

Often a camera's ability to take still HD photos is rated by giving the height (horizontal) only or giving a megapixel rating (multiplying the height by the width). For example, an HD camera with a resolution of 1280×720 could be listed as 720 or 720p. Its megapixel rating would be 0.92MP.

If you do not know a picture's resolution, you can determine whether the image is HD by using the Image Viewer utility on Raspbian. In the Raspbian GUI on your Raspberry Pi, click on the LXDE Programs Menu icon. (If you need help remembering where this icon is located, refer to Hour 2, "Understanding the Raspbian Linux Distribution.")

Within the LXDE menu, in the Accessories submenu is File Manager. When you open the File Manager, you can navigate to any photo or image files currently stored on or accessible by your Raspberry Pi. When the photo or image files are showing in the File Manager window, you can right-click an image file and have the Image Viewer open it. Figure 23.1 shows an example of an image file, `/home/pi/python_games/cat.png`.

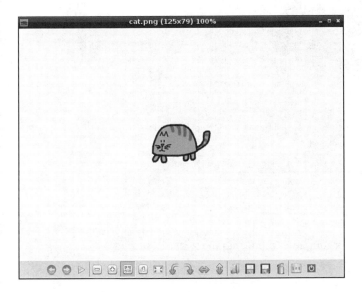

**FIGURE 23.1**
A low-resolution image.

In Figure 23.1, you can see that the Image Viewer's title bar shows the resolution of the image file: 125×79. This image of a cat is obviously not an HD image. Figure 23.2 shows an example of an HD image.

The photograph's resolution is 5616×3744, as you can see in the Image Viewer's title bar. Its resolution makes this still photo of an encased Raspberry Pi an ultra HD image.

The resolution of the pictures you want to present via your Raspberry Pi is up to you. Just remember that higher resolution photos generally provide a clearer image.

# Creating the Image Presentation Script

To create the HD image presentation script, you need to draw on the Python skills you've learned so far. For this script, you will be using several methods from the `PyGame` library.

**FIGURE 23.2**
An ultra HD photo.

The following are the basics for importing and initializing the PyGame library:

```
import pygame                    #Import PyGame library
from pygame.locals import *     #Load PyGame constants
pygame.init()                   #Initialize PyGame
```

This should look very familiar because the PyGame library is covered in Hour 19, "Game Programming." If you need a refresher on using PyGame, go back and review that hour.

## Setting Up the Presentation Screen

At this point, you need to determine what color you want to make the background screen. Your photos or images should guide your choice of the best background color to use. In general, you should use a monochrome color such as black, white, or gray. To generate the necessary background color, you need to set the RGB color settings. The RGB setting to achieve white is 255,255,255. Black is 0,0,0. The following example uses gray, with a setting of 125,125,125:

```
ScreenColor = Gray = 125,125,125
```

Notice that two variables are set up: ScreenColor and Gray. Using two variables not only makes the script more readable but gives you some flexibility in your script. If you plan to always use gray as your background color, you can use the Gray variable throughout the script. If you want to change the background color, you can use the variable ScreenColor in your script.

Flexibility is the name of the game with this Python script. The script you're creating now is very flexible. It will allow you to walk into a room and use whatever size presentation screen is available. It could be a 30-inch computer monitor sitting on your client's desk, a 75-inch television screen at your neighbor's house, or a 10-inch tablet at school. You will not need to edit your script to use the full screen size when you arrive at your presentation site, no matter what the screen size is! To accomplish this feat, when the screen is initialized using `pygame.display.set_mode`, you use the flag FULLSCREEN, as shown here:

```
ScreenFlag = FULLSCREEN | NOFRAME
PrezScreen = pygame.display.set_mode((0,0),ScreenFlag)
```

This causes the PyGame display screen to be set to the full size of the screen the script is encountering at that moment.

Notice that in addition to using FULLSCREEN, you also use the NOFRAME flag. This causes any frames around your screen to disappear during your presentation, allowing the entire available screen to be used for your HD images.

# Finding the Images

Now that you have your presentation screen set up, you need to find your images, so you need to know their file names. Once again, the idea is flexibility: You want to be able to add or delete photo files without needing to edit the Python script.

To have your script find your images for you, put two variables in the script that describe where the pictures are located and their file extensions. In this example, the variable `PictureDirectory` is set to point to the location of the stored photographs:

```
PictureDirectory = '/home/pi/pictures'
PictureFileExtension = '.jpg'
```

The variable `PictureFileExtension` is set to the photo's current file extension. If you have multiple file extensions, you can easily add additional variables, such as `PictureFileExtension1`.

After you set the location of the photos and their file extensions, list the files in that photo directory. To accomplish this, you need to import another module, the `os` module, as shown here:

```
import os                #Import OS module
```

You can use the `os.listdir` operation to list the contents of the directory that contains the photos. Each file located in the directory will be captured by the `Picture` variable. Thus, you can use a `for` loop to process each file as it is encountered:

```
for Picture in os.listdir(PictureDirectory):
```

You may have other files located in this directory, especially if you use a removable drive. (See the section "Storing Photos on a Removable Drive," later in this hour.) To grab and display only the desired images, you can have the script check each file for the appropriate file extension before loading it with the .endswith operation. You can use an if statement to conduct this operation. The for loop should now look as follows:

```
for Picture in os.listdir(PictureDirectory):
    if Picture.endswith(PictureFileExtension):
```

Now, all the pictures ending with the appropriate file extension will be processed. The rest of the loop simply loads each picture and then fills the screen with the designated screen color:

```
Picture=PictureDirectory + '/' + Picture
Picture=pygame.image.load(Picture).convert_alpha()
PictureLocation=Picture.get_rect()   #Current location
#Display HD Images to Screen ################3
PrezScreen.fill(ScreenColor)
PrezScreen.blit(Picture,PictureLocation)
pygame.display.update()
```

Also, you use the .blit operation to put the picture on the screen. Finally, the pygame.display.update operation causes all the pictures to be displayed.

DID YOU KNOW

### Movies

Many cameras can capture high-definition videos as well as photos. They use something called HDSLR (high-definition single-lens reflex) technology. These cameras save the videos in MPEG format. In the PyGame library, you can use the pygame.movie operation to display these MPEG-format videos.

Before moving on, you need to know about some additional flexibility concerning your images. Putting your photos on another storage device besides the SD card will allow you to not only store larger and higher-resolution photo files, it will also allow you to store more photos. This means more vacation photos to share. Your poor neighbors!

## Storing Photos on a Removable Drive

HD photos can take up a great deal of space. If you have a large number of photos to present, your SD card may not have the space needed to hold them all. Remember that the Raspbian operating system also resides on the SD card. One way to fix this problem is to use a removable drive to hold your presentation photos. However, using a removable drive introduces a new problem: How does your Python presentation script access the photos on the removable drive?

Before you start modifying the presentation script, you need to determine the "device file" name Raspbian will assign to your removable drive. Basically, a *device file* is a file name Raspbian uses to access a device, such as a removable hard drive. To determine the device file name, you follow these steps:

1. While in the GUI, plug your removable drive into an open USB port on your Raspberry Pi.

2. When the Removable Medium Is Inserted window appears, select Open in File Manager if it is not already selected and then click the OK button. The File Manager window opens, showing your files and some other important information (see Figure 23.3).

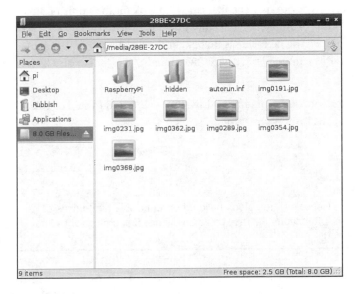

**FIGURE 23.3**
File Manager, showing a removable hard drive.

3. Look at the File Manager address bar for the directory that contains the files. Record the directory name for later use (The directory name in Figure 23.3 is /media/28BE-27DC).

4. Close the File Manager window.

5. Open LXTerminal in the GUI and type in the command ls directory, where directory is the name of the directory you recorded in step 3. Press Enter. You should see your picture files (and any other files located there) on your removable hard drive.

6. Now type mount and press Enter. You should see results similar to those shown in Listing 23.1. When you use the mount command, you can see the device file name that will be used for the removable drive when it is attached to the Raspberry Pi. (In Listing 23.1, the device file is /dev/sda1.)

**LISTING 23.1**    Using `mount` to Display the Device File Names

```
1: pi@raspberrypi ~ $ mount
2: /dev/root on / type ext4 (rw,noatime,data=ordered)
3: ...
4: /dev/mmcblk0p1 on /boot type vfat (rw,relatime,fmask=0022,dmask=0022,
5: codepage=cp437,iocharset=ascii,shortname=mixed,errors=remount-ro)
6: /dev/sda1 on /media/28BE-27DC type vfat (rw,nosuid,nodev,relatime,uid=1000,
7: ...
```

It looks like a bunch of just letters and numbers on the screen, but a couple clues show you the name of the removable drive's device file. The first clue is to look for the name of directory you recorded in step 3. In this example, the directory name is `/media/28B3-27DC`. Do you see it on line 6 of Listing 23.1? The next clue is to look for the word `/dev` on the same line as the directory name. In this example, the device file name is `/dev/sda1`, also listed on line 6.

7. Record the device file name for your removable hard drive that you found in step 6. You will need this information in your Python presentation script.

WATCH OUT!

**Changing Device File Names**

Device file names can change! This is especially true if you have other removable hard drives or devices attached to the USB ports on a Raspberry Pi. They can also change if you plug your removable hard drive into a different USB port.

8. Type `umount` device file name, where device file name is the device file name you recorded in step 6. (Yes, that command is `umount` and not `unmount`.) This command unmounts your removable hard drive from the Raspberry Pi. You can then safely remove it from the USB port.

9. Type `mkdir /home/pi/pictures` to create a directory for your removable hard drive's files.

Now that you have determined the device file name Raspbian will use to refer to your removable hard drive, you need to make some modifications to your presentation script. First, create a variable in your Python script to represent the device file name you determined in the steps above. Here's an example:

```
PictureDisk = '/dev/sda1'
```

The os module has a nice little function called .system. This function allows you to pass dash shell commands (which you learned about in Hour 2) from your Python script to the operating system.

In your script, you should make sure that the removable hard drive has not been automatically mounted. You can do this by issuing an umount command, using os.system, as shown here:

```
Command = "sudo umount " + PictureDisk
os.system(Command)
```

However, if the removable hard drive has not been mounted, this command generates an error message and then continues on with the presentation script. Getting this error message is not a problem. However, error messages certainly are not nice looking in a presentation! Not to worry: You can hide the potential error message by using a little dash shell command trick, as shown here:

```
Command = "sudo umount " + PictureDisk + " 2>/dev/null"
os.system(Command)
```

Now to mount your removable hard drive using the Python script, you use the os.system function again to pass a mount command to the operating system, like this:

```
Command = "sudo mount -t vfat " + PictureDisk + " " + PictureDirectory
os.system(Command)
```

Remember that the PictureDirectory variable was set earlier to /home/pi/pictures. The files on the removable hard drive will now be available in the /home/pi/pictures directory. (Don't worry if this concept is a little confusing to you. It is an advanced Linux concept!)

## WATCH OUT!

### Removable Drive Format

The majority of removable hard drives are formatted using VFAT. However, some have been formatted using NTFS. If your removable hard drive is NTFS, you need to change the -t vfat to -t ntfs in the mount command line of the script.

Now you have set up your presentation screen, found the image files, and even incorporated the use of a removable hard drive into your Python script. However, you need to resolve a few more issues before you are ready to conduct your presentation.

# Scaling the Photos

When you don't always know what the presentation screen size will be, you can end up with oversized photos on a screen that's too small. Figure 23.4 shows how an oversized photo might look on a small screen. In this photo, it appears that you are trying to show your audience a picture of a particular chip. However, you are really trying to show a picture of an entire Raspberry Pi.

**FIGURE 23.4**
A photo sized incorrectly for the display screen.

To ensure that your photos are properly sized, you need to determine the size of the current presentation screen. The .get_size operation helps with this. When the presentation screen is originally set up, you have your script determine the size of the current screen and assign it to a variable. Then you set up a variable called Scale, to be used to scale down any oversized pictures, as shown here:

```
PrezScreenSize = PrezScreen.get_size()
Scale=PrezScreenSize
```

Within your picture display for loop, you add the following if statement to check the size of the current picture. If the picture is larger than the current presentation screen, you have the script scale it down to screen size by using pygame.transform.scale, like this:

```
# If Picture is bigger than screen, scale it down.
if Picture.get_size() > PrezScreenSize:
        Picture = pygame.transform.scale(Picture,Scale)
```

This causes that one oversized photo in Figure 23.4 to now look like a great HD picture, as shown in Figure 23.5.

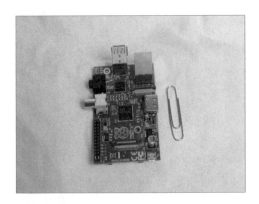

**FIGURE 23.5**
A photo sized correctly for the display screen.

Now that the photo is correctly sized, you can see the entire Raspberry Pi and not just a single chip. Also, the paper clip is now showing, giving the viewing audience a nice size comparison.

## Framing the Photos

To make your photo presentation just a little nicer, you can add a frame to all your photos. Just a minor adjustment to your Scale variable, as shown here, does the trick:

```
Scale=PrezScreenSize[0]-20,PrezScreenSize[1]-20
```

Now that you've added the frame, your photos will look as shown in Figure 23.6. Each one will have the screen's background color surrounding it.

Note that if you have photos of different sizes, the thickness of your frame will change. Also, if a photo is already smaller than the current presentation screen, the frame will have a different thickness than any displayed oversized photos that must be scaled.

## Centering the Photos

One problem you haven't seen yet is keeping your images in the middle of the presentation screen. When you use the functions in the PyGame library to display photos, by default, the photos are displayed in the upper-left corner of the screen. For larger pictures that are framed, the effect of being off-center is subtle. Notice in Figure 23.7 that the frame is missing from the upper and left sections of the display screen.

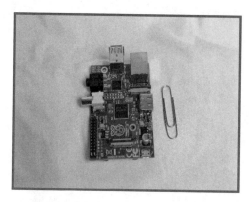

**FIGURE 23.6**
A "framed" photo.

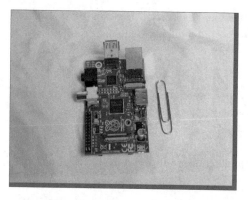

**FIGURE 23.7**
An uncentered photo.

To properly center your images, you need to add an additional variable to the place where the display screen is set up in your Python script. This variable, called CenterScreen, uses the .center method on the display screen to find exactly the center point of the current screen. Here's an example:

```
PrezScreenRect = PrezScreen.get_rect()
CenterScreen = PrezScreenRect.center
```

Within the display picture for loop, you modify the variable PictureLocation slightly. After the current location of the picture's rectangular area is obtained, the center of the picture's rectangle is set, using .center method:

```
PictureLocation=Picture.get_rect()   #Current location
PictureLocation.center=CenterScreen #Put picture in center of screen
```

Thus, when the picture is put on the screen using the following code, its center will be exactly the center of the presentation screen:

```
PrezScreen.blit(Picture,PictureLocation)
```

How cool is this?!

You have seen lots of bits and pieces of code to produce this presentation script. Listing 23.2 shows the entire script that has been put together so far.

### LISTING 23.2   The `script2301.py` Presentation Script

```
#script2301.py - HD Presentation
#Written by Blum and Bresnahan
#
###########################################
#
##### Import Modules & Variables ######
import os                  #Import OS module
import pygame              #Import PyGame library
import sys                 #Import System module
#
from pygame.locals import *    #Load PyGame constants
#
pygame.init()                      #Initialize PyGame
#
# Set up Picture Variables ####################
#
PictureDirectory = '/home/pi/pictures'
PictureFileExtension = '.jpg'
PictureDisk = '/dev/sda1'
#
# Mount the Picture Drive ####################
#
Command = "sudo umount " + PictureDisk + " 2>/dev/null"
os.system(Command)
Command = "sudo mount -t vfat " + PictureDisk + " " + PictureDirectory
os.system(Command)
#
# Set up Presentation Screen #################
#
ScreenColor = Gray = 125,125,125
```

```
#
ScreenFlag = FULLSCREEN | NOFRAME
PrezScreen = pygame.display.set_mode((0,0),ScreenFlag)
#
PrezScreenRect = PrezScreen.get_rect()
CenterScreen = PrezScreenRect.center
#
PrezScreenSize = PrezScreen.get_size()
Scale=PrezScreenSize[0]-20,PrezScreenSize[1]-20
#
###### Run the Presentation  #######################################
#
while True:
    #
    #Get HD Pictures ###################################
    #
    for Picture in os.listdir(PictureDirectory):
        if Picture.endswith(PictureFileExtension):
            Picture=PictureDirectory + '/' + Picture
            Picture=pygame.image.load(Picture).convert_alpha()
            #
            # If Picture is bigger than screen, scale it down.
            if Picture.get_size() > PrezScreenSize:
                Picture = pygame.transform.scale(Picture,Scale)
            #
            PictureLocation=Picture.get_rect()   #Current location
            PictureLocation.center=CenterScreen #Put picture in center of screen
            #
            #Display HD Images to Screen ################
            PrezScreen.fill(ScreenColor)
            PrezScreen.blit(Picture,PictureLocation)
            pygame.display.update()
            pygame.time.delay(500)
            #
            # Quit with Mouse or Keyboard if Desired
            for Event in pygame.event.get():
                Command = "sudo umount " + PictureDisk
                os.system(Command)
                sys.exit()
```

This script works fine, but it runs slowly! Just getting the first picture to display can take a rather long time.

BY-THE-WAY

### While You Test

While testing your presentation script, use small, simple, non-HD image files, such as the `.png` files in the `/home/pi/python_games` directory. That way, you can get everything working correctly without having to deal with the slow loading of HD files. To do this, you just change the variable `PictureDirectory` to `/home/pi/python_games` and the variable `PictureFileExtension` to `.png`.

Unfortunately, when you load any HD image file, a Python script can really slow down. However, there are a few things you can do to speed up the script as well as give the appearance of speed to your presentation audience.

# Improving the Presentation Speed

To improve the speed of the Python HD image presentation script, here are some modifications you can make:

▶ Load only functions used in modules instead of loading entire modules.

▶ Remove any implemented delays.

▶ Add buffering to the screen.

▶ Do not convert images.

▶ Add a title screen.

▶ Add finer mouse and/or keyboard controls.

Each one of these changes might improve the speed of the presentation by only a second or even just a millisecond. However, each little bit will help improve the flow of your HD image presentation. The following sections describe how to implement these optimizations.

## Loading Only Functions Instead of Entire Modules

Loading only the functions used will speed up any Python script. When you load an entire module using the `import` statement, all the functions it contains are also loaded—and this can really slow down a script. A good tip is to create a chart of your Python script that shows the modules imported and the actual functions used. Table 23.2 shows how this might look. This table lists each module, along with each of the functions used from that module. This type of chart will be helpful as you make the necessary modifications to your script.

**TABLE 23.2**   Functions Used in Loaded Modules

| Module | Functions Used |
|--------|----------------|
| os | listdir, system |
| pygame | event, display, font, image, init, transform |
| sys | exit |

To load only the functions you use from each module, you modify your import statements, using the chart you've created as a guide. The import statements will now look similar to the following:

```
##### Import Functions & Variables #######
#
from os import listdir, system       #Import from OS module
#
                                      #Import from PyGame Library
from pygame import event, font, display, image, init, transform
#
from sys import exit                  #Import from System module
```

After you make this change, you need to change all the calls to the various functions. For example, you change pygame.init() to init(), and you change sys.exit() to exit(). Python no longer recognizes the entire pygame module because you no longer load it. Instead, it only recognizes the functions you load from that module.

Use the chart you created and step through the Python script, making all the necessary changes to the function calls. When you have completed these changes, you should test the Python image presentation script. You will be amazed at how much faster it starts up! This is a good activity for any Python script you write: Load up whole modules; tweak the script until you are happy with it; chart your modules and their functions; modify the script to load only the functions; and modify the function calls.

## Removing Any Implemented Delays

This optimization is an easy one! For smaller, non-HD images, you needed to include the pygame.time.delay operation to allow the images to "pause" on the screen. When loading the large HD images, this pause is not needed, so you simply remove the line below from the script:

```
pygame.time.delay(500)
```

You also need to be sure to remove the loading of the time function in the pygame module's import statement.

## Adding Buffering to the Screen

This speed fix will gain you only a millisecond or two, but implementing it is still worthwhile—and it's easy. You simply add an additional flag, DOUBLEBUF, to the flags for your presentation screen, as shown here:

```
ScreenFlag = FULLSCREEN | NOFRAME | DOUBLEBUF
PrezScreen = display.set_mode((0,0),ScreenFlag)
```

## Avoiding Converting Images

In Hour 19, you learned that for games, it is wise to use the .convert_alpha() operation to load images. This way, game images can be converted once to speed up the operation of the game. The opposite is true here because these pictures will be displayed to the screen only one time. To make this modification, you need to make the image.load function for each picture look as follows.

```
Picture = image.load(Picture)
```

Remember that this statement was previously written like this:

```
Picture = pygame.image.load(Picture).convert_alpha()
```

You remove pygame here because you are no longer loading the entire pygame module. You remove .convert_alpha() to improve the speed of the image loading.

Making this change improves the speed of image loading by about 3 to 5 seconds. Making this change also gives you the exact pixel format displayed on the screen that is in the image file. So you get two improvements in one!

Even with these two improvements to image.load, this particular command will still be the slowest one in the entire script due to the large size of HD image files. However, many performance enhancements are in the works for the Raspberry Pi, so this may not be a problem in the future.

WATCH OUT!

## Preloading Images

Since loading images takes so long, it would make sense to preload the images into a Python list before you begin displaying them to the screen. However, if you try to do this, you will most likely run out of memory. It would work, though, if you had non-HD images or just a few pictures. But trying to load up several HD photos will cause your Python script to run out of memory, suddenly quit, and leave you with the message "Killed" displayed on the screen.

## Adding a Title Screen

The next best thing to do to improve the speed of the presentation script is to give the audience the illusion of speed.

Sending text to the presentation screen happens relatively quickly. Thus, by adding a title screen at the beginning of the script, you can give that first picture time to load. Also, this will prevent your audience from staring at a blank screen for 20 to 30 seconds.

To incorporate a title screen, you set up some variables and text to be used within the presentation, as follows:

```
# Set up Presentation Text ######################
#
# Color #
#
RazPiRed = 210,40,82
#
# Font #
#
DefaultFont='/usr/share/fonts/truetype/freefont/FreeSans.ttf'
PrezFont=font.Font(DefaultFont,60)
#
# Text #
#
IntroText1="Our Trip to the"
IntroText2="Raspberry Capital of the World"
IntroText1=PrezFont.render(IntroText1,True,RazPiRed)
IntroText2=PrezFont.render(IntroText2,True,RazPiRed)
```

Notice that a color called `RazPiRed` is being used for the text color to provide a nice contrast to the presentation screen's gray background.

Now you can place the text prior to the `for` loop in the presentation script, as follows:

```
# Introduction Screen #############################
#
PrezScreen.fill(ScreenColor)
#
# Put Intro Text Line 1 above Center of Screen
IntroText1Location = IntroText1.get_rect()
IntroText1Location.center = AboveCenterScreen
PrezScreen.blit(IntroText1,IntroText1Location)
#
# Put Intro Text Line 2 at Center of Screen
IntroText2Location = IntroText2.get_rect()
IntroText2Location.center = CenterScreen
PrezScreen.blit(IntroText2,IntroText2Location)
#
```

```
display.update()
#
#Get HD Pictures ################################
#
for Picture in listdir(PictureDirectory):
```

## Adding Finer Mouse and/or Keyboard Controls

This last fix will provide you with another speed illusion. Many people give business presentations while holding a remote or using a mouse to control the flow of the images shown on the screen. Adding an event loop immediately after a picture is loaded allows you to incorporate this type of control, as shown here:

```
Picture = image.load(Picture)
#
Continue = 0
# Show next Picture with Mouse
while Continue == 0:
        for Event in event.get():
                if Event.type == MOUSEBUTTONDOWN:
                        Continue = 1
```

This added event gives the illusion of picture display control by the presenter. You still have the same long load time, but by knowing the approximate time between pictures, you can talk through each image and then click the mouse after the approximate time. This gives the illusion of the pictures immediately loading when you click the mouse.

If you just want to show your friends and neighbors vacation pictures, you can leave out this optimization. In that case, the photos will continuously feed to the screen in a continuous loop.

## The Optimized Presentation

The HD image presentation script, with all of its "speed" modifications, has changed quite a bit. Listing 23.3 shows some of the new script, script2302.py, which is available to download at informit.com/title/9780789752055.

**LISTING 23.3**  The Optimized `script2302.py` Presentation Script

```
...
##### Import Functions & Variables #######
#
from os import listdir, system      #Import from OS module
#
                                    #Import from PyGame Library
from pygame import event, font, display, image, init, transform
#
from sys import exit                #Import from System module
```

```
#
from pygame.locals import *          #Load PyGame constants
#
init()                              #Initialize PyGame
#
# Set up Picture  Variables #####################
...
# Set up Presentation Text #####################
#
# Color #
#
RazPiRed = 210,40,82
#
# Font #
#
DefaultFont='/usr/share/fonts/truetype/freefont/FreeSans.ttf'
PrezFont=font.Font(DefaultFont,60)
#
# Text #
#
IntroText1="Our Trip to the"
IntroText2="Raspberry Capital of the World"
IntroText1=PrezFont.render(IntroText1,True,RazPiRed)
IntroText2=PrezFont.render(IntroText2,True,RazPiRed)
#
# Set up the Presentation Screen ###############
#
ScreenColor = Gray = 125,125,125
#
ScreenFlag = FULLSCREEN | NOFRAME | DOUBLEBUF
PrezScreen = display.set_mode((0,0),ScreenFlag)
#
PrezScreenRect = PrezScreen.get_rect()
CenterScreen = PrezScreenRect.center
AboveCenterScreen = CenterScreen[0],CenterScreen[1]-100
#
PrezScreenSize = PrezScreen.get_size()
Scale=PrezScreenSize[0]-20,PrezScreenSize[1]-20
#
###### Run the Presentation  ####################################
#
while True:
    # Introduction Screen ##########################
    #
    PrezScreen.fill(ScreenColor)
    #
    # Put Intro Text Line 1 above Center of Screen
    IntroText1Location = IntroText1.get_rect()
    IntroText1Location.center = AboveCenterScreen
```

```
PrezScreen.blit(IntroText1,IntroText1Location)
#
# Put Intro Text Line 2 at Center of Screen
IntroText2Location = IntroText2.get_rect()
IntroText2Location.center = CenterScreen
PrezScreen.blit(IntroText2,IntroText2Location)
#
display.update()
#
#Get HD Pictures ################################
#
for Picture in listdir(PictureDirectory):
    if Picture.endswith(PictureFileExtension):
        Picture = PictureDirectory + '/' + Picture
        #
        Picture = image.load(Picture)
        #
        Continue = 0
        # Show next Picture with Mouse
        while Continue == 0:
            for Event in event.get():
                if Event.type == MOUSEBUTTONDOWN:
                    Continue = 1
                if Event.type in (QUIT,KEYDOWN):
                    Command = "sudo umount " + PictureDisk
                    system(Command)
                    exit()
. . .
```

## Potential Script Modifications

Hopefully, as you read through the script in Listing 23.3, you thought of many improvements you could make. You have come a long way in learning Python! You may have noted improvements and changes such as these:

▶ Rewrite the script using `tkinter`, which is covered in Hour 18, "GUI Programming."

▶ Write an additional script which allows the creation of a configuration file that dictates where files are located and the file extensions of pictures. Modify the presentation script to use the information in the configuration file.

▶ Add to the script a dictionary that contains text to be displayed along with each image.

▶ Modify the script to determine the device file name on the fly, so it does not have to be determined beforehand.

Feel free to add as many changes as you desire. This is your HD image presentation script!

# Playing Music

You can use Python to create some creative scripts for playing your music—for free! After you create such a script, you can take your Pi over to someone else's place, hook it up to their television, and listen to your favorite music. The best part is that you're the one writing the script playing the music!

## Creating a Basic Music Script

To keep your music script simple, you will continue to use the PyGame library you've already learned about. PyGame does a decent job of handling music. You might think that the best way to handle music files would be to create a Sound object, as you did in Hour 19 for the Python game. That does, in fact, work, but loading the music files into Python this way goes very slowly. Thus, it's best to avoid using the Sound object to play music.

BY-THE-WAY

### Other Modules and Packages for Playing Music

There are several other modules and packages for Python that you can use to create scripts for playing music files. A rather detailed list of them is shown at http://wiki.python.org/moin/PythonInMusic.

Besides doing the basic PyGame initialization, you primarily need to use two methods in this script: pygame.mixer.music.load and pygame.mixer.music.play. Each music file must be loaded from the disk into the Python script, before it can be played. Here's an example of loading a music file:

```
pygame.mixer.music.load('/home/pi/music/BigBandMusic.ogg')
```

DID YOU KNOW

### Problems with MP3 Formats

Music comes in several standard file formats. Three of the most popular are MP3, WAV, and OGG. However, you need to be aware that the MP3 file format is a closed-source format, so the open-source world typically frowns on it.

Python and PyGame can handle MP3 file format, but be aware that MP3 files may not play on your Linux system and can even cause the system to crash. It is best to use either uncompressed music files, such as the WAV format, or open-source compressed files such as OGG files. You can convert your MP3 music files to supported formats by using either online conversion websites or locally installed software tools. You can find a list of many audio file conversion tools at http://wiki.python.org/moin/PythonInMusic.

After a music file is loaded, you use the `play` method for playing the file, as shown here:

```
pygame.mixer.music.play(0)
```

The number shown here, 0, is the number of times the music file will play. You might think that zero means it will play zero times, but actually, when the `play` method sees 0, it plays the file one time and then stops.

BY-THE-WAY
___

## Queuing It Up!

If you want to play only a couple songs, you can use the `queue` method. You simply load and play the first song, and the first song begins to play immediately. Then you load and queue a second song. The method to queue a song is `pygame.mixer.music.queue('filename')`. When the first song stops playing, the second song starts playing right away.

You can queue only one song at a time. If you queue a third song, before the second song starts playing, the second song is "wiped" from the queue list.

___

# Storing Your Music on a Removable Disk

Music files, especially if they are in uncompressed WAV file format, can take up a great deal of disk space. Once again, your SD card with Raspbian may not have the space needed to hold the music files you want to play. You can fix this problem by using a removable drive with your Python script.

Just as you used a removable hard drive in the HD image presentation script, you can use it in your music script. The only change needed is to set up variables which point to the disk and directory holding your music.

Be aware that you need to create a music directory before you run your script. To create a music directory, you open LXTerminal, type a command similar to `mkdir /home/pi/music`, and press Enter.

Unlike in the HD presentation script, you cannot simply unmount the drive at the end of the script. Playing music from a removable drive can introduce a few problems with keeping files open and can cause the drive to fail to unmount. However, you can clean up the `umount` commands from the HD presentation script and put them into a function (where they should have been in the first place). Here's what this looks like:

```
# Gracefully Exit Script Function ################
#def Graceful_Exit ():
    pygame.mixer.music.stop() #Stop any music.
    pygame.mixer.quit()        #Quit mixer
```

```
pygame.time.delay(3000)    #Allow things to shutdown
Command = "sudo umount " + MusicDisk
system(Command)    #Unmount disk
exit()
```

The method `pygame.mixer.music.stop` is called to stop any music from playing. Also, the mixer is shut down using `pygame.mixer.quit`. Finally, a delay is added, just to give everything time to shut down, before the `umount` command is issued. It's a little bit of overkill, but properly unmounting a removable drive with your music is worth it!

# Using a Music Playlist

While you could use the `os.listdir` method used earlier in this hour to load the music files, using a playlist will give you finer control (and more Python practice). You can create a simple playlist of the music files to play in a particular order by using either the nano text editor or the IDLE text editor.

You want the playlist file to have the file names of the songs, including their file extension. This way, you can play different types of music files, such as OGG or WAV. Also, each line of the playlist file should contain only a single music file name. No directory names are included because they will be handled in the Python script. The following is a simple example of a playlist that can be used with this script:

```
BigBandMusic.ogg
RBMusic.wav
MusicalSong.ogg
...
```

To open and read your playlist in the Python script, you use the `open` Python statement. (For a refresher on opening and reading files, see Hour 11, "Using Files.")

WATCH OUT!

## Ignoring the End

While PyGame is a wonderful utility for learning how to incorporate various features into your Python scripts, it can be a little flakey at times, especially when playing music. For example, PyGame often sporadically ignores the last music file loaded to play from a playlist. Therefore, you should put your least favorite song last in the playlist. When you learn how to properly code Python to play music, you can explore other music library options for Python, if you so desire.

The script opens the playlist and reads it all in, and then it saves the music file information into a list that the script can use over and over again. (If you need to review the concepts related to lists, see Hour 8, "Using Lists and Tuples.")

Each music file name is read from the playlist file and stored into a list called SongList. You need to strip off the newline character from the end of each read-in record. The .rstrip method will help you accomplish this. Use the following for loop to read in the music file names from the playlist after it is opened:

```
for Song in PlayList:        #Load PlayList into SongList
#
    Song = Song.rstrip('\n') #Strip off newline
    if Song != "":   #Avoid blank lines in PlayList
        Song = MusicDirectory + '/' + Song
        SongList.append(Song)
        MaxSongs = MaxSongs + 1
```

Notice that this example uses an if Python statement. This statement allows your script to check for any blank lines in the playlist file and discard them. It is very easy for blank lines to creep into a file like this one. This is especially true at the bottom of the file, if you accidently press the Enter key too many times.

The for loop appends any file names to the end of the SongList list and keeps a count of how many music files are loaded into the list. Instead of keeping a count, you could also wait until the SongList list is completely built. Then you can determine how many elements are in the list, using the Python statement len(SongList).

## Controlling the Playback

Now you have your playlist loaded, you have a removable drive with the music ready, and you know how to load and play the music. But just how do you control the playback of the music?

PyGame provides event handling that will work perfectly for controlling music playback. Using the .set_endevent method causes an event to queue up after a song has finished playing. This event is called an "end" event because it is sent to the queue when the song ends. The following is an example of an entire function that loads the music file, starts playing the music file, and sets an end event:

```
# Play The Music Function ########################
#
def Play_Music (SongList,SongNumber):
    pygame.mixer.music.load(SongList[SongNumber])
    pygame.mixer.music.play(0)
    pygame.mixer.music.set_endevent(USEREVENT)   #Send event when Music Stops
```

Notice that end event set is USEREVENT. This means that when the music stops playing, the event USEREVENT will be sent to the event queue.

Checking for the USEREVENT event should be handled in the main body of the Python script. You use a while loop to keep the music playing and a for loop to check for the song's end event:

```
while True:   #Keep playing the Music ###########
    for Event in event.get():
        #
            if Event.type == USEREVENT:
                if SongNumber < MaxSongs:
                    Play_Music(SongList,SongNumber)
                    SongNumber = SongNumber + 1
                if SongNumber >= MaxSongs:
                    SongNumber = 0 #Start over in PlayList
                    Play_Music(SongList,SongNumber)
                    SongNumber = SongNumber + 1
```

In this example, if the song's end event, USEREVENT, is found in the event queue, then a couple checks are made. If the song list has not been fully played (SongNumber < MaxSongs), the next song in the SongList is played. However, if all the songs have been played, then SongNumber is set back to 0 (the first file name in the SongList), and the playing of the list starts over.

BY-THE-WAY

### Getting Fancy

You have just seen a very simple way to handle playing music in Python. You can get very fancy with PyGame operations, though. For example, you can use .fadeout to slowly fade out music at its end and .set_volume to make certain songs (like your favorites) louder than others.

At this point, the Python music script plays endlessly. To add control for ending the script, you need to check for another event, such as pressing a key on the keyboard. You do this much the same way you controlled the HD image presentation script. Here's what it looks like:

```
if Event.type in (QUIT,KEYDOWN,MOUSEBUTTONDOWN):
    Graceful_Exit()
```

But wait! This actually doesn't work! For PyGame to properly handle events, the display screen must be initialized. Thus, you need to set up a simple display screen to gracefully control the end of your script. Here's an example:

```
MusicScreen = display.set_mode((0,0))
display.set_caption("Playing Music...")
```

Now, when your music plays, a screen pops up, with the caption "Playing Music" at the top. You can minimize that screen and listen to your music playlist. When you are done, you just maximize the screen and either click it with your mouse or press any key on the keyboard to stop the music.

Since the display screen is already initialized in your music script, you might think that you could add images to be displayed on the screen while the music plays—and that's a good idea!

You'll learn how to do that a bit later this hour, but first you need to do a few more things related to your music script, including reviewing the music script in its entirety. See Listing 23.4.

**LISTING 23.4**   The `script2303.py` Music Script

```
#script2303.py - Play Music from List
#Written by Blum and Bresnahan
#
###########################################
#
##### Import Functions & Variables #######
#
from os import system          #Import from OS module
#
                               #Import from PyGame Library
from pygame import display, event, init, mixer, time
#
from sys import exit           #Import from System module
#
from pygame.locals import *    #Load PyGame constants
#
init()                         #Initialize PyGame
#
# Load Music Play List Function ##################
#
# Read Playlist and Queue up Songs #
#
def Load_Music ():
#
      global SongList          #Make SongList global
      SongList = []            #Initialize SongList to Null
      #
      global SongNumber        #Make SongNumber global
      SongNumber = 0           #Initialize Song Number to 0
      #
      global MaxSongs          #Make MaxSongs global
      MaxSongs = 0             #Initialize Maximum Songs to 0
      #
      PlayList = MusicDirectory + '/' + 'playlist.txt'
      PlayList = open(PlayList, 'r')
      #
      for Song in PlayList:    #Load PlayList into SongList
      #
            Song = Song.rstrip('\n') #Strip off newline
            if Song != "":    #Avoid blank lines in PlayList
                  Song = MusicDirectory + '/' + Song
                  SongList.append(Song)
```

```
                MaxSongs = MaxSongs + 1
     PlayList.close()
#
# Play The Music Function ########################
#
def Play_Music (SongList,SongNumber):
     mixer.music.load(SongList[SongNumber])
     mixer.music.play(0)
     mixer.music.set_endevent(USEREVENT)     #Send event when Music Stops
#
# Gracefully Exit Script Function ################
#
def Graceful_Exit ():
     mixer.music.stop() #Stop any music.
     mixer.quit()        #Quit mixer
     time.delay(3000)   #Allow things to shutdown
     Command = "sudo umount " + MusicDisk
     system(Command)       #Unmount disk
     exit()
#
# Set up Music Variables ####################
#
MusicDirectory = '/home/pi/music'
MusicDisk = '/dev/sda1'
#
# Mount the Music Drive #####################
#
Command = "sudo umount " + MusicDisk + " 2>/dev/null"
system(Command)
Command = "sudo mount -t vfat " + MusicDisk + " " + MusicDirectory
system(Command)
#
# Queue up the Music ###########################
#
Load_Music()
Play_Music(SongList,SongNumber)
SongNumber = SongNumber + 1
#
#Set up Display for Event Handling #############
#
MusicScreen = display.set_mode((0,0))
display.set_caption("Playing Music...")
#
while True:  #Keep playing the Music ###########
     for Event in event.get():
     #
          if Event.type == USEREVENT:
               if SongNumber < MaxSongs:
```

```
            Play_Music(SongList,SongNumber)
            SongNumber = SongNumber + 1
    if SongNumber >= MaxSongs:
            SongNumber = 0 #Start over in PlayList
            Play_Music(SongList,SongNumber)
            SongNumber = SongNumber + 1
            #
    if Event.type in (QUIT,KEYDOWN,MOUSEBUTTONDOWN):
            Graceful_Exit()
```

Notice that this script imports only the needed functions. The names of the module operations in the script have been modified to reflect this.

## Making the Play List Random

If desired, you can make your script play music from the playlist randomly. Making this happen requires only a few minor changes. The first change is to import the randinit operation from the random module, as shown here:

```
from random import randint     #Import from Random module
```

The other two changes are small tweaks to an if statement within the main while loop of the music script. For the if SongNumber >= MaxSongs: statement, you need to replace SongNumber = 0 with the statement using the randinit method, as shown here:

```
while True:   #Keep playing the Music ############
     for Event in event.get():
...

            if SongNumber >= MaxSongs:
                 SongNumber = randint(0,MaxSongs - 1) #Pick random song
                 Play_Music(SongList,SongNumber)
                 SongNumber = MaxSongs #Keep songs random

...
```

Also, to keep the songs playing in a random fashion, instead of incrementing SongNumber after Play_Music is called, you set it back to being equal to MaxSongs.

## Creating a Special Presentation

By now you've probably figured out what the "special" presentation is all about: playing music along with displaying your HD images! There are many reasons you might want to do this. You may have a special business presentation that needs music behind it. You might want to see images display while your music is playing. Or, as in this example, you might be a teacher trying to encourage your school board to buy Raspberry Pis for the students and start up some classes teaching Python.

BY-THE-WAY

## Playing One Song Continuously

You might just want to play one loaded song, such as your company's marketing song, endlessly during a presentation. To do this, you use the Python statement `pygame.mixer.music.play(-1)`. The negative one (-1) tells Python to keep playing the song over and over again, until the script exits.

Basically, this project, shown in Listing 23.5, melds together the HD image presentation script and the music script. It assumes that both your HD images and your music will be on the same removable drive.

**LISTING 23.5**   The `script2305.py` Special Presentation

```
#script2305.py - Special HD Presentation with Sound
#Written by Blum and Bresnahan
#
###########################################
#
##### Import Functions & Variables #######
#
from os import listdir, system      #Import from OS module
#

                                     #Import from PyGame Library
from pygame import event, font, display, image, init, mixer, time, transform
#
from random import randint           #Import from Random module
#
from sys import exit                 #Import from System module
#
from pygame.locals import *          #Load PyGame constants
#
init()                               #Initialize PyGame
#
# Load Music Play List Function ################
#
# Read Playlist and Queue up Songs #
#
def Load_Music ():
#
        global SongList      #Make SongList global
        SongList = []        #Initialize SongList to Null
        #
        global SongNumber    #Make SongNumber global
        SongNumber = 0       #Initialize Song Number to 0
        #
        global MaxSongs      #Make MaxSongs global
```

```python
        MaxSongs = 0          #Initialize Maximum Songs to 0
        #
        PlayList = PictureDirectory + '/' + 'playlist.txt'
        PlayList = open(PlayList, 'r')
        #
        for Song in PlayList:   #Load PlayList into SongList
        #
            Song = Song.rstrip('\n') #Strip off newline
            if Song != "":   #Avoid blank lines in PlayList
                Song = PictureDirectory + '/' + Song
                SongList.append(Song)
                MaxSongs = MaxSongs + 1
        PlayList.close()
#
# Play The Music Function ########################
#
def Play_Music (SongList,SongNumber):
        mixer.music.load(SongList[SongNumber])
        mixer.music.play(0)
        mixer.music.set_endevent(USEREVENT)     #Send event when Music Stops
#
# Gracefully Exit Script Function #################
#
def Graceful_Exit ():
        mixer.music.stop() #Stop any music.
        mixer.quit()        #Quit mixer
        time.delay(3000)    #Allow things to shutdown
        Command = "sudo umount " + PictureDisk
        system(Command)     #Unmount disk
        exit()
#
# Set up Picture Variables ####################
#
PictureDirectory = '/home/pi/pictures'
PictureFileExtension = '.jpg'
PictureDisk = '/dev/sda1'
#
# Mount the Picture Drive ########################
#
Command = "sudo umount " + PictureDisk + " 2>/dev/null"
system(Command)
Command = "sudo mount -t vfat " + PictureDisk + " " + PictureDirectory
system(Command)
#
# Set up Presentation Text ####################
#
# Color #
#
```

```
RazPiRed = 210,40,82
#
# Font #
#
DefaultFont='/usr/share/fonts/truetype/freefont/FreeSans.ttf'
PrezFont=font.Font(DefaultFont,60)
#
# Text #
#
IntroText1="Why Our School Should"
IntroText2="Use Raspberry Pi's and Teach Python"
IntroText1=PrezFont.render(IntroText1,True,RazPiRed)
IntroText2=PrezFont.render(IntroText2,True,RazPiRed)
#
# Set up the Presentation Screen ###############
#
ScreenColor = Gray = 125,125,125
#
ScreenFlag = FULLSCREEN | NOFRAME | DOUBLEBUF
PrezScreen = display.set_mode((0,0),ScreenFlag)
#
PrezScreenRect = PrezScreen.get_rect()
CenterScreen = PrezScreenRect.center
AboveCenterScreen = CenterScreen[0],CenterScreen[1]-100
#
PrezScreenSize = PrezScreen.get_size()
Scale=PrezScreenSize[0]-20,PrezScreenSize[1]-20
#
###### Run the Presentation  #####################################
#
# Queue up the Music ############################
#
Load_Music()
Play_Music(SongList,SongNumber)
SongNumber = SongNumber + 1
#
while True:
    # Introduction Screen ############################
    #
    PrezScreen.fill(ScreenColor)
    #
    # Put Intro Text Line 1 above Center of Screen
    IntroText1Location = IntroText1.get_rect()
    IntroText1Location.center = AboveCenterScreen
    PrezScreen.blit(IntroText1,IntroText1Location)
    #
    # Put Intro Text Line 2 at Center of Screen
    IntroText2Location = IntroText2.get_rect()
```

```
IntroText2Location.center = CenterScreen
PrezScreen.blit(IntroText2,IntroText2Location)
#
display.update()
#
#Get HD Pictures ###################################
#
for Picture in listdir(PictureDirectory):
      if Picture.endswith(PictureFileExtension):
            Picture = PictureDirectory + '/' + Picture
            #
            Picture = image.load(Picture)
            #
            for Event in event.get():
                  #
                  if Event.type == USEREVENT:
                        if SongNumber < MaxSongs:
                              Play_Music(SongList,SongNumber)
                              SongNumber = SongNumber + 1
                        if SongNumber >= MaxSongs:
                              SongNumber = randint(0,MaxSongs - 1)
                              Play_Music(SongList,SongNumber)
                              SongNumber = MaxSongs #Keep it random
                  #
                  if Event.type in (QUIT,KEYDOWN):
                        Graceful_Exit()
            #
            # If Picture is bigger than screen, scale it down.
            if Picture.get_size() > PrezScreenSize:
                  Picture = transform.scale(Picture,Scale)
            #
            PictureLocation=Picture.get_rect()   #Current location
            PictureLocation.center=CenterScreen #Put in center
            #
            #Display HD Images to Screen ###############3
            PrezScreen.fill(ScreenColor)
            PrezScreen.blit(Picture,PictureLocation)
            display.update()
      #
      # Quit with Keyboard if Desired
      for Event in event.get():
            if Event.type in (QUIT,KEYDOWN):
                  Graceful_Exit()
```

Remember that you can also get a copy of these scripts from the publisher's website. That way, you do not have to retype an entire script into your favorite text editor to modify it for your own needs.

# Summary

In this hour, you learned how to create three Python projects: one that displays HD images to a presentation screen, one that plays music from a music playlist, and one that combines the two scripts into a special presentation.

Before you start thinking of all the modifications you can make to these projects, get ready. In the next hour, you are about to learn some really cool and rather advanced projects using Python on the Raspberry Pi!

# Q&A

**Q.** What is Wayland, and could it help speed up the HD image presentation script?

**A.** Wayland is a replacement under-the-hood program that is partially responsible for displaying windows in the GUI. It is promised to eventually be used in Raspbian. Will it speed up the HD image presentation script? Possibly.

**Q.** Where is the Raspberry capital of the world?

**A.** The Raspberry capital of the world is Hopkins, Minnesota. The city hosts an annual raspberry festival. The festival concerns fruit, and not computers.

# Workshop

## Quiz

1. You must use images that are already appropriately sized for the display screen. True or false?

2. What is the minimum resolution for an HD image?

3. Which `PyGame` operation handles the playing of music?

    **a.** `.pygame.music`

    **b.** `.pygame.music.play`

    **c.** `.pygame.mixer.music.play`

# Answers

1. False. When using the `PyGame` library, you can transform an image to an appropriate size. However, if you increase the size of an image, you might lose the image's original clarity.

2. An image is considered to be HD if it has at least a resolution of 1280×720 pixels.

3. Answer c is correct. The `pygame.mixer.music.play` operation handles the playing of a loaded music file.

# Working with Advanced Pi/Python Projects

**What You'll Learn in This Hour:**

▶ Working with the GPIO interface

▶ Exploring the Python RPi.GPIO module

▶ Using the GPIO for output

▶ Using the GPIO for input

One of the exciting features of the Raspberry Pi is the GPIO interface, which allows you to connect your Raspberry Pi to electronic circuits and then interact with the outside world. In this hour, you'll learn about the GPIO interface and how to use it to both accept input and send output to electronic circuits. This hour you'll use two popular Raspberry Pi electronic interface devices for the projects: the Pi Cobbler and the Gertboard.

## Exploring the GPIO Interface

One of the features included in the Raspberry Pi that you don't often see in consumer computers is the General Purpose Input/Output interface (called the GPIO interface for short). The GPIO interface is the key to getting your Raspberry Pi to interact with the outside world. You can use it to control all sorts of electronics, from temperature gauges to robots. In the sections that follow, you'll take a look at the Raspberry Pi's digital interface and what you need to interact with it.

## What Is the GPIO Interface?

The GPIO interface provides direct access to the Broadcom chip on the Raspberry Pi, which includes several built-in digital interface features:

▶ 17 digital input/output (I/O) pins

▶ A pulse-width modulation (PWM) output

▶ An Inter-Integrated Circuit (I2C) interface

- A Serial Peripheral Interface (SPI) connection

- A Universal Asynchronous Receiver/Transmitter (UART)

The 17 digital I/O pins allow you to read high or low digital signals from up to 17 separate devices or send up to 17 high or low digital signals to external devices—or some combination of the two. These signals can be used for controlling relays to turn circuits on or off or send signals to trigger devices (such as turn on your coffeemaker).

The PWM output is used to control the speed of electric motors. You can control the PWM signal to make a motor stop, start, speed up, or slow down.

The I2C and SPI interfaces provide a digital communications protocol for interfacing with integrated circuits. This protocol allows you to connect advanced microcontrollers, such as the Atmel ATmega microcontroller chip, made popular in the Arduino hobbyist unit.

Finally, the GPIO interface provides access to the UART pins on the Broadcom chip. The UART pins allow you to connect a serial device (such as a terminal or modem) to your Raspberry Pi.

## The GPIO Pin Layout

The GPIO interface is the series of 26 pins (in two rows of 13 pins each) that stick up at the upper-left corner of the Raspberry Pi circuit board. They directly interface with specific pins on the Broadcom integrated circuit chip and are assigned names based on the chip signal. Some pins have dual functions from the Broadcom chip, depending on how you program the chip. Table 24.1 shows the signal names and the alternate functions for each of the pins. When you start coding your Python scripts, you'll need to know which pin or signal you need to work with.

**TABLE 24.1**  The GPIO Pins

| Pin | Signal | Alternate Function |
| --- | --- | --- |
| 1 | 3.3V | |
| 2 | 5V | |
| 3 | GPIO 2 | I2C SDA |
| 4 | 5V | |
| 5 | GPIO 3 | I2C SCL |
| 6 | GND | |
| 7 | GPIO 4 | GPCLK0 |
| 8 | GPIO 14 | UART TX |
| 9 | GND | |
| 10 | GPIO 15 | UART RX |

| Pin | Signal | Alternate Function |
|-----|--------|--------------------|
| 11 | GPIO 17 | |
| 12 | GPIO 18 | PWM |
| 13 | GPIO 27 | |
| 14 | GND | |
| 15 | GPIO 22 | |
| 16 | GPIO 23 | |
| 17 | 3.3V | |
| 18 | GPIO 24 | |
| 19 | GPIO 10 | SPI MOSI |
| 20 | GND | |
| 21 | GPIO 9 | SPI MISO |
| 22 | GPIO 25 | |
| 23 | GPIO 11 | SPI SCLK |
| 24 | GPIO 8 | SPI CE0 |
| 25 | GND | |
| 26 | GPIO 7 | SPI CE1 |

WATCH OUT!

### GPIO Pins Versus Signals

The GPIO signals are numbered after the pin number on the Broadcom chip. Unfortunately, they don't correlate the actual pins used in the GPIO interface. (For example, GPIO signal 2 is on pin 3 of the GPIO interface.) You must be careful when referencing the pin connections. Make sure you know whether you're working with pin numbers or signal numbers. The code in this hour uses signal numbers because this is the method most hardware interface devices use.

# Connecting to the GPIO

There are three common ways to connect to the GPIO pins on the Raspberry Pi motherboard:

▶ Directly plug wires into them.

▶ Use the Pi Cobbler breakout box.

▶ Use the Gertboard experimental device.

### Connecting to the GPIO

Although you can plug wires directly into the GPIO pins on the motherboard, doing so is somewhat of a risky adventure. If you accidentally short out the wrong pins, you risk damaging your entire Raspberry Pi unit! It's much safer (especially for beginners) to use either the Pi Cobbler or the Gertboard.

Let's take a closer look at how to connect to the GIO using the Pi Cobbler and the Gertboard.

## Connecting to the GPIO via the Pi Cobbler

The Pi Cobbler is an inexpensive breakout box that connects to the GPIO pins using a standard 26-pin ribbon cable. It then breaks out the pins into a form that you can plug into a standard breadboard socket (see Figure 24.1).

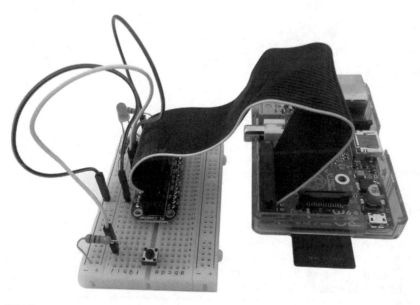

**FIGURE 24.1**
The Pi Cobbler breakout box connected to a Raspberry Pi via a ribbon cable.

The Pi Cobbler unit labels the breakout pins using the GPIO signal names, so you can easily identify which pin is which signal. Once you plug the Pi Cobbler interface into the breadboard, you can wire up your projects directly on the breadboard.

### Plugging in the Pi Cobbler Ribbon Cable

Be careful when connecting the Pi Cobbler ribbon cable to the GPIO interface on the Raspberry Pi. In the model that I purchased, the ribbon cable points to the inside of the Raspberry Pi circuit board (see Figure 24.1), not the outside, as you might assume.

## Connecting to the GPIO via the Gertboard

For more advanced Raspberry Pi experimenting, the Gertboard has it all. Created by Gert van Loo and sold through various electronics distributors around the world, it's a full circuit board of handy components that plugs directly into the Raspberry Pi GPIO pins (see Figure 24.2). If you purchased a case for your Raspberry Pi, you may have to remove the top of the case to plug in the Gertboard.

**FIGURE 24.2**
The Gertboard plugged in to a Raspberry Pi.

The Gertboard contains circuits for experimenting with many common features of the Raspberry Pi:

- ▶ 12 buffered I/O ports

- ▶ 12 LEDs for displaying logic levels

- ▶ Three push-button switches for input

- ▶ Six open collector relays for turning higher-voltage circuits on and off

- ▶ An 18v, 2A motor controller

- ▶ A two-channel analog-to-digital converter

- ▶ A two-channel digital-to-analog converter

- ▶ An Atmel ATmega microcontroller (just like the Arduino)

The Gertboard is designed as a modular board with pins that interface to all the onboard components. Setting up a circuit is as easy as connecting wires between the pins on the board. The Gertboard kit comes with a set of jumpers (short clips that connect two adjacent pins) and a set of straps (longer wires that connect two pins).

To use the Gertboard, you need to become familiar with the pin layout on the board. Each group of pins is identified by a J number that you see written on the circuit board. Table 24.2 shows what each J block of pins is used for.

**TABLE 24.2**   **The Gertboard Pin Block Layout**

| Block | Description |
| --- | --- |
| J1 | Interfaces the Raspberry Pi GPIO pins (on the bottom of the board) |
| J2 | Provides pins for each of the GPIO signals |
| J3 | Provides I/O buffer pins |
| J7 | Applies 3.3V power to the Gertboard circuits |
| J25 | Interfaces with the ATmega microcontroller |

The three-pin J7 block is crucial. You must place a jumper between the middle pin and the upper pin, labeled 3V3, to power on the Gertboard. Without it, none of your projects will work.

Also inside the circuit are two sets of 12 pins labeled B1 through B12. One set is for using the buffered inputs, and the other set is for using the buffered outputs. Consult the Gertboard manual for a complete description of how to use the buffered input and output pins.

# Using the `RPi.GPIO` **Module**

To interface your Python programs with the GPIO signals, you have to use the `RPi.GPIO` module. The `RPi.GPIO` module uses direct memory access to provide an interface to control the GPIO signals. The following sections walk through the basics of the `RPi.GPIO` module.

## Installing `RPi.GPIO`

At this writing, the Raspberry Pi installs the Python v2 version of the `RPi.GPIO` module by default (called `python-rpi.gpio`). To use Python 3 programs, you have to install the Python v3 version of the module from the software repository, like this:

```
sudo apt-get update
sudo apt-get install python3-rpi.gpio
```

After you install the module, you can test to make sure it's installed, as in the following example:

```
pi@raspberrypi ~ $ python3
Python 3.2.3 (default, Mar  1 2013, 11:53:50)
[GCC 4.6.3] on linux2
Type "help", "copyright", "credits" or "license" for more information.
>>> import RPi.GPIO as GPIO
>>>
```

Because of the long module name, it has become somewhat of a default standard to use the GPIO alias when importing the `RPi.GPIO` module. This hour uses that convention.

When you have the `RPi.GPIO` module installed for Python v3, you're ready to start experimenting!

## Startup Methods

You need to know only a handful of basic methods in order to access the GPIO pins. Before you can start interacting with the interface, you have to use the `setmode()` method to set how the library will reference the GPIO pins:

```
GPIO.setmode(option)
```

As mentioned earlier this hour, to confuse things, there are two ways to reference the GPIO signals in the option placeholder:

▶ Using the pin number on the GPIO interface

▶ Using the GPIO signal number from the Broadcom chip

The GPIO.BOARD option, which you use like this, tells the library to reference signals based on the pin number on the GPIO interface:

```
GPIO.setmode(GPIO.BOARD)
```

The other option is to use the Broadcom chip signal number, specified by the GPIO.BCM value, as shown here:

```
GPIO.setmode(GPIO.BCM)
```

For example, GPIO signal 18 is on pin 12 of the GPIO interface. If you use the GPIO.BCM mode, you reference it using the number 18, but if you use the GPIO.BOARD mode, you reference it using the number 12.

This hour uses GPIO.BCM mode because it's easier to see in both the Pi Cobbler and the Gertboard.

After you select the mode, you must define which GPIO signals to use in your program and whether they will be used for input or output. You do that with the setup() method, as shown in the following syntax:

```
GPIO.setup(channel, direction)
```

For the direction parameter, you can use constants defined in the library: GPIO.IN and GPIO. OUT. For example, to set GPIO signal 18 to use for output, you'd write this:

```
GPIO.setup(18, GPIO.OUT)
```

Now the GPIO 18 signal is ready to use for output from your Python program. The next step is to actually control what you output.

# Controlling GPIO Output

The GPIO pins allow you to send a digital output signal to an external device. The following sections walk through how to control the output signal from your Python program.

## Setting Up the Hardware to View the GPIO Output

Before you can dive into coding, you need to set up the hardware environment for the project. You can use either the Pi Cobbler breakout box with your own components or the Gertboard, which contains all the components you'll need for the project. The following sections show the instructions for both methods.

## Setting Up the Pi Cobbler for Output

Unfortunately, to use the Pi Cobbler for this project, you need to collect a few more pieces of hardware:

- ▶ A breadboard

- ▶ A 1,000-ohm resistor

- ▶ An LED

- ▶ A piece of wire for connecting the breadboard sections

### Build the Pi Cobbler Circuit

Follow these steps to set up your Pi Cobbler to test the GPIO output:

WATCH OUT!

#### Working with Power

It's always a good idea to wire your project with the Raspberry Pi turned off. If the Raspberry Pi is turned on, the pins on the Pi Cobbler interface are live, and can be accidentally shorted out!

1. Connect one end of the Pi Cobbler ribbon cable to the GPIO interface and then connect the other end to the Pi Cobbler breakout box.

2. Connect the Pi Cobbler breakout box on the breadboard, making sure the two rows of pins straddle the middle of the breadboard so they don't connect to each other.

3. Connect a wire from one of the GND pins on the Pi Cobbler to a common location on the breadboard. (Most breadboards have two common rails that run the length of the breadboard for the ground and power supply.)

4. Place the 1,000-ohm resistor in the breadboard path for the Pi Cobbler pin labeled #18 and an empty area on the breadboard. (Pin #18 is the GPIO 18 signal pin.)

5. Place the LED so that the long lead connects to the 1,000-ohm resistor and the other lead connects to the ground rail on the breadboard.

Figure 24.3 shows a diagram of what the circuit should look like when you're finished with these steps.

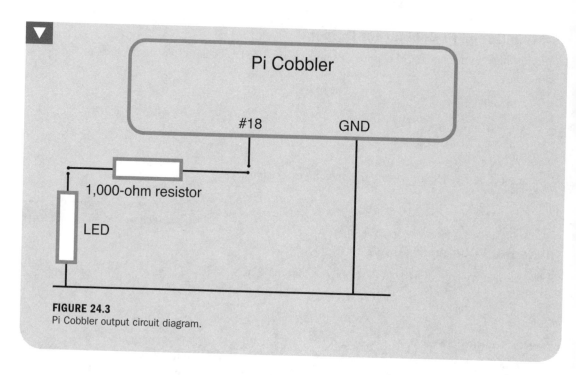

**FIGURE 24.3**
Pi Cobbler output circuit diagram.

With this circuit, when the GPIO 18 signal goes HIGH, the LED light ups, and when it goes LOW, the LED goes out.

### Setting Up the Gertboard for Output

The beauty of the Gertboard is that it already has all the components on the board for you, so all you need to do is connect some jumpers and wires.

### Build the Gertboard Circuit

The Gertboard makes developing circuits a snap! Here are the steps you need to follow:

1.  Connect a jumper to the 3.3V power supply side in the J7 block (the middle pin to the top pin).

2.  Connect a wire from the GP18 pin in the J2 block to the B12 pin in the J3 block. This redirects the GPIO 18 signal to the I/O buffer 12 area on the Gertboard.

3.  Connect a jumper between the two B12 output pins, directly above the U5 chip.

This circuit uses the D12 LED in the row of LEDs at the top of the Gertboard for the output LED. When the GPIO 18 signal goes HIGH, the LED lights up, and when it goes LOW, the LED goes out.

Now you're ready to start testing the GPIO output!

## Testing the GPIO Output

You should test the GPIO output before you start coding. To do this, you can run a test directly from the Python v3 command prompt to turn the LED on and off, using the GPIO output signal.

Because the RPi.GPIO module accesses the GPIO pins using direct memory access, you must run commands at the Python v3 command prompt as the root user account using the sudo program, as shown in this example:

```
pi@raspberrypi ~ $ sudo python3
Python 3.2.3 (default, Mar  1 2013, 11:53:50)
[GCC 4.6.3] on linux2
Type "help", "copyright", "credits" or "license" for more information.
>>>
```

You need to set the GPIO.BCM mode and set up the GPIO pin 18 signal for output:

```
>>> import RPi.GPIO as GPIO
>>> GPIO.setmode(GPIO.BCM)
>>> GPIO.setup(18, GPIO.OUT)
>>>
```

Now you can turn the LED on and off by using these two commands:

```
>>> GPIO.output(18, GPIO.LOW)
>>> GPIO.output(18, GPIO.HIGH)
```

Toggle back and forth a few times and watch the LED turn on and off. Then you use the cleanup() method to return the GPIO ports back to a neutral setting, like this:

```
>>> GPIO.cleanup()
>>>
```

If this doesn't work, you need to make sure that you started the Python v3 command prompt using the sudo command. If ensuring that you use sudo doesn't help, you may have to double-check your wiring to make sure it's all okay.

WATCH OUT!

### Resetting the GPIO Interface

It's always a good idea to use the `cleanup()` method when you're done with the GPIO signals. It places all the GPIO pins in a LOW status, so no extraneous signals are present on the interface. If you do not use the `cleanup()` function, the `RPi.GPIO` module produces a warning message if you try to set up a GPIO signal that is already assigned a signal value.

## Blinking the LED

Now you're ready to start writing some Python code. Listing 24.1 shows the `script2401.py` program, which toggles the GPIO 18 signal LED 10 times, causing the LED to blink 10 times. Just open your editor and enter the code shown in the Listing.

**LISTING 24.1**   The `script2401.py` Program Code

```
#!/usr/bin/python3

import RPi.GPIO as GPIO
import time
GPIO.setmode(GPIO.BCM)
GPIO.setup(18, GPIO.OUT)
GPIO.output(18, GPIO.LOW)
blinks = 0
print('Start of blinking...')
while (blinks < 10):
    GPIO.output(18, GPIO.HIGH)
    time.sleep(1.0)
    GPIO.output(18, GPIO.LOW)
    time.sleep(1.0)
    blinks = blinks + 1
GPIO.output(18, GPIO.LOW)
GPIO.cleanup()
print('End of blinking')
```

After you save this code, you need to use the chmod command to change the permissions so you can run the code from the command line. Because the script accesses direct memory, you need to use the sudo command to run it, as shown here:

```
pi@raspberrypi ~ $ chmod +x script2401.py
pi@raspberrypi ~ $ sudo ./script2401.py
Start of blinking...
End of blinking
pi@raspberrypi ~ $
```

When the program is running, you should see the LED blink on and off. Congratulations! You just programmed your first digital output signal!

# Creating a Fancy Blinker

You had to write a lot of code just to get the LED to blink. Fortunately, the GPIO has a feature that can help make that easier.

PWM is a technique used in the digital world mainly to control the speed of motors using a digital signal. You can apply it to your blinking project as well. With PWM, you control the amount of time the HIGH/LOW signals repeat (called the *frequency*) and the amount of time the signal stays in the HIGH state (called the *duty cycle*).

It just so happens that the Broadcom GPIO signal 18 doubles as a PWM signal. You can set the GPIO 18 signal to PWM mode by using the GPIO.PWM() method, as shown here:

```
blink = GPIO.PWM(channel, frequency)
```

After you set up the GPIO 18 signal, you can start and stop it by using the start() and stop() methods, as shown here:

```
blink.start(50)
blink.stop()
```

The start() method specifies the duty cycle (from 1 to 100). After you start the PWM signal, your program can go off and do other things. The GPIO 18 continues to send the PWM signal until you stop it.

Listing 24.2 shows the script2402.py program, which demonstrates using PWM to blink the LED.

**LISTING 24.2   The script2402.py Program Code**

```
1:  #!/usr/bin/python3
2:
3:  import RPi.GPIO as GPIO
4:  GPIO.setmode(GPIO.BCM)
5:  GPIO.setup(18, GPIO.OUT)
6:  blink = GPIO.PWM(18, 1)
7:  try:
8:      blink.start(50)
9:      while True:
10:         pass
11: except KeyboardInterrupt:
12:     blink.stop()
13: GPIO.cleanup()
```

The code starts the PWM signal on GPIO 18, at 1Hz (line 6), and then it goes into an endless while loop doing nothing (using the `pass` command on line 10). You set the loop in a `try` code block to catch the Ctrl+C keyboard interrupt to stop things.

After you start the program (using `sudo`), the LED should blink once per second (because of the 1Hz frequency in the `PWM()` method) until you press Ctrl+C.

# Detecting GPIO Input

Using the GPIO pins to detect input signals is a little bit trickier than using them for output. The following sections walk through a couple different ways to handle digital input signals on the GPIO pins. First, you need to set up the hardware you need for this project.

## Setting Up the Hardware for Detecting Input

In this project, you'll simulate a house with two doorbells: one for the front door, and one for the back door. When someone is ringing one of the doorbells, the project will tell you which one, and it will give you the opportunity to do some cool things with that information.

The following sections describe how to set up the hardware for the Pi Cobbler and Gertboard environments.

### Setting Up the Pi Cobbler for Input

For the doorbells, you need two push-button switches. You can use any type of switch you can find, as long as it conducts when pushed and breaks the connection when released. You can get specialty miniature push buttons that plug directly in to a breadboard, or you can use larger buttons and connect them to your breadboard using wires.

▼ TRY IT YOURSELF

### Connect the Pi Cobbler Circuit

For this project, you need to start with the circuit you created for controlling GPIO output in the previous section. In addition to that setup, you need just four additional pieces of hardware: two push-button switches and two 1,000-ohm resisters. When you have all the hardware you need, follow these steps:

1. Connect one side of each push-button switch to the ground signal, using a 1,000-ohm resistor.

2. Connect the other side of one push button to the GPIO 24 pin on the Pi Cobbler (marked #24) using a piece of wire.

3. Connect the other side of the other push button to the GPIO 25 pin on the Pi Cobbler (marked #25) using a piece of wire.

Figure 24.4 shows a diagram of what your final circuit should look like.

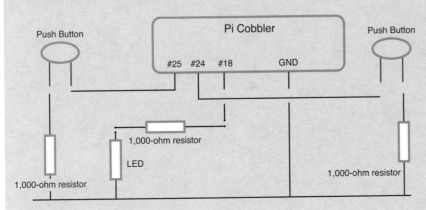

**FIGURE 24.4**
The Pi Cobbler input circuit diagram.

Remember to keep the LED and resistor plugged into the GPIO 18 pin because you'll use that in this project as well.

## Setting Up the Gertboard for Input

For the doorbells, you'll use two of the three built-in push button switches on the Gertboard.

### Set Up the Gertboard for Input

To set up the Gertboard for input, keep the B12 output buffer set to the B12 LED and plugged into the GP18 pin that you used for the output test. Also make sure you have the 3V3 jumper on the J7 block for power. Then follow these steps to set up the switches for input:

1. Connect a wire between GP24 in the J2 block and B2 in the J3 block.

2. Connect a wire between GP25 in the J2 block and B1 in the 3 block.

That's it! Your setup is complete, and you're all set to start coding!

# Working with Input Signals

On the surface, working with input signals is a breeze in the RPi.GPIO library. You just set up the GPIO pin for input and then read the pin status using the input() method, like this:

```
pi@raspberrypi ~ $ sudo python3
Python 3.2.3 (default, Mar  1 2013, 11:53:50)
[GCC 4.6.3] on linux2
Type "help", "copyright", "credits" or "license" for more information.
>>> import RPi.GPIO as GPIO
>>> GPIO.setmode(GPIO.BCM)
>>> GPIO.setup(24, GPIO.IN)
>>> print(GPIO.input(24))
0
>>> print(GPIO.input(24))
1
>>> print(GPIO.input(24))
0
>>>
```

After you enter this code, try holding down the push button you connected to the GPIO 24 pin as you run the print() methods. You should get different values of 0 or 1, depending on whether the button is pushed in.

However, there's a hidden problem with this setup, and you may have already run into it with your testing. Pushing the button connects the GPIO 24 pin to ground, forcing the LOW value (which is displayed as a 0). However, when the button isn't pressed, the GPIO 18 pin isn't connected to anything. That means the pin could be in either a HIGH or LOW state, and it may even switch back and forth without your doing anything. This is called *flapping*.

To avoid flapping, you need to set the default value of the pin for when the button isn't pressed. This is called a pull-up (when you set the default to a HIGH signal) or pull-down (when you set the default to a LOW signal). There are two ways to implement a pull-up or pull-down:

- ▶ **Hardware**—Connect the GPIO 18 pin to either the 3.3V voltage pin for a pull-up (using a 10,000- to 50,000-ohm resistor to limit the current) or to a GND pin (using a 1,000-ohm resistor) for a pull-down.

- ▶ **Software**—The RPi.GPIO library provides the option of defining a pull-up or pull-down for the pin internally, using an option in the setup() method:

  ```
  GPIO.setup(18, GPIO.IN, pull_up_down=GPIO.PUD_UP)
  ```

Adding this line forces the GPIO 18 pin to always be in a HIGH status if the pin is not connected directly to ground.

If you're using the Pi Cobbler, you can use either the hardware or software pull-up or pull-down method. However, the Gertboard doesn't provide for the hardware feature, so in this hour, you'll stick with using a software pull-up on your input lines, and then you'll use the push-button switch to connect the pin to the GND signal to trigger the LOW value.

Now you're ready to move on to some coding!

## Input Polling

The most basic method for watching for a switch is called *polling*. The Python code checks the current value of a GPIO input pin at a regular interval. The GPIO input changing value means the switch was pressed. Listing 24.3 shows the script2403.py program, which demonstrates this feature.

**LISTING 24.3**   The script2403.py Program Code

```
1:  #!/usr/bin/python3
2:
3:  import RPi.GPIO as GPIO
4:  import time
5:
6:  GPIO.setmode(GPIO.BCM)
7:  GPIO.setup(18, GPIO.OUT)
8:  GPIO.setup(24, GPIO.IN, pull_up_down=GPIO.PUD_UP)
9:  GPIO.setup(25, GPIO.IN, pull_up_down=GPIO.PUD_UP)
10: GPIO.output(18, GPIO.LOW)
11:
12: try:
13:     while True:
14:         if (GPIO.input(24) == GPIO.LOW):
15:             print('Back door')
16:             GPIO.output(18, GPIO.HIGH)
17:         elif (GPIO.input(25) == GPIO.LOW):
18:             print('Front door')
19:             GPIO.output(18, GPIO.HIGH)
20:         else:
21:             GPIO.output(18, GPIO.LOW)
22:         time.sleep(0.1)
23: except KeyboardInterrupt:
24:     GPIO.cleanup()
25: print('End of test')
```

The code sets up the GPIO 18 pin for output (line 7) and then the GPIO 24 and GPIO 25 pins for input (for the back and front doorbells, respectively; lines 8 and 9). Then the code goes into a loop, polling the status of the GPIO 24 and GPIO 25 pins in each iteration. If the GPIO 24 pin is

LOW, the code prints a message that the back doorbell is ringing and lights the LED. If the GPIO 25 pin is LOW, the code prints a message that the front doorbell is ringing and lights the LED.

BY-THE-WAY

### Doorbell Emailer

You can add any code you like to the `if-then` code block when a doorbell ring is detected. For example, you can use the email feature from Hour 20, "Using the Network," to send a customized email message to yourself each time a doorbell rings.

Polling is a simple way of detecting an input value, but there are other ways. The next section explores them.

# Input Events

Polling is a somewhat tricky way to determine when a switch has been pressed. You have to manually read the input value in each iteration and then determine whether the value has changed.

Most of the time, you're not as interested in the value of the input at any specific moment as you are in when the value changes. *Rising* occurs when the input changes from LOW to HIGH, and *falling* happens when the input changes from HIGH to LOW.

A couple different methods in the RPi.GPIO module allow you to detect rising and falling events on an input pin.

## Synchronous Events

The `wait_for_edge()` method stops your program until it detects either a rising or falling event on the input signal. If you just want your program to pause and wait for the event, this is the method to use. Listing 24.4 shows the `script2404.py` program, which demonstrates how to use the `wait_for_edge()` method to wait for a change in the input.

**LISTING 24.4**   The `script2404.py` Program Code

```
1:   #!/usr/bin/python3
2:
3:   import RPi.GPIO as GPIO
4:
5:   GPIO.setmode(GPIO.BCM)
6:   GPIO.setup(24, GPIO.IN, pull_up_down=GPIO.PUD_UP)
7:   GPIO.wait_for_edge(24, GPIO.FALLING)
8:   print('The button was pressed')
9:   GPIO.cleanup()
```

This script listens for the GPIO 24 signal. The program pauses at line 7 and does nothing until it detects a falling input value. (Remember: You're tying the input channel HIGH, so when you press the button, the signal goes from HIGH to LOW.) When the event occurs, the program is released and continues processing.

The downside to this is that you can wait for only one event at a time. If someone rings the front doorbell while you're waiting for the back doorbell to ring, you'll miss the event. The next method solves this problem.

## Asynchronous Events

You don't have to stop the entire program and wait for an event to occur. Instead, you can use asynchronous events. With asynchronous events, you can define multiple events for the program to listen for. Each event points to a method inside your code that runs when the event is triggered.

You use the add_event_detect() method to define the event and the method to trigger, like this:

```
GPIO.add_event_detect(channel, event, callback=method)
```

You can register as many events as you need in your program to monitor as many channels as you need. Listing 24.5 shows the script2405.py program, which demonstrates how to use this feature.

### LISTING 24.5   The script2405.py Program Code

```
 1:  #!/usr/bin/python3
 2:
 3:  import RPi.GPIO as GPIO
 4:  import time
 5:
 6:  GPIO.setmode(GPIO.BCM)
 7:  GPIO.setup(18, GPIO.OUT)
 8:  GPIO.output(18, GPIO.LOW)
 9:  GPIO.setup(24, GPIO.IN, pull_up_down=GPIO.PUD_UP)
10: GPIO.setup(25, GPIO.IN, pull_up_down=GPIO.PUD_UP)
11:
12: def backdoor(channel):
13:     GPIO.output(18, GPIO.HIGH)
14:     print('Back door')
15:     time.sleep(0.1)
16:     GPIO.output(18, GPIO.LOW)
17:
18: def frontdoor(channel):
19:     GPIO.output(18, GPIO.HIGH)
20:     print('Front door')
```

```
21:    time.sleep(0.1)
22:    GPIO.output(18, GPIO.LOW)
23:
24: GPIO.add_event_detect(24, GPIO.FALLING, callback=backdoor)
25: GPIO.add_event_detect(25, GPIO.FALLING, callback=frontdoor)
26:
27: try:
28:     while True:
29:         pass
30: except KeyboardInterrupt:
31:     GPIO.cleanup()
32: print('End of program')
```

The `script2405.py` code registers two events—one for each button. In this project, the code goes into a loop and does nothing while it waits for a button to be pressed (lines 27 through 31). You can easily incorporate other features in the loop, such as checking the temperature. (See Hour 20 for a refresher on using the `urllib` module to read temperatures from a webpage.)

BY-THE-WAY

### Reducing Switch Bounce

You may have noticed when testing the input project that sometimes using push-button switches can be a bit touchy (such as triggering two separate contacts with one button push). This is commonly called *switch bounce*. You can reduce switch bounce by adding a capacitor across the switch inputs. You can also control switch bounce by using software: The `add_event_detect()` method has a `bouncetime` parameter that you can add to set a timeout feature that helps with the switch bounce problem.

Now that you know the basics of working with input and output from the GPIO interface, you can create many applications. You can mix and match which pins you use for input and output, creating complex projects that detect input and send output based on the inputs.

# Summary

This hour explores the GPIO interface on the Raspberry Pi. You worked on a project that outputs a digital signal to a GPIO pin, as well as a project that outputs a PWM signal you can use to control motors. You also worked a project to read input values from the GPIO pins, which allows you to detect switch presses. You can use these concepts to control any type of electronic circuit, from reading temperatures to running robots!